Perspectives on Process Safety from Around the World 2024

AA008994

Held at the 2024 AIChE Spring Meeting and 20th Global Congress on Process Safety

New Orleans, Louisiana, USA
24-28 March 2024

ISBN: 979-8-3313-1701-0

Printed from e-media with permission by:

Curran Associates, Inc.
57 Morehouse Lane
Red Hook, NY 12571

Some format issues inherent in the e-media version may also appear in this print version.

AIChE
120 Wall Street, FL 23
New York, NY 10005-4020

Phone: (800) 242-4363
Fax: (203) 775-5177

www.aiche.org

Additional copies of this publication are available from:

Curran Associates, Inc.
57 Morehouse Lane
Red Hook, NY 12571 USA
Phone: 845-758-0400
Fax: 845-758-2633
Email: curran@proceedings.com
Web: www.proceedings.com

TABLE OF CONTENTS

Overview of Risk-Based Decision-Making Methodologies and Scenario Prioritization in Bow Tie Analysis: Unveiling Benefits and Challenges

Deyse Casais Camara
Petrobras
deyse.camara@petrobras.com.br

André Dantas
Petrobras
andrerd@petrobras.com.br

Fábio Leandro Rossi
Petrobras
flrossi@petrobras.com.br

Fábio Ventura Machado
Petrobras
fabioventura@petrobras.com.br

Keywords: Risk-Based Decision-Making; Bow Tie Analysis; Scenario Prioritization

Abstract

This study provides an overview of methodologies for risk-based decision-making and scenario prioritization within the context of Bow Tie analysis. The research holds particularly significant for large process facilities managing multiple Bow Ties, associated barriers, and crucial elements, as they require structured methodologies that support decision-making in cases of barrier degradation or unavailability. The choice among the existing methodologies for a company should consider aspects such as the company's safety culture, technological infrastructure, corporate standards, maintenance and inspection software, human resources. Our critical analysis explores into both traditional and innovative approaches for decision-making, with a specific focus on the Bow Tie method and its qualitative and quantitative applications. This includes the qualitative approach relying on robust verification, monitoring activities, and robust critical analysis, as well as the Bow Tie - LOPA method and the AHP method. We emphasize the fundamental benefits of these methodologies, particularly their capacity to prioritize and track multiple consequences within a single scenario, thereby enhancing their efficacy in implementing crucial safeguards. Furthermore, we address the challenges inherent in implementing these methodologies, such as the need for reliable data, complexity of modeling cause-and-effect relationships, software requirements to implement these methodologies for multiple Bow Ties, and the human resources needed for data processing. We highlight the significance of robust data collection and analysis strategies to enhance the accuracy and reliability of risk analyses. In addition to these methodologies, we discuss the increasing relevance of integrating Big Data and Artificial

Intelligence technologies into the process of risk-based decision-making. We conclude by emphasizing the continuous significance of research and development in this domain, with a focus on enhancing process safety and strengthening practices in industrial risk management.

1 Introduction

The Bow Tie analysis is a risk diagram that illustrates how various threats can lead to the loss of control over a hazard, allowing this unsafe condition to develop into a series of unwanted consequences. The diagram can display all deployed barriers and degradation controls. This intuitive model details barriers and degradation controls, offering a clear and comprehensible representation of the risk assessment and management process, crucial for diverse audiences, from operational staff to regulators (1).

The intrinsic value of Bow Tie analysis lies in its ability to effectively support risk management and communication. Through its graphical representation - including threats, consequences, and control measures - the method is accessible to a wide range of stakeholders, fostering a comprehensive understanding of both prevention barriers, which prevent the main event, and mitigation barriers, essential for reducing the severity of the consequences, should they occur (1).

Methodologies for constructing a Bow Tie are diverse, reflecting the complexity of industrial processes. Traditionally, the development of a Bow Tie begins with the identification of hazards and the top event, followed by the analysis of consequences and causes. Essentially, each Bow Tie is enriched by protection barriers, representing the implemented control measures. Additionally, degradation factors and associated degradation controls can be added to the Bow Tie diagram. The entire construction of this diagram can be carried out through a workshop with the participation of a multidisciplinary team (1). Another way to develop a Bow Tie diagram is through the combination of Fault Tree Analysis and Event Tree Analysis, allowing for a more detailed and in-depth risk analysis (2) (3) (4).

In recent years, Bow Tie analysis has evolved to incorporate quantitative methods, such as Layer of Protection Analysis (LOPA), and advanced data analysis techniques, enhancing its relevance and applicability in a broader spectrum of industrial risks (5) (6) (7) (8) (9). This evolution reflects the increasing demand for risk analysis approaches that are not only robust but also adaptable and capable of addressing operational changes and emerging risk scenarios.

In the context of industrial risk management, Bow Tie analysis stands out as a fundamental tool, both for the clear visualization of risks and for identifying areas where safety barriers can be improved, offering valuable insights to strengthen risk management measures (7) (9). This methodology is crucial for decision-making in environments where operational safety cannot be compromised (7) (9).

The importance of innovative methodologies in risk-based decision-making becomes prominent in the face of the complexity of modern industrial systems. The integration of new techniques and technologies in Bow Tie analysis not only enhances the accuracy and effectiveness in risk management but also ensures that organizations are prepared to proactively respond to constantly evolving risk scenarios (7) (9).

This paper aims to discuss the evolution of methodologies and applicability of Bow Tie analysis in industrial risk management, highlighting the impact of methodological innovations in the field of process safety for the enhancement of risk analysis in the decision-making process, with a special focus on the continuous need for research and development to face emerging challenges in industrial safety.

2 Traditional Approaches in Bow Tie Analysis

2.1 Conventional Techniques in Bow Tie Analysis

Bow Tie analysis is a widely recognized tool in industrial risk management, known for its effective graphical representation that illustrates the causes of hazardous events, the control measures implemented, and the potential consequences. Its bidirectional approach plays a crucial role in identifying risks and planning both preventive and mitigative measures (10).

- **Risk Modeling:** At the heart of Bow Tie analysis is the visual modeling of risks. The main event, or top event, is placed at the center of the diagram, with potential causes on one side and possible consequences on the other. This bidirectional approach not only aids in identifying existing risks but also in planning proactive measures to prevent and mitigate these risks (10).
- **Safety Barriers and Controls:** A critical aspect of traditional Bow Tie analysis is the emphasis on safety barriers. These barriers, which can be physical, technical, or procedural, are represented along the paths leading to the potential consequences. This focus on barriers aids organizations in understanding and strengthening defenses against hazardous events.
- **Risk Communication:** A key advantages of Bow Tie analysis is its ability to facilitate risk communication across various levels of an organization. The Bow Tie model's straightforward graphical representation makes the technical aspects of risks accessible to a wide audience, including individuals without technical training in safety. (5) (11).

2.2. Limitations and Challenges of Traditional Approaches

Conventional Bow Tie techniques, while effective in various aspects, face significant limitations:

- **Lack of Quantification:** Traditional Bow Tie approaches are predominantly qualitative, which may be inadequate for detailed analyses in modern industrial environments where accuracy in assessing the probability and impact of risks is crucial (12) (5) (13).
- **Operational Dynamics and Risk Changes:** The static nature of traditional models often does not reflect the operational dynamics and rapid changes in risks. For instance, in

industries undergoing rapid technological changes, Bow Tie analysis may not be agile enough to capture emerging new risks (7) (14).

- **Complexity in Diverse Scenarios:** In situations with multiple sources of hazard, Bow Tie analysis can become complex and challenging to interpret, especially in industries with a wide range of risk variables (9).
- **Maintenance and Data Updating:** The need to keep Bow Tie diagrams updated is a constant challenge, especially in industries subject to frequent changes (10).
- **Need for Deeper Analyses:** For scenarios requiring a more detailed understanding of risks, traditional Bow Tie approaches may be insufficient, highlighting the need for complementary methodologies (5).
- **Dependence on Experts:** Effective construction and interpretation of Bow Tie heavily rely on expert knowledge, which can be a limitation in organizations with limited resources for training and professional development.

Traditional approaches in Bow Tie analysis, while fundamental in understanding and communicating industrial risks, face significant challenges that limit their effectiveness in dynamic and complex environments. The limitations and challenges, as highlighted by scholars like Xu et al. (2018) (9) and Marszal (2023) (5), underline the need for the continuous evolution of this methodology. Integrating quantitative and adaptive approaches, such as LOPA, and the application of advanced methods like fuzzy logic and Bayesian analysis, are examples of how Bow Tie analysis is adapting to maintain its relevance in contemporary industrial risk management.

3 Fuzzy Logic Integration in Bow Tie Analysis

In the dynamic field of risk management, the intersection of fuzzy logic and Bow Tie analysis offers an approach for addressing complex challenges. This chapter explores how this integration enhances barrier management and optimizes the decision-making process in the context of risks.

Bow Tie analysis, effectively connecting Fault Tree Analysis (FTA) and Event Tree Analysis (ETA), is widely used in the industry to map and understand risks. The inclusion of fuzzy logic in this analysis significantly improves the accuracy and relevance of the modeled risk scenarios. It allows risk assessors to consider variability and uncertainty in the estimates of events and the effectiveness of safety barriers.

Fuzzy logic, proposed by Zadeh (1965), brings a new dimension to risk analysis. It stands out for its ability to transform qualitative uncertainties into quantitative analyses, allowing for a better interpretation of ambiguous data and the creation of models that reflect the complexity of the real world (15; 16). In the maritime industry, where risks are often accompanied by uncertain data and subjective judgments, fuzzy logic provides a means of capturing the essence of these ambiguities.

As illustrated by the study of Ferdous et al. (16), the application of fuzzy logic is carried out in four distinct steps: using fuzzy numbers to define probabilities of input events; aggregation of these numbers; determination of the probabilities of critical and outcome events; and finally, defuzzification.

One of the main challenges in quantifying Bow Tie analysis is calculating the probability of basic events present in fault trees. Traditionally, the probability of a basic event is derived from sources such as reliability manuals or field data. Subsequently, the probability of the main event is mathematically determined using logical gates, applying methods such as logical gate analysis or minimal cut sets. However, experts' opinions, often vague and imprecise, are better handled through fuzzy logic, efficiently translating these subjective views into quantifiable probabilities (17).

According to research conducted by Ferdous et al. (16), it's possible to categorize sources of uncertainty into three main dimensions, namely, data uncertainty, model uncertainty, and quality uncertainty. Quality uncertainty involves a thorough assessment of risks, including the identification and description of relationships in fault and event trees. Methods like HAZOP, HAZID, and FMEA can effectively reduce this type of uncertainty (16). In the context of bow tie analysis, both fuzzy theory-based approaches and evidence theory-based approaches have been considered to deal with data and model uncertainties (18) (10) (15) (19).

The application of fuzzy logic in Bow Tie analysis enables a quantitative approach in environments where data uncertainty and imprecision are predominant. By integrating fuzzy logic with Bow Tie analysis, there is a significant improvement in risk modeling, by:

- **Modeling Uncertainties:** The main contribution of fuzzy logic to Bow Tie analysis is the ability to convert qualitative and subjective assessments into precise quantifications. This transformation is fundamental, as risk quantification is often challenged by the lack of exact data and the subjective nature of risk estimates (17) (20).
- **Prioritization of Risk Scenarios:** By applying fuzzy sets and inference rules, it is possible to structure and prioritize risk scenarios based on probabilities and impacts. This approach allows for a more effective identification of critical points where mitigation actions are most needed (20).
- **Improvement in Barrier Management:** Fuzzy analysis offers a refined way to assess the reliability and effectiveness of barriers, which is crucial for the maintenance and continuous improvement of safety measures (21).

The incorporation of fuzzy logic into Bow Tie analysis significantly improves the basis for making strategic decisions, thereby enhancing how effectively risks can be mitigated (19). This approach is not only advantageous due to its robust support in decision-making but also due to its adaptability and flexibility. These qualities make it applicable in a variety of risk contexts, establishing it as a valuable tool for safety assessments (18) (15) (22; 19). On the other side, On the flip side, there are notable challenges in implementing this method. The primary difficulty lies in the technical intricacies of applying fuzzy logic, posing a barrier for organizations that don't possess specialized skills in mathematics and statistics. Furthermore, a persistent challenge is the subjective nature of data inputs. Relying on expert opinions and judgments can lead to biases in assessing risks, impacting the overall quality of the analysis.

The Fussell-Vesely (FV) methodology is another crucial approach in Bow Tie analysis, used to evaluate the contribution of basic events in Bow Tie analysis studies. This approach is highlighted for its ability to quantify the criticality of basic events in the occurrence of a main event (10) (17).

Key Advantages:

1. **Ability to Capture Variability:** The FV is recognized for its effectiveness in capturing variability in the probability of basic events, offering a more detailed analysis of risk. This aspect is crucial for understanding the dynamics of complex risks (17).
2. **Ease of Computation and Application:** This methodology is valued for its ease of calculations and practical applicability in different risk analysis contexts (10).
3. **Classification of Threats by Criticality:** The application of FV in Bow Tie analysis allows for classifying threats according to their criticality, aiding in identifying priority areas for mitigation actions.

The Fussell-Vesely methodology, therefore, represents a valuable approach in Bow Tie analysis for risk management, standing out for its efficiency in quantifying and classifying threats, which is fundamental in safety decision-making and in implementing effective risk mitigation strategies (10) (17).

Another method was employed to identify the basic event that most substantially contributed to the occurrence of accident scenarios. This approach involved sensitivity analysis to determine the input events that most significantly contribute to the output events in the Bow Tie method analysis, providing an assessment for mitigating the percentage of risk reduction in a system (16).

Therefore, the integration of fuzzy logic in Bow Tie analysis, complemented by the FV methodology, or another sensitivity analysis methodology represents a valuable and effective approach in risk management, standing out for its ability to quantify and classify threats, which is fundamental in safety decision-making and in implementing effective risk mitigation strategies.

3.1 Cloud Model and CREAM Approach in Risk Analysis

The Cloud Model, an advanced tool for expressing and processing uncertainties, offers a unique perspective in risk management. Developed as an extension of fuzzy logic, it combines elements of fuzzy set theory with probability theory to better capture uncertainty and randomness (23). In a recent study on the risk assessment of explosion and fire in oil tankers, the Cloud Model was employed to quantify the uncertainties associated with static electricity, demonstrating its effectiveness in modeling complex risks in industrial environments (24).

The CREAM (Cognitive Reliability and Error Analysis Method), on the other hand, focuses on the human element of risks. This method, developed by Hollnagel (1998), is used to assess cognitive reliability and human errors in complex systems. The application of CREAM in conjunction with fuzzy logic in Bow Tie analysis allows for a more comprehensive assessment of risks, considering both technical and human aspects (25).

A notable example of this integrated approach's application is found in the maritime industry. In a study on the risks associated with static electricity on oil tankers, combining Bow Tie analysis with fuzzy logic and CREAM provided valuable insights into risk management, helping to identify key areas of vulnerability and the most effective mitigation strategies (25).

The reviewed studies indicate that the integration of fuzzy logic and Bow Tie analysis, enriched by the Cloud Model and CREAM, offers a robust and flexible approach to risk analysis. This methodology not only enhances the accuracy of risk assessments but also provides a deeper understanding of the interactions between different risk factors. Furthermore, the studies highlight the importance of considering both technical and human aspects in risk management, emphasizing the need for a holistic approach (25) (24).

The combination of fuzzy logic with Bow Tie analysis represents a significant advancement in risk management. It provides a holistic and nuanced approach to addressing safety, crucial in environments where decisions have far-reaching consequences (21).

Although the results to date are promising, the studies also suggest several areas for future research. This includes exploring new ways of integrating Bow Tie analysis with other risk assessment models, expanding the use of fuzzy logic to address different types of uncertainties, and developing tools and techniques to facilitate the practical implementation of this integrated approach.

4 Relationship between LOPA and Bow Tie Analysis in Barrier Management and Decision Making

In this chapter, we explore the relationship between two key methodologies: Layer of Protection Analysis (LOPA) and Bow Tie analysis. LOPA, a semi-quantitative risk assessment technique, is often used to determine the need for additional safety measures in industrial processes (26). It typically uses order of magnitude categories for the frequency of the initiating event, the severity of the consequences, and the probability of failure of independent protection layers (IPLs) to estimate the risk of a scenario (26). In contrast, Bow Tie analysis provides a visual representation of the paths between the causes of unwanted events and their consequences, facilitating the understanding of risk scenarios.

The integration of LOPA with Bow Tie analysis enriches the process safety decision-making process. This combined approach allows for a more detailed analysis of risk scenarios, covering a variety of potential causes and consequences (27). The effectiveness of this integration is particularly evident in complex scenarios, where the isolated analysis of each method may not be sufficient to capture all the nuances of the risk.

In traditional LOPA, an accident scenario is defined as a single cause-consequence pair, analyzing only one path leading to a greater hazard. However, for more complex scenarios, LOPA must be used multiple times for each separate initiating event. The bow tie model overcomes this limitation, encompassing the entire accident scenario, including all initiating events and their possible consequences (6).

In the study conducted by Markowski, A. S., & Kotynia, A. (2011), a combination of methodologies: bow tie, LOPA, and fuzzy logic was applied. According to Markowski, A. S., & Kotynia, A. (2011), the application of fuzzy logic helps to refine uncertainties and lack of knowledge, especially for the frequencies and probability of initiating events and safety functions, allowing the use of uncertain qualitative data provided by experts and converting them into precisely determined outputs. The use of fuzzy logic in risk analysis provides more realistic risk index values and is effective in precisely determining additional safety measures, essential for achieving safety objectives. The proposed hybrid approach, combining the bow tie model, fuzzy logic, and LOPA, enhances the management of protection layers but requires the application of computer-assisted analyses, which may conflict with the simplicity of traditional LOPA (6).

One of the advantages of combining the bow tie and LOPA methodologies is the ability to evaluate mitigating safeguards. Unlike LOPA, which may struggle to assess mitigating safeguards, the quantitative analysis of bow tie can track not only the consequences when mitigating safeguards fail but also when they function (27). One of the conclusions of his study was that the quantitative analysis of bow tie is a powerful tool that can help process safety professionals assess risks more comprehensively and accurately, contributing to more effective risk management.

However, there are inherent challenges and limitations to this combined approach. The complexity in modeling cause-and-effect relationships and the need for reliable and accurate data are some of the main concerns (27). Additionally, effective integration of these methodologies requires a deep understanding of both processes and may be more time-consuming than using each method separately.

Valuable insights into the practical implementation of the combination of LOPA and Bow Tie are provided by some case studies (27) (7) (9). For example, the application in a distillation unit illustrates how this integrated approach can significantly improve process safety decision-making, providing a deeper understanding of the risks involved.

In conclusion, the combination of LOPA and Bow Tie analysis has the potential to transform risk-based decision-making.

5 Integrating Bayesian Networks into Bow Tie Analysis for Enhanced Risk Management

The Bow-tie model, traditionally used in risk analysis, presents limitations, especially in terms of dynamism and complex interactions. Its inability to adapt to changes in operational conditions and to capture the evolution of risks over time highlights the need for more flexible and adaptive approaches, such as Bayesian Networks (BNs). These graphical structures represent random variables and their conditional dependencies, allowing for the modeling of uncertainties and the updating of probabilities as new information becomes available (28; 21; 3; 4).

The integration of Bayesian Networks (BNs) with the Bow-tie model enhances its capabilities by introducing quantification and adaptability. Within a BN, events and safety barriers are converted into probabilistic nodes, enabling a dynamic risk assessment that accounts for variations in operational conditions and barrier effectiveness. An algorithm inspired by Khakzad et al.'s model was employed to map the Bow-tie framework into BNs, encompassing basic, intermediate, and top events, as well as safety barriers and consequences (29).

The software GeNIe (28; 3; 30; 31; 32; 14; 33) and Hugin (21; 13) have been utilized in the construction of Bayesian Networks (BNs), showcasing the methodology's applicability in diverse situations, and highlighting the necessity for software assistance in both analyses and the development of Bayesian networks.

The proposed steps for constructing dynamic risk analysis include (28):

1. Construction of the Bow-tie model from the fault tree and event tree.
2. Identification of possible barriers.
3. Mapping BNs from the Bow ties.
4. Performing dynamic risk analysis, evaluating different states of barriers and their performance in risk assessment.

Some important conclusions from this work are:

- The probability of occurrence of hazardous events and phenomena is dynamically updated as new information is provided.
- Analyzing the influence strength of adjacent nodes allows for the implementation of risk mitigation with the addition of safety barriers.
- Assessing the performance of safety barriers is essential to identify key barriers in various accident risks, supporting risk management decision-making and the implementation of preventive strategies.

In the context of risk-based decision-making, the combination of Bow tie and BNs becomes a robust tool for identifying and prioritizing risks, facilitating efficient resource allocation for risk management. This approach is essential in the industry for quick identification and response to risks, significantly impacting safety and operational efficiency (29; 23)

Wu et al.'s (2023) proposed method allows for the dynamic graphical representation of risk variations when the overall safety measures of the system are modified or when safety barriers are in different states. Moreover, the study rigorously distinguishes between safety barrier failures and events, fully considering the prevention and control effects of safety barriers in use and the safety strategy implemented on the event (28).

The implementation of dynamic risk assessment clarifies which events can be prevented and controlled and to what extent consequences are mitigated. Additionally, the ability to adjust safety barrier nodes in the Bayesian network diagram without the need to rebuild the entire model is a highlighted advantage. The proposed method and its findings are relevant in improving on-site safety management, providing insights into the impact of specific events on the overall event chain, tracking critical causes, and assessing risk levels under different safety barrier states. This enables the formulation of targeted prevention and control strategies and identifies the most likely paths of event development, benefiting on-site managers.

Various sensitivity analysis techniques have been applied, including the rate of variation (RoV) of probabilities and Birnbaum's importance (BIM), to identify the most critical basic events in a Bayesian Network (29). The effectiveness of sensitivity analysis in determining how the output of the model responds to the input in the Bayesian Network has been highlighted (30). Sensitivity analysis in Bayesian Networks is recognized as an efficient and straightforward method for emphasizing critical aspects of the analyzed risk scenario.

The application of this integrated approach in various industries is evidenced in the literature. In the petrochemical industry, for example, the efficacy of this methodology in identifying risks and managing safety barriers is demonstrated (28). In the energy industry, the integration of BNs with the Bow-tie model has been used to improve risk analysis and decision-making in complex and dynamic scenarios (34).

The complexity of constructing BN models and the need for high-quality data represent challenges for this integrated approach. Additionally, the lack of standardization in risk analysis methodologies can lead to variations in application and results. However, the integration of Bow tie and BNs in risk analysis has the potential to significantly transform risk management in the industry, offering a more dynamic, adaptable, and quantitative approach.

In summary, the integration of Bayesian Networks and Bow tie in risk analysis represents a significant advancement in industrial risk management, promising significant improvements in safety and operational efficiency. As this methodology evolves, it is likely to find increasingly broad applications in various industries, contributing to a safer and more resilient operational environment.

5.1 Integration of Bow Tie, Fuzzy Logic, and Bayesian Networks in Risk Analysis

The dynamism and complexity of industrial processes require risk analysis methodologies that are adaptable and robust. The integration of Bow Tie techniques, fuzzy logic, and Bayesian Networks emerges as a promising approach, overcoming the limitations of traditional risk analysis methods. This methodology combines the effective visualization of Bow Tie analysis with the probabilistic modeling and adaptability of Bayesian Networks, enhanced by the flexibility of fuzzy logic to handle uncertainties and imprecise data (23; 32).

A crucial advancement in this methodology in integrating Bow Tie with Bayesian networks is the incorporation of fuzzy logic into these analyses, allowing for the use of imprecise or subjective data, which is common in industrial environments. Fuzzy logic transforms uncertain estimates into quantifiable probabilities, significantly broadening the applicability and accuracy of risk analysis, especially in scenarios with limited or difficult-to-quantify data (31; 32).

Key advantages of incorporating the Fuzzy methodology into Bow Tie analysis and Bayesian Networks in risk management:

1. **Flexibility and Adaptation to Uncertainties:**
    - Fuzzy logic provides greater flexibility in analysis, allowing expert opinions and imprecise data to be effectively integrated into the risk model.

2. **Dynamic Risk Analysis:**
    - The fuzzy approach, integrated with BNs, allows for a dynamic update of probabilities, crucial for industrial environments where operational conditions change rapidly.
3. **Detail and Sensitivity in Analysis:**
    - The incorporation of fuzzy logic enhances sensitivity analysis and the identification of critical events, allowing for a more accurate assessment of risks and the most effective mitigation measures.

Summary of the main steps of the methodology:

1. **Application of Bow Tie Analysis:**
    - It begins with the identification of potential causes of a hazardous event (e.g., gas leak) and its possible consequences. Here, the combination of Fault Tree Analysis (FTA) and Event Tree Analysis (ETA) is used for a complete visualization of the risk scenario.
2. **Conversion to Bayesian Network:**
    - The Bow Tie model is transformed into a Bayesian Network. This step transposes the identified events and their interrelations into a format that allows for probabilistic analysis and bidirectional inferences.
3. **Incorporation of Fuzzy Logic:**
    - The integration of fuzzy logic allows the inclusion of uncertain data and estimates, such as expert opinions or unclear probabilities, converting them into quantifiable data within the model.
4. **Development of Fuzzy Bayesian Network (FBN):**
    - The resulting FBN is a fusion of the probabilistic modeling of the BN with the adaptability of fuzzy logic. This network can represent and analyzing the complex interactions and uncertainties present in industrial environments.
5. **Dynamic Updating and Sensitivity Analysis:**
    - The FBN allows dynamic updates of probabilities based on new information or evidence. Additionally, sensitivity analysis is performed to identify the most critical events and their consequences in the model.

The integration of Bow Tie, fuzzy logic, and Bayesian Networks represents a significant advancement in industrial risk analysis, offering a more comprehensive, adaptable, and quantitative approach. This approach is particularly valuable in risk management, where rapid identification and response to risks can have a substantial impact on safety and operational efficiency. Research validates the effectiveness and versatility of this integrated methodology, marking a significant advancement in industrial risk management practices (14) (32) (31).

6 Application and Impact of AHP in the Analysis and Prioritization of Risk Scenarios

The Analytic Hierarchy Process (AHP) is a structured and mathematical technique essential for complex decision-making. In the context of industrial risk management, AHP is a valuable tool for breaking down multifaceted problems into simpler subproblems, promoting clearer understanding and evaluation of various aspects of risk (35) (2).

The importance of AHP in industrial risk management is threefold:

- **Facilitation of Complex Decision-Making:** By breaking down a complex problem into smaller components, AHP facilitates understanding and detailed analysis of risk factors.
- **Quantification of Subjective Elements:** AHP converts subjective judgments into numerical scales, providing a solid basis for comparing different risk factors.
- **Effective Prioritization:** AHP assists in identifying and prioritizing risks, contributing to more efficient resource allocation in mitigating these risks.

Several studies demonstrate the efficacy of the Analytic Hierarchy Process (AHP) in risk analysis (2) (30). One study applied AHP to quantify probabilities in Bow Tie diagrams and integrated these with Bayesian Networks, leading to a marked improvement in the accuracy and usefulness of the diagrams for risk analysis in the oil and gas industry. In a similar vein, another study combined AHP with Bayesian Networks to evaluate the probability of success in Escape, Evacuation, and Rescue (EER) operations on offshore platforms, underscoring the method's relevance in the offshore exploration sector (30).

The incorporation of AHP in industrial risk management, especially when combined with Bow Tie and Bayesian Networks, offers a more comprehensive and quantitative approach to informed decision-making. Its application across various industrial sectors, from offshore exploration to the chemical industry, demonstrates its versatility and value in improving safety and operational efficiency.

7 Challenges and Benefits of Integrated Methodologies

7.1 Challenges in Implementing Advanced Methodologies

The integration of advanced methodologies into Bow Tie analysis brings complex challenges:

1. **Technical Complexity:** The implementation of techniques such as fuzzy logic and Bayesian Networks requires specialized knowledge in mathematics and statistics, which can be a barrier for organizations without such expertise (29) (34).
2. **Need for Specialized Software:** The construction of Bayesian Networks often requires specialized software, such as GeNIe and Hugin, used in various studies for risk modeling (29) (28). This need can represent a barrier, both in terms of costs and necessary technical skills.

3. **Integration of Data and Models:** Combining different models and data for an integrated analysis is a complex and challenging process, requiring a deep understanding of each methodology and how they interact (28).
4. **Maintenance and Updating:** Maintaining integrated models up-to-date and aligned with changes in operational conditions is a continuous and intensive process (33).
5. **Organizational Acceptance:** Adapting to new methodologies can be seen as complex or challenging, affecting the organizational acceptance of these advanced techniques (21) (13).

7.2 Benefits of Integrating Multiple Techniques in Bow Tie Analysis

Despite the challenges, the integration of multiple techniques in Bow Tie analysis offers significant benefits:

1. **In-Depth Risk Analysis:** The combination of qualitative and quantitative methods allows for a more complete analysis and a richer understanding of risks (27).
2. **Improvement in Decision-Making:** This integrated approach provides a more solid foundation for strategic decisions, enhancing risk management and mitigation (7).
3. **Flexibility and Adaptability:** The integration of different techniques increases the versatility of the analysis, allowing its application in a variety of contexts and types of risks. For example, in the petrochemical industry, the integration of fuzzy logic with Bow Tie analysis has been used to assess risks in refining operations (16). In the energy industry, the combination of Bow Tie with Bayesian Networks has improved safety analysis in nuclear facilities (34). These examples demonstrate the applicability of this approach in contexts with varied levels of complexity and types of risks.
4. **Dynamic Response to Changes:** The ability to update and modify analyses based on new information makes risk management more adaptable and dynamic (29). For example, in a situation where a safety barrier becomes unavailable in a chemical plant, the use of integrated Bayesian Networks allows for quickly recalculating the risk, assessing how it fits into the company's acceptability criteria (28). This dynamic response capability is essential in industrial environments, where conditions can change suddenly, requiring an immediate review of safety plans (36).

7.3 Overview

The integration of advanced methodologies in Bow Tie analysis presents notable challenges, including the need for specialized software and advanced technical knowledge. However, the benefits in terms of analysis precision, effectiveness in decision-making, and adaptability are substantial, outweighing the difficulties. Continuing to innovate and research in these integrated methodologies is essential to overcome emerging challenges in industrial safety and improve the effectiveness of risk management.

8 The Role of Artificial Intelligence and Big Data

Artificial Intelligence (AI) has emerged as a transformative force in the technological and industrial landscape, dating back to its roots in the Dartmouth Summer Research Project on Artificial Intelligence in 1956 by John McCarthy. This concept was initially inspired by Alan Turing's seminal work in 1950 on "Computing and Intelligence," aiming to mimic human problem-solving abilities in computers. Since then, AI has evolved to encompass tasks that require perception, cognition, planning, learning, communication, and physical action similar to humans, reflecting a significant expansion in its scope and applicability.

Within the AI spectrum, "generative AI" represents an innovative sub-discipline, focused on generating content in response to prompts. Tools like ChatGPT and DALL-E exemplify the transformative potential of generative AI, capable of revolutionizing the execution of various tasks. However, the full impact and associated risks of this technology are still under investigation and understanding.

In the domain of health, safety, and environment (HSE), it has been highlighted that the main benefits of AI include saving time and effort, scaling specialized knowledge, recovering lost lessons, and accelerating the time to value. The application of AI in HSE analysis notably transforms unstructured data into valuable insights. A key example of this is AI's ability to automatically generate Bow Tie diagrams from safety alerts, thereby enhancing the efficiency in identifying causes, controls, risks, and activities linked to incidents, and enriching strategic decision-making in HSE (37).

The crucial debate about human-AI interaction in the workplace has been addressed, highlighting that while AI-powered intelligent systems have the potential to autonomously solve problems, their operation under human supervision is essential for ensuring safe and reliable chemical manufacturing. This synergy between humans and AI is akin to the evolution of advanced process control, where computer algorithms provide optimal recommendations based on established operational parameters, and humans uphold operational discipline and manage abnormal conditions (38).

This overview of AI, from its origins to its current applications and the vital role of human-machine collaboration, reflects the ongoing evolution and increasing importance of AI in various disciplines. The integration of AI in HSE barrier management exemplifies its transformative potential and the need for a balanced approach that values both technological innovation and human oversight and expertise.

8.1 Reflections on the Future and Exploring New Possibilities with the Incorporation of Artificial Intelligence in Risk-Based Decision Making and Scenario Prioritization in Bow Tie Analysis

Bow Tie analysis, an established methodology for risk-based decision making and scenario prioritization, plays a key role in identifying and controlling risks in complex industrial processes. In this environment, the integration of Big Data and Artificial Intelligence (AI) technologies offers significant opportunities to enhance these methodologies, especially in large process facilities managing multiple Bow Tie Analyses.

AI can play a crucial role in various stages of Bow Tie analysis, such as data collection and analysis. One of the primary advantages of AI is its ability to process and analyze large volumes of data rapidly, essential for identifying hidden patterns and predicting risk scenarios. This is particularly valuable in facilities dealing with a substantial number of Bow Ties, associated barriers, and critical elements, where the need for structured decision-making methodologies is most pressing.

8.2 Applications of AI in Bow Tie Analysis

1. **Identification of Risks and Prioritization of Scenarios:** AI can analyze historical and real-time data to identify emerging risks and prioritize scenarios based on likelihood and potential impact. This enhances the ability to predict and mitigate risks before they materialize.
2. **Data Analysis for Decision Making:** Using advanced algorithms, AI can process large data sets to extract relevant insights, crucial in informed decision-making. It can identify trends and correlations that may not be evident to human analysts.
3. **Modeling Cause and Effect Relationships:** The complexity of modeling cause and effect relationships in industrial systems is one of the primary difficulties in Bow Tie analysis. AI, with its advanced machine learning capabilities, can help build more accurate and comprehensive models.
4. **Continuous Improvement:** AI algorithms can continuously learn from new data and outcomes of previous decisions, constantly refining decision-making, and risk analysis processes.

8.3 Challenges and Future Considerations

While AI offers significant benefits, its implementation comes with challenges. Data quality and reliability are crucial for the success of AI application. Additionally, the complexity of AI models requires a deep understanding and specialized skills to develop, implement, and maintain these systems. Collaboration between AI experts and process safety professionals is essential to ensure that AI systems are effective and aligned with the practical needs of the industry.

In conclusion, the integration of AI in Bow Tie analysis represents a significant advancement in industrial risk management. As we continue to explore the potential of AI, it is crucial that this technology is applied in a way that complements and enhances existing process safety practices, leading to safer and more efficient industrial operation.

9 Conclusion and Future Perspectives

The trajectory of Bow Tie analysis in risk management has been one of constant adaptation and enhancement. Originally a tool for visualization and communication of hazards and controls, it has evolved into a sophisticated methodology, integral to risk-based decision-making and scenario prioritization. The transition from a predominantly qualitative approach to one that embraces quantitative elements, such as LOPA, fuzzy logic, Bayesian Networks, and the AHP method, represents a fundamental shift. This evolution reflects the increasing complexity of industrial operations and the need for more data-driven approaches to risk analysis.

In risk management, especially in the process of risk-based decision-making and scenario prioritization, Bow Tie analysis has established itself as a vital tool. It facilitates a comprehensive view of potential risks, from identification to mitigation strategies. The methodology's flexibility to integrate with other advanced techniques has expanded its adaptability to various industrial contexts, offering a more dynamic and responsive approach to risk management.

Looking to the future, the field of risk analysis and industrial safety is poised for further transformation, primarily influenced by technological advancements and data analysis. The integration of Artificial Intelligence (AI) and Big Data into Bow Tie analysis is expected to be a significant trend. AI's ability to process large volumes of data and image processing can lead to more accurate risk predictions, timely identification of emerging hazards, and more efficient resource allocation for risk mitigation.

The potential of these technologies to provide real-time monitoring and dynamic recalibration of risk scenarios presents a new frontier in industrial safety. They could enable organizations to quickly adjust their risk management strategies in response to operational changes or unforeseen events, thus enhancing operational safety and efficiency.

Another area of future growth lies in the development of more user-friendly software and tools that simplify the complexity of advanced risk analysis techniques. This would make sophisticated risk assessment methodologies more accessible to a wider range of industries, including smaller organizations with limited specialized knowledge.

In conclusion, as the field of risk analysis and industrial safety continues to evolve, it is likely to see a greater convergence of traditional methodologies with cutting-edge technologies. This synergy promises to enhance the effectiveness of risk management practices, making industrial environments safer and more resilient to the ever-changing landscape of risks and hazards. Continuous research and development in this area are essential to realize the full potential of these emerging tools and methodologies in shaping the future of industrial safety.

10 References

1. **CCPS, Energy Institute.** *Bow Ties in Risk Management: A Concept Book for Process Safety.* New York, NY, London, UK : American Institute of Chemical Engineers and John Wiley & Sons, Inc, 2018.

2. **Badreddine, A., & Amor, N. B.** A Bayesian approach to construct bow tie diagrams for risk evaluation. *Process Safety and Environmental Protection.* 91(3), 159-171., 2013.

3. **Zarei, E., Azadeh, A., Khakzad, N., Aliabadi, M. M., & Mohammadfam, I.** Dynamic safety assessment of natural gas stations using Bayesian network. *Journal of hazardous materials.* 321, 830-840., 2017.

4. **Zerrouki, H.** Risk assessment of a liquefied natural gas process facility using bow-tie and Bayesian networks. *Process Safety Progress.* 41(3), 480-491, 2022.

5. **Marszal, E. M.** Expanding from LOPA to quantitative bow tie analysis. *Process Safety Progress.* 42(2), 258-268., 2023.

6. **Markowski, A. S., & Kotynia, A.** "Bow-tie" model in layer of protection analysis. *Process Safety and Environmental Protection.* 89(4), 205-213., 2011.

7. **Xu, Q., Xu, K., Li, L., & Yao, X.** Safety assessment of petrochemical enterprise using the cloud model, PHA–LOPA and the bow-tie model. *Royal Society open science.* 5(7), 180212., 2018.

8. **Yazdi, M.** The application of bow-tie method in hydrogen sulfide risk management using layer of protection analysis (LOPA). *Journal of failure analysis and prevention.* 17, 291-303., 2017.

9. **Xu, Q., Xu, K., Yao, X., Zhang, J., & Wang, B.** Sand casting safety assessment for foundry enterprises: fault tree analysis, Heinrich accident triangle, HAZOP–LOPA, bow tie model. *Royal Society open science.* 5(10), 180915., 2018.

10. **Almeida, R. S., da Silva, F. V., & Vianna, S. S.** Combining the bow-tie method and fuzzy logic using Mamdani inference model. *Process Safety and Environmental Protection.* 169, 159-168., 2023.

11. **Pitblado, R. M., Tahilramani, R., & Joseph, J. Y.** Integrated Risk Management: Using Intranet-Based Tools To Effectively Communicate Critical Risk Information From Bow Ties And Hazard & Risk Registers. *In SPE International Conference and Exhibition on Health, Safety, Environment, and Sustainability.* pp.SPE-127201, 2010.

12. **Agarwal, A., Ayello, F., Raghunathan, V., Demay, V., & Sridhar, N.** Quantitative assessment of failure probability of underground natural gas storage wells using an integrated bow-tie Bayesian network approach. *Journal of Failure Analysis and Prevention.* 21,107-123, 2021.

13. **Khakzad, N., Khan, F., & Amyotte, P.** Quantitative risk analysis of offshore drilling operations: A Bayesian approach. *Safety science.* 57, 108-117., 2013.

14. **Ahmadi, O., Mortazavi, S. B., Mahabadi, H. A., & Hosseinpouri, M.** Development of a dynamic quantitative risk assessment methodology using fuzzy DEMATEL-BN and leading indicators. *Process Safety and Environmental Protection.* 142, 15-44., 2020.

15. **Akyuz, E., Arslan, O., & Turan, O.** Application of fuzzy logic to fault tree and event tree analysis of the risk for cargo liquefaction on board ship. *Applied Ocean Research.* 101, 102238., 2020.

16. **Ferdous, R., Khan, F., Sadiq, R., Amyotte, P., & Veitch, B.** Analyzing system safety and risks under uncertainty using a bow-tie diagram: An innovative approach. *Process Safety and Environmental Protection.* 91(1-2), 1-18., 2013.

17. **Das, S., Garg, A., Maiti, J., Krishna, O. B., Thakkar, J. J., & Gangwar, R. K.** A comprehensive methodology for quantification of Bow-tie under type II fuzzy data. *Applied Soft Computing.* 103, 107148, 2021.

18. **Arici, S. S., Akyuz, E., & Arslan, O.** Application of fuzzy bow-tie risk analysis to maritime transportation: The case of ship collision during the STS operation. *Ocean Engineering.* 217, 107960., 2020.

19. **Shahriar, A., Sadiq, R., & Tesfamariam, S.** Risk analysis for oil & gas pipelines: A sustainability assessment approach using fuzzy based bow-tie analysis. *Journal of loss prevention in the process Industries.* 25(3), 505-523., 2012.

20. **Aqlan, F., & Ali, E. M.** Integrating lean principles and fuzzy bow-tie analysis for risk assessment in chemical industry. *Journal of Loss Prevention in the process Industries.* 29, 39-48., 2014.

21. **Khakzad, N., Khan, F., & Amyotte, P.** Dynamic safety analysis of process systems by mapping bow-tie into Bayesian network. *Process Safety and Environmental Protection.* 91(1-2), 46-53., 2013.

22. **Kaptan, M.** Risk assessment of ship anchorage handling operations using the fuzzy bow-tie method. *Ocean Engineering.* 236, 109500., 2021.

23. **Li, Y., Xu, D., & Shuai, J.** Real-time risk analysis of road tanker containing flammable liquid based on fuzzy Bayesian network. *Process Safety and Environmental Protection.* 134, 36-46, 2020.

24. **Xie, S., Dong, S., Chen, Y., Peng, Y., & Li, X.** A novel risk evaluation method for fire and explosion accidents in oil depots using bow-tie analysis and risk matrix analysis method based on cloud model theory. *Reliability Engineering & System Safety.* 215,107791, 2021.

25. **Elidolu, G., Akyuz, E., Arslan, O., & Arslanoğlu, Y.** Quantitative failure analysis for static electricity-related explosion and fire accidents on tanker vessels under fuzzy bow-tie CREAM approach. *Engineering Failure Analysis.* 131, 105917, 2022.

26. **CCPS.** *Layer of Protection Analysis. Simplified Process Risk Assessment.* . New York : . American Institute of Chemical Engineers and John Wiley & Sons, Inc, 2001.

27. **Marszal, E. M.** Expanding from LOPA to quantitative bow tie analysis. *Process Safety Progress.* 42(2), 258-268., 2023.

28. **Wu, X., Huang, H., Xie, J., Lu, M., Wang, S., Li, W., ... & Sun, X.** A novel dynamic risk assessment method for the petrochemical industry using bow-tie analysis and Bayesian network

analysis method based on the methodological framework of ARAMIS project. *Reliability Engineering & System Safety.* 237, 109397, 2023.

29. **Borgheipour, H., Tehrani, G. M., Eskandari, T., Mohammadieh, O. C., & Mohammadfam, I.** Dynamic risk analysis of hydrogen gas leakage using Bow-tie technique and Bayesian network. *International Journal of Environmental Science and Technology.* 1-12, 2021.

30. **Ping, P., Wang, K., Kong, D., & Chen, G.** Estimating probability of success of escape, evacuation, and rescue (EER) on the offshore platform by integrating Bayesian Network and Fuzzy AHP. *Journal of loss prevention in the process industries.* 54, 57, 2018.

31. **Mirzaei, M.** Fire and explosion Risk Analysis Using Bow-Tie Method and Fuzzy-Bayesian Network in Process Industries. *Iran Occupational Health.* 18(1), 139-154, 2021.

32. **Aliabadi, M. M., Pourhasan, A., & Mohammadfam, I.** Risk modelling of a hydrogen gasholder using Fuzzy Bayesian Network (FBN). *International Journal of Hydrogen Energy.* 45(1), 1177-1186., 2020.

33. **Abimbola, M., Khan, F., Khakzad, N., & Butt, S.** Safety and risk analysis of managed pressure drilling operation using Bayesian network. *Safety science.* 76, 133-144., 2015.

34. **Yan, F., Xu, K., Yao, X., & Li, Y.** Fuzzy Bayesian network-bow-tie analysis of gas leakage during biomass gasification. . *PLoS One.* 11(7), e0160045, 2016.

35. **Badreddine, A., & Amor, N. B.** A dynamic barriers implementation in Bayesian-based bow tie diagrams for risk analysis. *In ACS/IEEE International Conference on Computer Systems and Applications-AICCSA.* (pp. 1-8). IEEE., 2010.

36. **Vairo, T., Benvenuto, A. C., Tedeschi, A., Reverberi, A., & Fabiano, B.** Make Bow-tie Dynamic by Rethinking It as a Hierarchical Bayesian Network. Dynamic Risk Assessment of an Lng Bunkering Operation. *Chemical Engineering Transactions.* 91, 277-282, 2022.

37. **Garvin, T., & Kimbleton, S.** Artificial intelligence as ally in hazard analysis. *Process Safety Progress.* 40(3), 43-49., 2021.

38. **Chiang, L. H., Braun, B., Wang, Z., & Castillo, I.** Towards artificial intelligence at scale in the chemical industry. *AIChE Journal.* 68(6), e17644., 2022.

Implementation methodology of third part interference frequency reduction using ISO 13.623 aiming Social Risk reduction

Vitor Cersosimo Esposito
TAG – Transportadora Associada de Gás
Avenida República do Chile, 330 - Torre Leste/ 23° andar – Centro
Vitor.esposito@ntag.com.br

Keywords: Asset Integrity and Reliability, Hazard Identification and Risk Analysis and Measurement and Metrics

Abstract

Pipelines are a fundamental part of the energy and infrastructure matrix of countries and within this context, safer and more efficient transport technologies are developed and implemented. Besides this, environmental licensing and regulatory premises are established, being the Quantitative Risk Analysis (QRA) one of the most important, considering the regions surrounding the pipeline, regarding communities and the environment. This paper establishes the implementation of the methodology used to reduce the frequency of occurrence of third-party action. Derived from ISO 13.623 (Pipeline transportation systems), the methodology generates a new global failure frequency and, as a result, a reduction in social risk is achieved as it is calculated based on the mitigating measures applied, following the premises from the Guidance Standards of CETESB P.4.261, which addresses the terms of reference and methods for deciding this type of study. Based on QRA results, it is possible to have a better technical assessment of the risks and assertiveness in the analysis of its impacts by the licensing and regulatory agencies. The application of this methodology is essential for the safety of operations given the projected population density in the coming years, helping pipeline companies in the analysis and characterization of risk in surrounding populous regions.

1 Introduction

Pipelines are critical elements in the gas transportation industry because they facilitate the transport of large volumes over long distances and link gas production (*upstream*) to distribution markets (*downstream*). In order to obtain the licenses and authorizations needed for the construction and operation of pipelines, the elaboration of a Risk Analysis Study (Estudo de Análise de Risco – EAR in Portuguese) is necessary since it is demanded by licensing and regulatory agencies. The core of this study is a risk quantitative analysis and its main outcome being the calculation of

Social Risk. Social Risk represents the risk for a group of people that lives in the vicinity of a pipeline for a defined period of time (CETESB, 2011).

The calculation of Social Risk is directly related to the materialization of the hazards associated with pipelines and their operation. Its expression is represented by the F-N curve, where F represents the accumulated frequency of occurrence of scenarios with a number of fatalities N or more. Therefore, there are two fundamental factors for its calculation. The first factor is the frequency of occurrence of events, derived from an accumulation of external or internal factors related to the undertaking. The second factor is the number of people in the vicinity of the pipelines exposed to possible accidental hypotheses.

This work aims to outline the methodology employed in the EAR to minimize the calculation of Social Risk using the ISO 13.623 as a reference. The approach envolves the implementation of mitigating measures, generating a reduction in the frequency of third-party action, which is the largest contribution to pipeline failure (EGIG, 2020).

2 Methodology

2.1 Risk Analysis Study

Natural gas transportation systems are composed of different assets, which pipelines and surface facilities[1] are included. Understating the risk associated with these facilities is essencial for optimizing resources in emergency response plans and existing risk management programs within the pipeline network It also supports analyzes and impact studies for environmental licensing processes and compliance with licenses conditions. The tool employed to achieve these goals is the Risk Analysis Study (Estudo de Análise de Risco – EAR in Portuguese).

The EAR is performed aiming to anticipate and assess potencial impacts that results from accidental events from an enterprise to the company, as well as the surrounding population, presenting specific analysis objectives:

- Identification and analysis of potential accidental scenarios related to the products involved in the activities, which may cause harm to people, property, continuity of operations, environment and image;

- Focus on events of liberation of flammable products with the potential to cause significant impacts on populations near the gas pipeline, and;

- Generation of technical fundations for the development or revision of the Emergency Response Plan (PRE - Plano de Respostas a Emergências in Portuguese) and the Risk Management Program (PGR – Plano de Gerenciamento de Riscos in Portuguese) for the gas pipeline.

---

[1] Examples of surface facilities: Delivery Point, Receiving Point, Compression Station

In order to achieve these results, the sequential steps for the elaboration of the EAR involve identifying dangerous conditions using the Preliminary Risk Analysis (APR – Análise Preliminar de Riscos in Portuguese), selection of the scenarios with high impact and subsequent analysis of its consequences. During the latest step it is possible to obtain the calculated Social Risks for the selected sections and assess within the tolerability parameters of the Guidance Standard for the Preparation of Risk Analysis Studies of CETESB P.4.261(Norma de Orientação para Elaboração de Estudos de Análise de Riscos da CETESB P.4.261 in Portuguese).

The levels of tolerability used for technical parametrization are: (i) Tolerable: Risks at an acceptable level that must have residual risk treated in a specific management program; (ii) ALARP: Risks as low as possible, with implemented mitigating measures within technical and financial feasibility. Other risks are continuously managed within a specific management program; (iii) Intolerable: Unacceptable risks, where the project becomes unfeasible or requires technical changes for safe implementation/operation.

2.2 Consequence Analysis

To calculate the risk and analyze consequences, specific technical documentation serves as references, such as Terms of Reference of licencing agencies that are indicated for each project, and the Guidance Standard for the Preparation of Risk Analysis Studies of CETESB P.4.261 as a general reference.

In order to analyze the project risk , areas of population density are identified and high impact scenarios determined in the APR are selected. After this step the calculation of Social Risk, which is a measurement that refers to a determined number of people that are exposed to the physical events that result from accidental scenarios, is performed. It is presented through F-N curves, which are obtained by plotting the accumulated frequency data of the studied events and their quantitative effects. Figure 1 illustrates the F-N curve with the criteria of Tolerability for Social Risk given by CETESB Standard (CETESB, 2011).

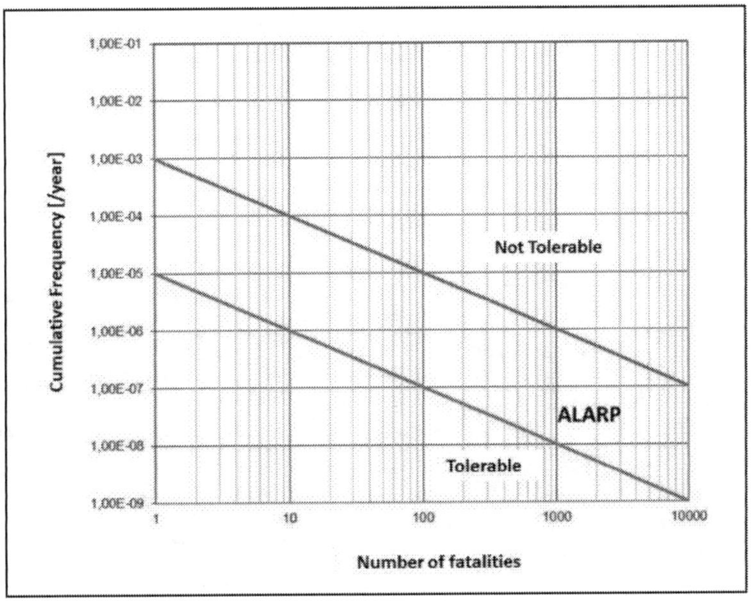

Figure 1 - Criteria of Tolerability for Social Risk

Source: CETESB, 2011

As mentioned before, Figure 1Figure 1 illustrates the tolerability limits of the F-N curve, where F represents the quantification of consolidated frequency of occurrences with the potential to result in fatal victims, portrayed Figure 1 by the N axis.

The consolidated frequency, a central aspect of the methodology used for this study, is obtained through the sum of the frequencies of accidental scenarios with the same number of fatalities in their severity, maintaining all other conditions such as wind direction, sum of day and night periods, among other environmental aspects.

2.3 Frequency of Accidental Scenarios for Natural Gas Transportation Pipelines

As previouslly mentioned, for the calculation of Social Risk is necessary the consolidated frequency of accidental scenarios. To gather this data, the database of failures associated to gas pipelines (European Gas Pipeline Incident Data Group - EGIG) is used. From this database the detailed frequencies for the simulations are obtained, either from the size of the failure/leaking, or from the causes of the failures, as presented in Table 1 and Table 2.

Table 1. Failure Frequency Caused by Leak Size – 2010 to 2019

Causes	Failure frequency (oc./1000km.year)			
	Rupture	**Fissure/ Crack**	**Hole**	**Unknown**
External interference	0,006	0,015	0,015	-
Corrosion	-	-	0,033	0,001
Construction Error/Material Failure	0,001	0,001	0,017	0,001
Incorrect Hot-Tap	-	0,001	0,001	-
Ground movement	0,007	0,005	0,008	0,001
Others / unknown	0,001	0,001	0,014	0,001

SOURCE: EGIG, 2020

Table 2. Distribution of Accident Frequency by Leak in the Period from 2015 to 2019

Size of the leaking	Failure frequency (oc./1000km.year)
Unknown	0,003
Hole	0,088
Crack	0,022
Rupture	0,013

SOURCE: EGIG, 2020

Therefore, the failure frequencies obtained are:

- Hole = 8,8 E-05 oc/Km.year;
- Crack = 2,2 E-05 oc/Km.year;
- Rupture = 1,3 E-05 oc/Km.year.

2.4 Frequency reduction of Accident Scenarios for Natural Gas Transport Pipelines using ISO 13.623

As highlighted in the item 2.3 the calculation of the Social Risk relies on the distribution of the frequency of the accidental scenarios. In cases where demographic changes are presented, specifically, variation in population density in pipeline vicinity, the social risk obtained may fall outside the acceptable tolerability parameters, making it necessary to implement mitigating measurements to reduce these uprising risks.

Within the principle of calculation it is feasible to act on the consequences, with the primary measurement being to change the Maximum Operating Pressure. However, it might not be viable due to minimum reduction possibilities or impractability of transport, as well as potencial breach of existing contracts. Another measurement is to act in the frequency of occurrence of accidents causes . Thus, the methodology outlined in ISO 13.623 will be presented below, providing guidelines on behalf of frequency reduction.

The Table 3 presents the reduction rate for some mitigating measures designed to prevent damage resulting from the action of third parties or external interference to the pipeline, within natural gas transport operations.

Table 3. Frequency Reduction Rate of Mitigant Actions Related to Third Party Activities

Mitigation measure	**Failure Frequency Reduction Rate**
Ground cover of 1,5m	20%
Ground cover of 2,0m	40%
Ground cover of 3,0m	80%
Reinforced concrete barrier or steel encapsulation	**85%**
Buried tape	40%
Increase in duct strip landmarks, Reduction in spacing between duct strip landmarks less than or equal to 25m	25%
Pipeline patrolling twice a month	20%

SOURCE: ADAPTED FROM ISO 13.623

As can be seen, the most significant reduction factor is obtained from the implementation of a physical barrier, concrete or steel, as a form of mitigating measure. Table 4 and Table 5 provide the revised frequencies of accident scenarios after application of the proposed reduction measurements.

Table 4. Frequency of Failure Caused by Leak Size – 2010 to 2019 – Reduction in Frequency of Third Party Action

Causes	Failure frequency (oc./1000km.year)			
	Rupture	**Fissure/ Crack**	**Hole**	**Unknown**
External interference	**0,0009**	**0,00225**	**0,00225**	-
Corrosion	-	-	0,033	0,001
Construction Error/Material Failure	0,001	0,001	0,017	0,001
Incorrect Hot-Tap	-	0,001	0,001	-
Ground movement	0,007	0,005	0,008	0,001
Others / unknown	0,001	0,001	0,014	0,001
Total Frequency	**0,0099**	**0,01025**	**0,0755**	**0,004**

SOURCE: ADAPTED FROM EGIG

Through the analysis of Table 4, it is possible to estimate that between 2010 and 2019, approximately, 75.77% of the types of causes of failure in natural gas pipelines result in holes; 10.29% generate Cracks; 9.93% result in pipeline rupture and 4.01% have an unknown leakage class after reducing the frequency due to external interference.

Table 5. Distribution of the Frequency of Accidents According to the Dimension of Leak with Use of a Mitigating Measure – Concrete Plates

Leaking Size	**Size estimation**	**Share of total frequency by class of Leak (%)**	**Total Frequency -Natural Gas Pipeline (oc/km year)**	**Frequency by class of leak (oc/km year) considering concrete**
Hole	5% Pipeline Diameter	75,77	9,965 E-05	7,55 E-05
Crack	20% Pipeline Diameter	10,29		1,03 E-05
Rupture	100% Pipeline Diameter	9,93		9,90 E-06

SOURCE: DESENVOLOPED BY THE AUTHOR

Therefore, the failure frequencies obtained are:

- Hole= 7,55 E-05 oc/Km.year;
- Crack = 1,03 E-05 oc/Km.year;
- Rupture = 9,90 E-06 oc/Km.year.

3 Results

Using the methodology presented by ISO 13.623, the reduced frequencies of occurrences of accidental scenarios can be obtained. From this, it is possible to analyze the direct impact on the reduction of social risk. The following figures were generated for an EAR of a gas pipeline belonging to TAG and are part of the Company's technical analysis framework:

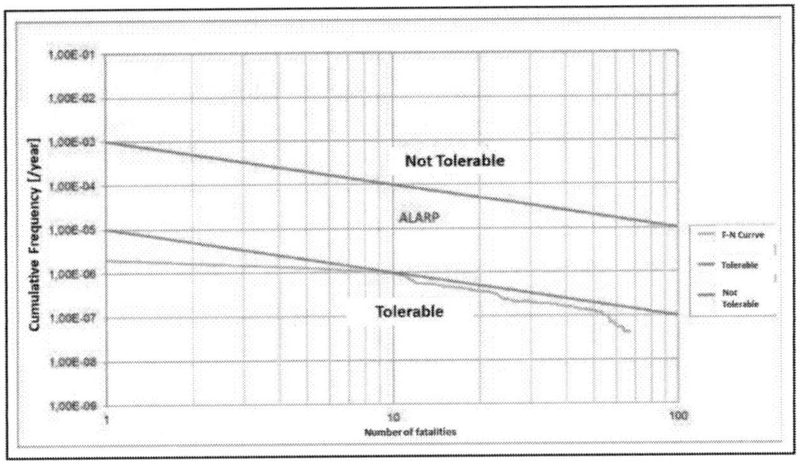

Figure 2 - Social Risk Without Mitigating Measures at Point 01

Source: Created by the Author

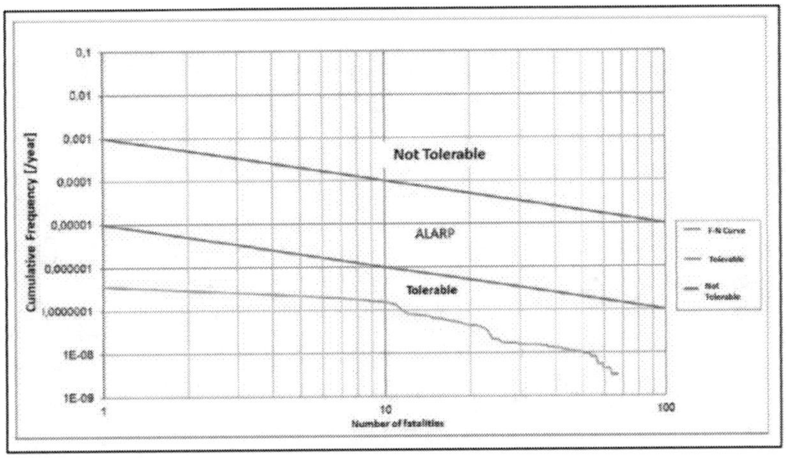

Figure 2 - Social Risk with the Adoption of Mitigating Measures at Point 01 – (Concrete Plates)

Source: Created by the Author

Based on the F-N curves shown in Figure 2 and Figure 3, presented above, it can be seen that the Social Risk calculated before applying the mitigating measures was very close to the tolerability limit at the ALARP level at several points along the curve. However, after applying the mitigating actions defined by ISO 13.623, the result obtained demonstrates that the F-N curve is completely below the parameter established by the CETESB Standard, thus showing the reduction in all points to the Tolerable level for the section.

Conclusions

Risk analysis and implementation of mitigation measures play a crucial role in ensuring safety and structural reliability of pipelines, especially in densely populated regions. Prioritizing structural reliability of pipelines and minimizing the probability of incidents resulting from external interference are imperative to protect both assets and surrounding communities.

It is notable that the approach of adding multiple security barriers, as permitted by ISO 13.623, provides even greater protection against potential risks. This methodology is widely accepted globally and has found support in regulatory and licensing agencies in Brazil.

In this context, effectively using these practices not only preserves the operational continuity of the pipelines, but also contributes to strengthening the country's infrastructure, especially at a crucial time of opening up the gas market. Therefore, by adopting proactive approaches to risk management and applying security measures, it is possible to guarantee greater security and reliability in the analysis and mitigation of potential risks.

4 References

[1] CETES. CETESB P4.261 - Norma de Orientação para Elaboração de Estudos de Análise de Riscos; 2011.

[2] ISO 13623:2017. Petroleum and Natural Gas Industries — Pipeline Transportation Systems. Geneva, Switzerland: International Organization for Standardization; 2017.

[3] EGIG. *European Gas Pipeline Incident Data Group; 2020.*

5 Acknowledgments

Acknowledgments are due to TAG – Transportadora Associada de Gás S.A., the company that operates in the transportation of natural gas and holds 47% of the gas transportation infrastructure in Brazil for sponsoring the project.

Acknowledgments to Environpact for the support and capability in the development of the risk analysis for TAG assets.

Dynamic Barriers Monitoring in the Mining Industry and Its Contribution to Process Safety

Hellen Mello
Vale SA, Corporative Process Safety and Operational Risk Management
Praia de Botafogo, 186, Botafogo, 22250-145, Rio de Janeiro, RJ, Brazil
hellen.mello@vale.com

Marcelle Lobato
Vale SA, Integrated Enterprise Risk Management

Sylette Arruda
Vale SA, Corporate IT Innovation Management

Bruno Menegatti Esperandio
Vale SA, Corporative Process Safety and Operational Risk Management

Jorge Mascarenhas
Vale SA, Corporative Process Safety and Operational Risk Management

Fabio Teixeira
Vale SA, Integrated Enterprise Risk Management

Thiago Costa
Vale SA, Technical Operational Excellence Management

Silvana Campos
Vale SA, Technical Operational Excellence Management

Rubens Silva
Vale SA, Technical Operational Excellence Management

Leonardo Martins
Vale SA, Internal Controls and SOx Management

Keywords: Barrier Management, Maintenance, Dynamic, Manage Risks, Asset Integrity and Reliability, Process Safety

Abstract

Different dynamic barrier monitoring methods are increasingly being adopted across the industries for enhanced facilities safety. Considering the effort challenges in manually analyzing maintenance and inspection plans during barrier performance audits, in addition to the potential for errors and inaccuracies, Continuous Control Monitoring (CCM) was developed aiming to automate and assertively execute those checks. The strategy involves integrating online risk and maintenance management tools, enabling periodic and automatic verifications of asset maintenance conditions, and providing timely signaling of vulnerabilities with precision, data detail and traceability. Additionally, the prioritization of maintenance activities and leadership notifications in cases of non-compliance also became automated, on a risk-based approach. Significant improvements in assertiveness, optimization, and compliance in both barrier and asset management processes could be noticed. Process safety as a whole benefits from the automatic and timely identification of possible non-conformities in the maintenance of assets, supporting corrective measures, risk classification analysis, and the adoption of appropriate governance procedures.

1 Introduction

The development of methods for dynamic monitoring of barriers is increasingly a reality in different industrial sectors. The solid systematization of routines through computerized solutions enables the development of systemic interfaces that can provide a number of benefits to the facilities' safety, such as those to be pointed out by this study. Vale S.A., like other major players in the mining industry, follows ICMM's (International Council on Mining and Metals) guidelines[1][2] for critical controls management, which are closely aligned with the RBPS (Risk Based Process Safety) model proposed by the CCPS (Center for Chemical Process Safety).

The company's integrated management system, the VPS (Vale Production System), is comprised of 3 pillars: management, leadership and technical, within which the main process safety elements and requirements are established. The entire risk management is based on the lines of defense model, where the first line (operations) is responsible for directly managing the risks, identifying, assessing, treating, preventing, monitoring and mitigating the events in an integrated manner. The second line of defense comprises the enterprise risk management and specialist teams, who develop and support the implementation of policies, methodologies, technical standards and tools, as well as promoting the integrated communication and disseminating the company's safety culture. Lastly, the third line verifies the effectiveness of risk management, internal controls and compliance, through total independent evaluations, inspections and investigations carried out by the internal audit and whistleblower channel.

One of the specialist areas of the second line of defense concerns process safety and has stablished the HIRA (Hazard Identification and Risk Analysis) program[3] for the major operational risks, herein called MUEs (Material Unwanted Events). The cycle begins with the application of the bowtie methodology for the evaluation of incidents of high severity of impact to people, the environment and/or operational disruption. For each cause and consequence included in the diagram, proper preventive and mitigative barriers are identified[4], then classified as critical or not. The criteria for categorization considers (i) the effectiveness of the control to interrupt the chain of events, avoiding the incident occurrence or reducing its impacts, (ii) its independency, (iii) if it is effective for multiple risks, among other aspects[1]. Afterwards, the performance criteria for each critical control are defined, as well as any site-specificities and accountabilities. Based on all this information, the periodic verification and reporting of the barriers' performance are carried out, leading to a proper response to reestablishment when inadequate. Finally, the management loop closes returning to the initial step with the review of the safety studies.

As commonly acknowledged and stated in the Guidelines for Risk Based Process Safety[5] of CCPS, safety systems must be available when required and capable of functioning at a specified performance level. This is typically attained through an integrated combination of ITPM (Inspection, Testing, and Preventive Maintenance) tasks and quality assurance measures[5], making maintenance a key aspect of critical controls management[6]. During the barriers' performance audit though, an activity that can require a relatively long time when carried out manually is the analysis of maintenance and inspection plans adherence. Some cases may involve hundreds or thousands of items, making the verification procedure extensive and susceptible to failure which ultimately may lead to an inaccurate assessment and management of the related risks. In this context, CCM (Continuous Control Monitoring) was developed as a way to automatically and assertively perform these checks, as well as constantly update the maintenance status related to each barrier mapped by the process safety studies. It additionally provides a risk-based approach for maintenance management, optimizing the prioritization of activities besides resulting in either more or less rigorous modifications to the program, based on the outcomes of risk analysis and decision-making criteria.[6]

According to Hosseinnia et al[7], accident investigations in process industry indicate that inadequate barrier management has been one of the main causes of many major accidents. Moreover, barriers degrade over time and at different rates, which may gradually and unnoticeably drift the system towards a state of high risk. It is advised that such issues can be addressed by the dynamic barrier management concept through the combination of all available information, such as inspection, preventive maintenance, audit, sensors, process control, and near misses or incident records, with the main goal to infer barrier status in near real-time and evaluate its impact on risk level. An approach based on systems engineering was considered suitable for managing complex problems by considering the big picture and to provide a practical roadmap for dynamic barrier management.

Yuan et al.[8] in turn suggests an integrated approach aimed at enhancing dynamic risk-informed safety barrier management within chemical process industries. A simulation tool was created to merge dynamic risk assessment and safety barrier management, offering multiple functionalities to aid decision-making in optimizing safety barriers, evaluating their risk reduction performance by incorporating periodic proof test data, continuous condition-monitoring, and accident precursor data to update risk profiles, considering safety barrier degradation. It utilizes a combination of cost-effectiveness analysis and optimization algorithms to identify optimal strategies for establishing and improving safety barriers.

Similarly, a framework for maritime operations decision support using dynamic barriers monitoring is propositioned by Karolius et al[9]. It involves identifying hazards and corresponding barriers, equipped with sensors to provide real-time information on their status. A parametric model enables effective prioritization of barrier performance information and real-time monitoring of its status is considered to enhance the capability of operators to predict and address specific outcomes. This seamless synergy is facilitated by ship connectivity, leveraging advanced analytics to process numerous inputs continuously.

In addition, the systematic review of literature journals by Qiao et al.[10] suggests three key issues of dynamic barrier management for further study: (i) the need for addressing benchmark determination, (ii) balancing costs and gains in obtaining real-time data, and (iii) understanding the interaction among different types of barrier elements in dynamic barrier assessment activities. Determining a reasonable benchmark for dynamic barrier assessment may be challenging, especially because the standards designed for a single barrier element and the barrier system can be different. Thus, the relationship among different benchmarks needs to be well understood. Also, obtaining and integrating real-time data and information is crucial and while current technical developments aim to enable real-time data acquisition, the challenge lies in balancing the associated costs with the gains, as companies may be reluctant to implement risk control activities at high expenses. Lastly, it argues that the objective of barrier management is supported by integrating multiple barriers, including technical, operational, and organizational elements. Therefore, understanding the interaction between these elements would be vital for effective safety management.

2 Continuous Control Monitoring - CCM

The adopted strategy for CCM is based on the integration between risk and maintenance online management tools, which enables data exchange through the systems, thus supporting the determination of the protection layers' status.

The project was executed in two phases. The first one developed the interface between the systems in which the data was intended to flow from the maintenance management tool (MMT) to the risk one (RMT) only. Therefore, assets info such as tag, description and criticality, started to be constantly updated in the RMT, alongside with the links between those assets and the critical barriers already registered in the risk management system. Once specified to which assets each safeguard is related, periodic and automatic verifications of the maintenance condition initiated, allowing timely signaling of vulnerabilities in the RMT with precision, detailing and traceability. Additionally, the connection between assets and protection layers of high-potential hazardous events turned into a central provision to the definition of prioritization of maintenance activities and the automatic escalation of leadership notification in cases of non-compliances (previously implemented in MMT), optimizing the maintenance management in a risk-based approach.

The controls and assets associations are defined by local engineering teams, with the proper knowledge on the operational equipment at each site, supported by the corresponding risk management staff. At this first stage of the project, the selection of applicable safeguards for the engineering analysis (MUEs' operative critical barriers) and the upload of the results in MMT used to be manually carried out in a monthly basis.

During the second phase of CCM implementation, the reverse path of data flow was also settled and the info exchange began to occur from both systems. The automatic and continuous screening of suitable safeguards was operationalized, avoiding eventual human factors and enabling the timely availability of the material for the engineering analysis. Moreover, the controls and assets correlations commenced to be registered directly in RMT by engineering, allowing better tracking of changes and approvals by the system's log and providing the personnel more independence and flexibility to register the information at any time. Therefore, RMT started to feed MMT with refreshed data regarding barriers and risks, assets and controls' links and safeguards applicability to these associations. Ultimately, the assets' criticality classification became automated in MMT, given all those info received in nearly real-time from RMT, allowing both the triggering and inhibiting of assets prioritization in the maintenance managing system depending on the criteria compliance status.

The automatic checks of barriers already associated with assets currently occur at a pre-determined frequency of twice a week and verify: (i) the existence of a maintenance plan, (ii) the absence of leadership notification about delays for more than fifteen days, (iii) whether the last maintenance order was executed and (iv) whether it is within the deadline considering the maximum tolerance registered in the system. For the test result to be positive, all rules must be met on all assets linked to the barrier. As already exposed, the process relies on constant data exchange between systems, which keeps updated the databases of assets and maintenance plans, notes and orders, besides risks, controls and associations between the elements. Additionally, pending analysis and KPIs (Key Performance Indicators) results were made available online through specific reports, with data accuracy and efficiency.

3 Results and discussion

The main foundations of CCM approach are the identification and analysis of risks and the relationships between barriers and assets data. Such connections are still being established in the system by engineering, but currently 56% of critical controls related to operational risks with high potential impact have already been made available for automatic verification, as can be observed in Figure 1, with the main results showed in Figure 2.

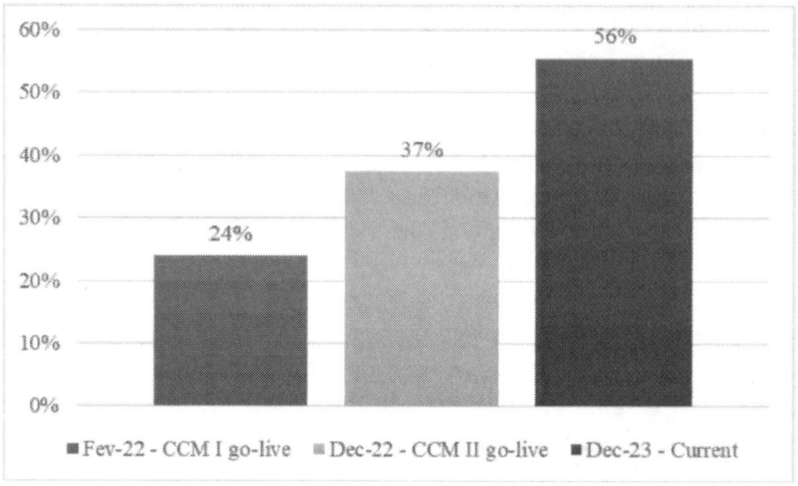

Figure 1. Evolution of MUEs' operative critical barriers association with assets.

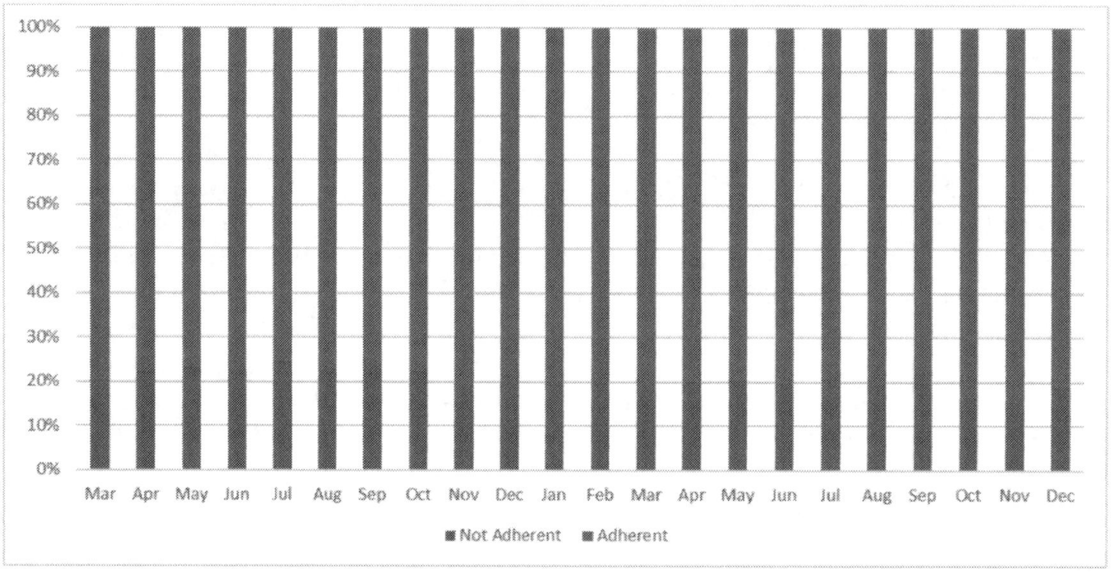

Figure 2. Results of automatic maintenance verifications for MUEs' operative critical barriers (2022 and 2023).

Despite the percentage of non-adherent protection layers seems to be kept at around 20%, not only it is possible to observe a downward trend but it is also worthy to highlight that it is a work in progress and the number of associations, therefore the volume of assets being assessed, are both increasing throughout the two years, as already presented in Figure 1.

Nowadays, the tool performs around 12,000 checks of operational barriers' maintenance status monthly. Despite being an expressive volume, impractical to achieve manually, it is yet lower than expected (see Figure 3 and Figure 4). A few gaps in the systems' interface are being investigated in order to solve the present absence of automatic verifications record for several controls, an issue presumed to be in the IT (Information Technology) context only.

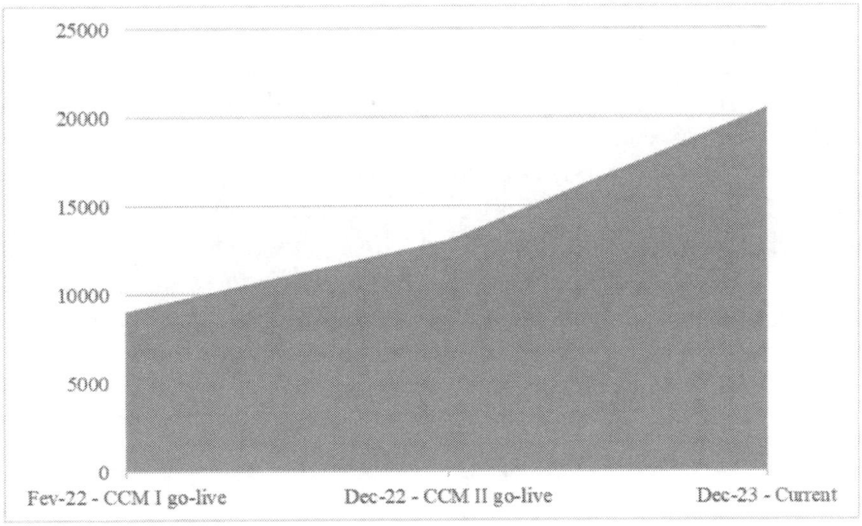

Figure 3. Expected number of automatic verifications of MUEs' operative critical barriers (in each indicated month).

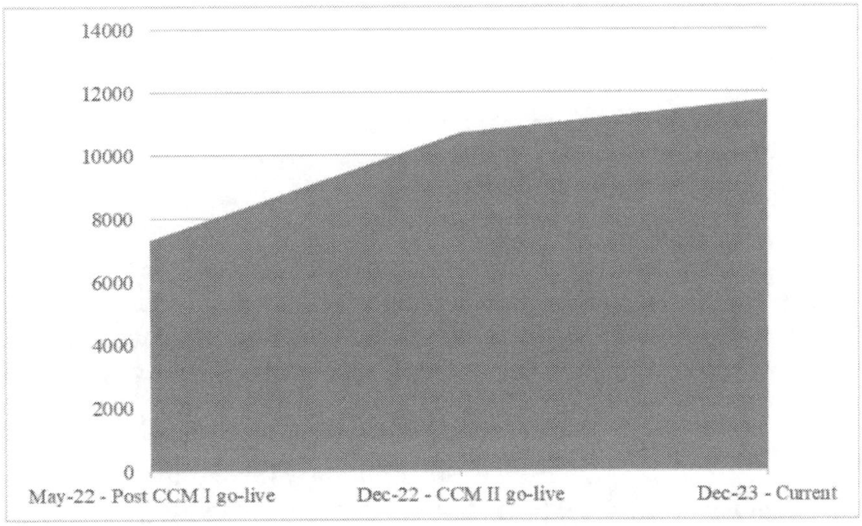

Figure 4. Actual number of automatic verifications of MUEs' operative critical barriers (in each indicated month).

Another opportunity of improvement is related to possible inconsistencies that may arise in the database due to criteria changing, updates in the risk register (which could change the control applicability if, e.g. the risk is eliminated or reduced to levels beneath the MUE's ones) or even errors. At the moment, most of those cases needs to be manually identified and/or addressed, requiring extra effort and attention. A constant cleanup of the database should be settled to eliminate or at least reduce any deviations that may arise.

Nonetheless, the implementation of the herein discussed automation brought significant gains in assertiveness, optimization and compliance in the barrier and asset management processes, especially from the point of view of integrity and reliability. In the second phase of the project, a more appropriate tool was made available for the definition of links between barriers and assets, with the required workflow for approval and log of all actions in the system. And once those connections are registered, the maintenance verifications of critical controls are accurately performed by CCM with findings properly indicated in portlets and reports, as configured in the first phase. Furthermore, saving time with ready automatic results, a better use of resources is attainable, with focus on more challenging tasks. Ultimately, the risk management as a whole benefits from the automatic and timely identification of possible non-conformities in the maintenance of assets, as it signals the need for corrective measures, as well as supporting the analysis of risk classification and the adoption of appropriate governance.

4 Conclusion

The present study aimed to delineate benefits of the dynamic monitoring of barriers to process safety and performance through a practical case implemented in mining industry, more specifically Vale S.A. The adopted approach is closely related to the risk management cycle, specifically for high-potential hazardous events and corresponding critical controls.

During the barriers' performance audit, the analysis of maintenance and inspection plans adherence can require a relatively high effort when carried out manually, besides being susceptible to failure which finally may lead to an inaccurate assessment and management of the related risks. In an attempt to optimize and bring more accuracy to the activity, CCM was developed as a way to automatically and assertively execute these checks, as well as constantly update the maintenance status related to each critical barrier mapped by the process safety studies. The strategy is based on the integration between online risk and maintenance management tools, enabling data exchange between systems, and periodic and automatic verifications of the maintenance condition of the applicable assets.

The results of CCM's checks allow timely signaling of vulnerabilities in the system with precision, data detail and traceability. Additionally, the association between assets and barriers of high severity risks, which is the basis of the systems' interface, was also used to support and automatically determine the prioritization of maintenance activities and escalation of leadership notification in cases of non-compliances.

This correlation between the systems' data is a working in progress, but more than 50% of critical controls related to operational risks with high potential impact have already been made available for automatic verification of the CCM. Nowadays, the tool performs around 12,000 operational critical controls checks monthly, a volume impractical to achieve manually.

The implementation of this automation brought significant gains in assertiveness, optimization and compliance in the barrier and asset management processes. The risk management as a whole benefits from the automatic and timely identification of possible non-conformities in the maintenance of assets, as it signals the need for corrective measures, as well as supporting the analysis of risk classification and the adoption of appropriate governance.

As the most immediate next steps towards continuous improvement, it is essential to solve the gaps in the tools' integration that are preventing the record of automatic verifications for several controls, as well as promote the constant cleanup of the database eliminating or at least reducing inconsistencies that may arise. Moreover, deepen the maintenance automatic analysis to also consider non-systematic orders, verifying their origin, deadlines, possible related MOCs (Management of Changes) etc. would expand CCM's scope of work in a very reasonable way.

Thinking ahead, an interesting advancement of the current approach would be to similarly include in the automatic evaluation further aspects related to the barrier management such as failure and triggering histories, bypass managing, setpoints configuration, personnel training, and incident records. In this manner, a broaden picture of the controls' situation would be assessed and all the advantages already mentioned would be extrapolated to those additional features.

Finally, a thorough analysis of the costs associated with the controls dynamic monitoring implementation and support compared to the gains obtained from it is another possible theme to be explored. It could assist the consolidation of the concept, as proposed by Qiao et al.[10], and facilitate in-depth studies and improved future methodologies.

5 References

[1] International Council on Mining & Metals (ICMM) (2015). Critical Control Management: Implementation Guide. https://www.icmm.com/website/publications/pdfs/health-and-safety/2015/guidance_ccm-good-practice.pdf?cb=60007. Accessed January 9th, 2024.

[2] International Council on Mining & Metals (ICMM) (2015). Health and Safety Critical Control Management: Good Practice Guide. https://www.icmm.com/website/publications/pdfs/health-and-safety/2015/guidance_ccm-implementation.pdf?cb=60006. Accessed January 9th, 2024.

[3] Esperandio, B. M. Hazard Identification and Risk Analysis at Mining Industry - Vale´s Experience. *AIChE Spring Meeting and GCPS* 2022.

[4] Center for Chemical Process Safety (CCPS), Energy Institute (EI). *Bow Ties in Risk Management: A Concept Book for Process Safety*. New Jersey John Wiley & Sons; 2018.

[5] Center for Chemical Process Safety (CCPS). *Guidelines for Risk Based Process Safety*. New Jersey John Wiley & Sons; 2007.

[6] Petroleum Safety Authority (PSA) Norway (2017). Principles for Barrier Management in the Petroleum Industry: Barrier Memorandum 2017. https://www.havtil.no/contentassets/43fc402b97e64a7cbabdf91c64b349cb/barriers-memorandum-2017-eng.pdf . Accessed January 28th, 2024.

[7] Hosseinnia B., Haskins C., Reniers G., Paltrinieri N. A Guideline for the Dynamic Barrier Management Framework Based on System Thinking. *Chemical Engineering Transactions* 2019. 77, 103-108. DOI: 10.3303/CET1977018.

[8] Yuan S., Reniers G., Yang M. Dynamic-risk-informed safety barrier management: An application to cost-effective barrier optimization based on data from multiple sources. *J. Loss Prev. Process. Ind.* 2023. 83, 105034. DOI: https://doi.org/10.1016/j.jlp.2023.105034.

[9] Karolius K. B., Psarros G. A., Astrup O. C., Liang Q., Van Welter C., Vassalos D. Maritime Operational Risk Management using Dynamic Barriers. *Ships and Offshore Structures* 2021. DOI: https://doi.org/10.1080/17445302.2021.1894028.

[10] Qiao W., Huang E., Guo H., Liu Y. and Ma X. Barriers Involved in the Safety Management Systems: A Systematic Review of Literature. *Int. J. Environ. Res. Public Health* 2022. 19, 9512. DOI: https://doi.org/10.3390/ijerph19159512.

MAIN CHALLENGES AND DEVIATIONS IN THE OPERATIONAL SAFETY MANAGEMENT OF OFFSHORE OIL AND GAS PRODUCTION UNITS IN THE DESIGN AND COMMISSIONING PHASE, IN THE VIEW OF THE BRAZILIAN REGULATOR

Thiago da Silva Ormonde
Agency for Petroleum and Natural Gas (ANP)
Av. Rio Branco, 65 – Rio de janeiro-RJ - Brazil
tormonde@anp.gov.br

Keywords: safety audits, commissioning phase, design phase, construction phase, main deviations, safety regulation, fpso, lifecycle.

Abstract

Since 2009, the Agency for Petroleum and Natural Gas (ANP) has carried out inspections of offshore production units during their operational phase to verify the effectiveness of the operators' safety management systems.

During these operational audits, serious and critical deviations (high severity) are identified, sometimes even leading to the shutdown of some platforms by the regulator.

Given the identification that many of these deviations identified in the audits were related to safety issues that were not adequately managed and resolved during the design and commissioning phase, the ANP understood that it should initiate an approach that was capable of previously identifying these weaknesses in the management systems in the initial phase of the units' life cycle. This preliminary audit was called "shipyard audit".

This work aimed to critically analyse the deviations identified in shipyard audits (carried out in the design and commissioning phase), in order to identify the main challenges in safety management faced by operators, that affect start-up and safety of the operations of offshore production units.

1 Technical safety regulation for Brazilian offshore (SGSO)

The operational safety audits carried out by the Brazilian regulator (ANP) on offshore units are governed by the agency's technical regulation called SGSO (Operational Safety Management System), created based on ANP Resolution Nº 43/2007.

This technical regulation establishes that installation operator must have an operational safety management system based on 17 (seventeen) management practices. The SGSO is not a prescriptive technical regulation, but rather a risk-oriented and performance-based regulation.

Therefore, what the Brazilian regulator expects in safety audits is an adequate demonstration of risk management on the part of the operator rather than mere compliance with prescriptive standards.

Table 1. Management practices from technical regulation (SGSO from ANP

Management Practices	Title
Nº 1	SAFETY CULTURE, COMMITMENT AND MANAGEMENT RESPONSIBILITY
Nº 2	STAFF INVOLVEMENT
Nº 3	PERSONNEL QUALIFICATION, TRAINING AND PERFORMANCE
Nº 4	WORK ENVIRONMENT AND HUMAN FACTORS
Nº 5	SELECTION, CONTROL AND MANAGEMENT OF CONTRACTORS
Nº 6	CONTINUOUS PERFORMANCE MONITORING AND IMPROVEMENT
Nº 7	AUDITS
Nº 8	INFORMATION AND DOCUMENTATION MANAGEMENT
Nº 9	INCIDENT INVESTIGATION
Nº 10	DESIGN, CONSTRUCTION, INSTALLATION AND DEACTIVATION
Nº 11	SAFETY CRITICAL ELEMENTS
Nº 12	RISK IDENTIFICATION AND ANALYSIS
Nº 13	MECHANICAL INTEGRITY
Nº 14	PLANNING AND MANAGEMENT OF MAJOR EMERGENCIES
Nº 15	OPERATIONAL PROCEDURES
Nº 16	MANAGEMENT OF CHANGES
Nº 17	SAFE WORK PRACTICES AND CONTROL PROCEDURES IN SPECIAL ACTIVITIES

2 Motivation for carrying out "shipyard" audits

The audits carried out by the ANP during the operational phase of offshore production units have consistently identify critical deviations that led to the shutdown of several units in the Brazilian offshore.

Below, we summarize some of the main deviations identified during the operation phase of these units that motivated the implementation of shipyard audits, as well as the continuation of this approach of auditing units even in an earlier phase of their life cycle, such as project and commissioning.

It is important to highlight that to this day, deviations are still being identified in the operation phase that feed back into ANP audits at the shipyard.

For analytical purposes, each deviation will be presented correlated to the type of platform (FPSO or fixed) and the unit's time in operation. For a correct reading of the table, the following parameter was established:

- Operational life stage (up to 3 years of operation): **Beginning;**
- Operational life stage (from 4 to 15 years of operation): **Middle;**
- Operational life stage (over 15 years of operation): **Final.**

Table 2 Main deviations identified in the operation phase of offshore production units

Deviation	How many times has it been identified	Platform type	Lifecycle stage
Failure to systematically assess risks			
Lack of assessment of the consequences of the H2S dispersion scenario	1	FPSO	Middle
Lack of assessment of the consequences of different explosion scenarios	1	FPSO	Beginning
	2	FPSO	Final
Lack of implementation of risk analysis recommendation			
Lack of implementation of non-tolerable scenario recommendation, identified in APR.	1	FPSO	Início
Lack of implementation of recommendations to adapt the F&G detection plant to the minimum acceptable requirements, through the installation of dozens of F&G detectors.	3	FPSO	Middle
Lack of implementation of recommendations for installing Passive Fire Protection on safety critical elements or on main structures to avoid collapse.	2	FPSO	Beginning
	3	FPSO	Middle
	2	FPSO	Final
	1	Fixed	Final
Design failure (system undersizing or commissioning failure)			
Undersized CO2 Fire Suppression System	1	FPSO	Middle
F&G detection system with low reliability, due to incorrect positioning of several detectors	2	FPSO	Middle
	2	Fixed	Final
Undersized open drainage system	1	FPSO	Beginning
	1	FPSO	Middle
	1	FPSO	Final
Lack of critical system established in the Safety Philosophy	1	FPSO	Middle
	1	Fixed	Final

3 Main difficulties faced by operators in managing operational safety, in the design and commissioning phase, in the regulator's view

Since 2015, a total of 24 audits have been carried out during the design and commissioning phase, involving 8 different installation operators. These audits were carried out before start-up, with the majority taking place at shipyards, while others were carried out on offshore location (also before start-up).

Below is a list of operators whose units have undergone audits during the design and commissioning phase (shipyard audit).

Table 2. Installation operators audited in the design and commissioning phase

Operator of Installation
Altera
Equinor
MISC
Modec
Perenco
Petrobras
SBM Offshore
Yinson

Therefore, in this topic, based on a preliminary analysis of these 22 shipyard audit reports from 2015 to 2023, containing 219 deviations identified by the ANP, we arrived at a substantial datas about safety management during this stage of the units′ safety life cycle.

Firstly, it is important to make a correlation between the identified deviations and the management practices outlined in the regulatory framework (SGSO).

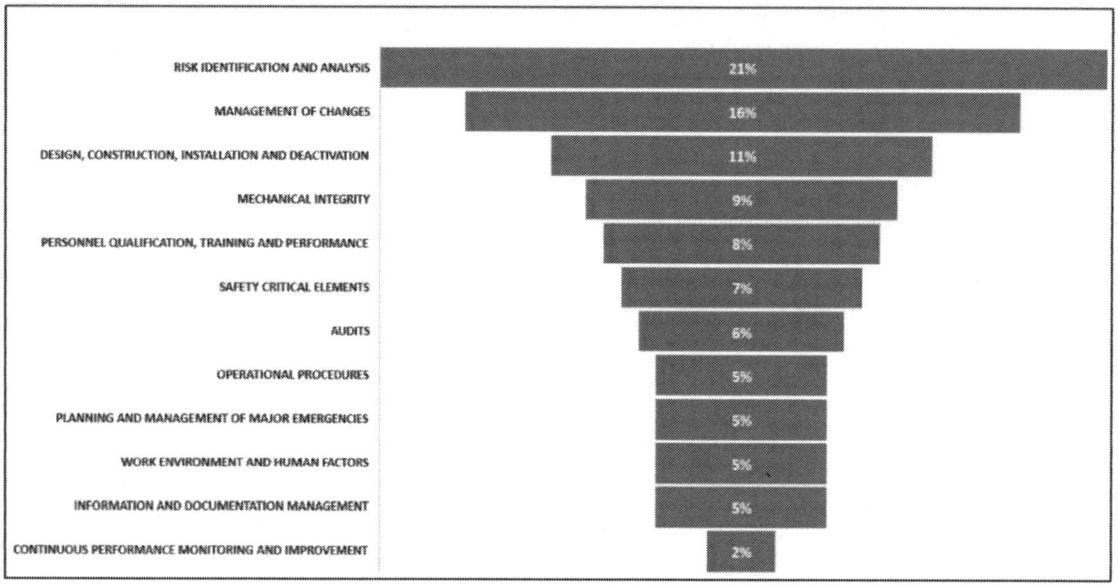

Figure 1. Deviations identified by the ANP classified by the SGSO management practices (MP)

Based on the provided data, it is possible to verify that failures in risk analysis and identification, management of changes and design/construction (MP 12, 16 and 10) correspond to approximately half of the deviations identified during site audits, totaling 48% of all deviations.

Considering the above, from this point onwards, the main deviations identified in shipyard audits in each of these management practices will be detailed.

All selected deviations were categorized as "high severity" by the regulatory body and required the implementation of corrective and preventive actions by the operators, before authorization for the start of operations on the units was granted by the regulator.

3.1 Main deviations identified by the ANP in site audits – Management practice 12 (risk identification and analysis)

Regarding risk identification and analysis, a matter that concentrates the largest number of deviations identified in shipyard audits, the deviations were divided into 4 (four) main topics.

In the following graph, it is possible to see that the greatest challenge faced by operators are in managing recommendations and adequately evaluating the identified scenarios.

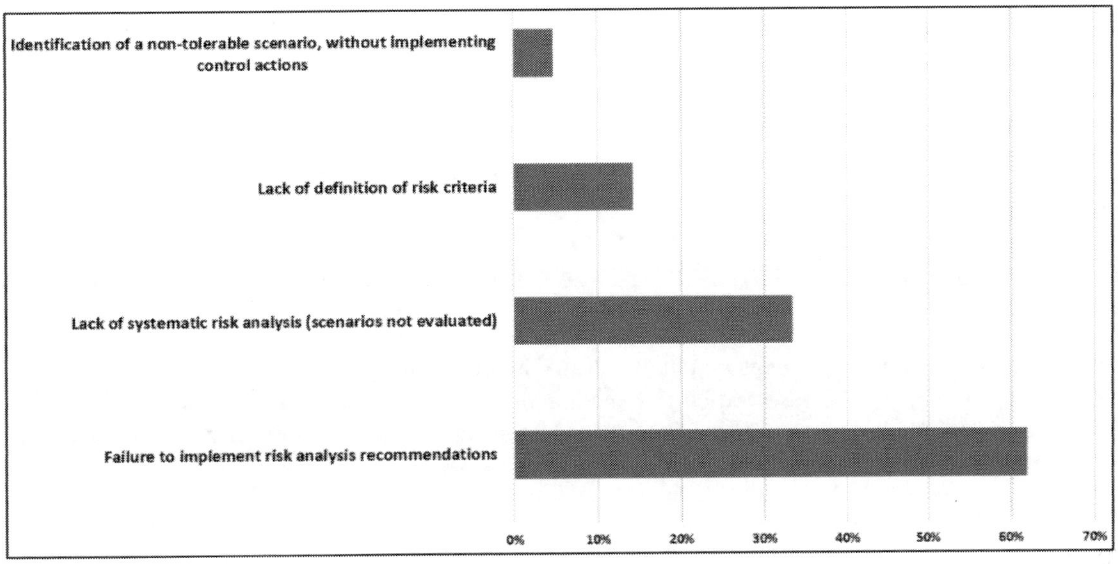

Figure 2. Percentage of units with deviations in the management of risk analysis

Listed below are some of themain deviations identified by the regulator (all of high severity).

3.1.1 Failure to implement risk analysis recommendations

Operators are expected to implement all risk analysis recommendations before unit start-up, as the tranfer of recommendations between life cycle phases (from shipyard phase to operation phase) brings greater risks to the operational phase.

Furthermore, the challenges of implementing recommendations from the design phase during the operational phase are much greater, as the number of personnel on board an offshore unit is significantly reduced, compared to the workforce dedicated to the shipyard phase. Additionally,, there is still the logistical challenge of making any physical changes to the process plant when the operation phase (offshore) begins.

3.1.2 Lack of systematic risk analysis (scenarios not evaluated)

In one-third of the audited units, failures were identified due to the lack of risk assessment of some identified scenarios, which were mandatory to be evaluated according to the specific requirements of the operators' management systems.

In some units, it was evident that simulations assessing the physical effects of overpressures (resulting from explosion studies) were not carried out for all topside modules where explosion risks were qualitatively identified. Consequently, these risks were not systematically evaluated in breach of the operator´s own management system requirements.

Another case observed in some units concerns the lack of simulations and analyses regarding the physical effects of pool fires and jet fires scenarios, due to a screening decision that was not based on best practices.

3.1.3 Lack of definition of risk criteria

In two different operators, failures were identified in the definition of risk criteria. In one case the deviation was related to the lack of definition of the criteria for the F/N curve (fatalities).

In another case, it was evident that the Operator failed todefine a minimum coverage criterion for the fire and gas detection systems.

3.1.4 Acceptance of study results with a non-tolerable scenario, without implementing control actions

Although this deviation only occurred once, due to the high criticality associated with it, it is important to mention it. It was verified that the operator's management system was not able to mitigate non-tolerable scenarios identified in their studies without any implementation of actions to reduce the risk to the acceptable level (moderate or low risk).

This deviation was related to the frequency of vessel collisions (calculated in the Ship Collision study) which was approximately 5 (five) times above the Operator's risk criteria (4.9E-03 versus 1E-03) without any measure being taken to address the intolerable risk.

3.2 Main deviations identified by the ANP in site audits – Management practice 16 (management of changes)

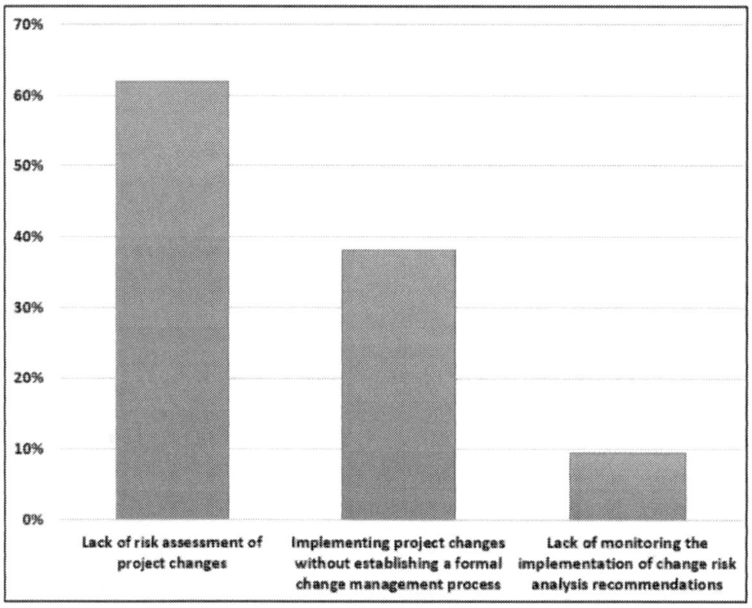

Figure 3. Percentage of units with deviations in management of changes

Occupying the second position in the number of deviations identified, failures related to project change management were categorized into 3 (three) different topics.

3.3.1 Lack of risk assessment of project changes

The most frequent type of deviation identified by the regulator is the failure to analyze the risks associated with changes made to the project.

It turns out that, even with the opening of a formal management of change process in the management system, operators often do not assess the safety risks of these changes. Instead only assess the financial and logistical impact on the project. In other words, the use of the management of change process is more focused on evaluating the impact on the project budget and costs and not on process risks.

This deviation was particulary common in the initial years of shipyard audits.

3.3.2 Implementing project changes without establishing a formal management of changes process

This type of deviation identified by the regulator concerns the absence of formal change management in the operator's management system, that is, it involves changes made to the project without any type of management.

It is known that from a moment in the design of a platform there is a milestone called "design freeze", after which any change to the project requires a formal evaluation and management process for the proposed change. However, this deviation reveals itself as one of the main failures among operators.

3.3.3 Lack of monitoring the implementation of change risk analysis recommendations

It is known that within a risk analysis of a change, recommendations may be generated.

However, as these recommendations are generated in the change management system and not in the same management system for recommendations from studies and formal risk analyses of the unit, in many cases the operators do not manage these recommendations (missing the closeout process).

3.3 Main deviations identified by the ANP in shipyard audits – Management practice 10 (design, construction, installation and deactivation)

3.3.1 Failure to meet design criteria (either by not following the provisions of the Design Safety Philosophy or by failure to commission safety critical systems/equipment)

Regarding project management, construction, assembly and commissioning, almost all of the platforms audited (82%) had deviations in these aspects.

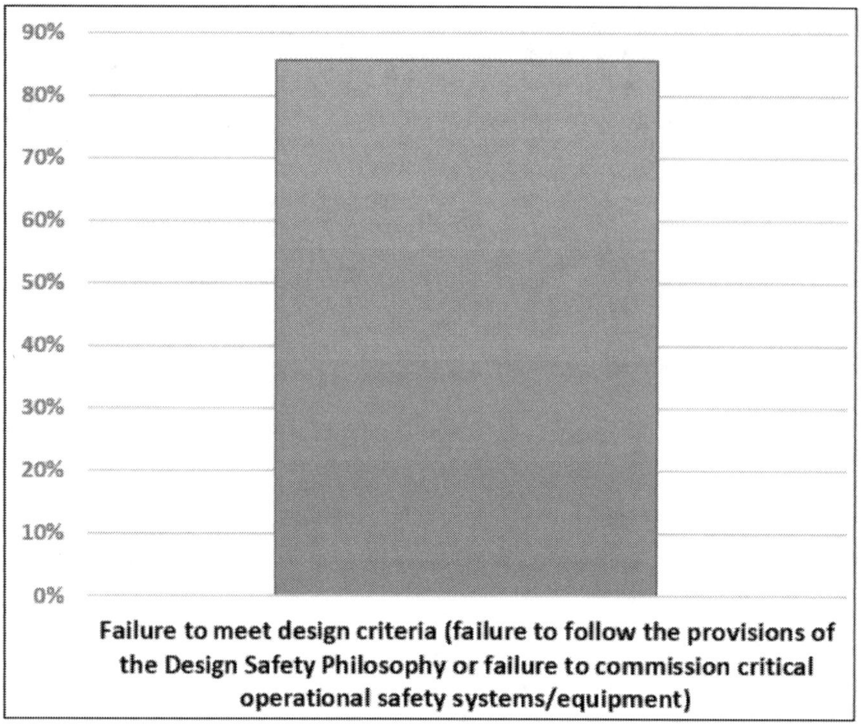

Figure 4. Percentage of units with deviations in project safety management and/or commissioning of critical systems

Below are some of the main deviations identified related to failures in the design or commissioning of the units.

- Lack of testing on safety-critical systems – This type of deviation significantly compromises the safety of operations, since by not testing a safety-critical system it is not possible to guarantee that it is meeting the required performance standard.

  One of the most serious cases identified by the regulator was the absence of a full discharge test for the fixed CO2 fire suppression system (for several compartments on the platform). After this deviation was noted in several FPSOs during shipyard audits, the test results identified serious degradations in the CO2 systems, such as undersizing of the system and excessive pressurization of the protected room (leading to critical failure of the room's sealing).

  Therefore, if these tests had not been carried out, operators would only discover these critical failures when this critical system (CO2) was demanded, resulting in the worst type of failure: failure on demand.

- Lack or undersizing of Safety Critical System – This type of deviation, although infrequent, is highly critical from a risk perspective.

One of the most serious cases ever seen was the absence of a drainage system for the well arrival area of an FPSO (turret area), which directly contradicted theISO 13702 standard and with the installation Operator's own design requirements.

It should be noted that the absence of a drainage system, precisely in the turret area, would result in an escalation of pool fire scenarios, as the burning oil would not be contained on that floor in the event of a loss of containment.

In another FPSO, the undersizing of the deluge system for a process vessel was evident. The water demand was undersized by around 40% of that required by the NFPA 15 standard.

- Transfer of a Safety Critical System from the commissioning phase to the operation phase with pending safety issues – This deviation is recurrent among audited units. Operators often face challenges in accurately classifying commissioning, construction and assembly pending items that are impediments to system operation. In this way, some systems are transferred to operation with impeding punch items that compromise the performance required for the system.

- Approval of safety-critical system performance test (during commissioning), however, without meeting the acceptance criteria – One deviation identified by the regulator in some FPSOs was the approval of the deluge system performance test with a flow rate below that required by the hydraulic calculation, without complying with the provisions of the calculation based on NFPA 15.

  The regulator has also identified the approval of a CO2 fire fighting system performance test without the system being capable of reaching the concentrations required by NFPA 12.

- Serious and systemic failure in the commissioning of the Fire & Gas system – This deviation was recurrent at the beginning of shipyard audits, especially in the first units used in the pre-salt region.

  In several of these units, serious deviations were identified in the fire and gas detection system, such as: (i) allocation of the detection plant (hundreds of detectors) in a position different from that recommended in the project study; (ii) Approval of the fire and gas detection system in the management system and its delivery for operationswith elements pending installation (detectors, junction boxes and cables).

- Failure to prepare the cause and effect matrix, with significant divergences in relation to the Safety Design Philosophy - In some FPSOs, cause and effect matrices (C&E matrix) were identified, prepared in disagreement with the design philosophy. For example, some matrices failure to account for the opening of the blowdown valves in the event of a confirmed fire in the process plant.

- Failure to lock valves in right position (LO/LC) – In several units, failures in controlling valve locking have already been evidenced, in disagreement with the engineering documentation.

4 Conclusion

From the critical analysis of the deviations identified in the shipyard audits carried out in the design and commissioning phase by the ANP, it is possible to conclude that failures in the management of identification and risk analysis, management of changes and design/construction/commissioning correspond to approximately half (48%) of the deviations identified.

This fact shows that operators have had recurring difficulties managing the execution and implementation of safety studies, management of changes and the construction, assembly and commissioning stages of safety critical systems. In other words, this demonstrates that from the initial phases of the installations' life cycle, operators already encounter difficulties in managing these management practices, just as these challenges are perpetuated during the operational life of offshore production platforms.

It is also interesting to note that the shutdown of offshore units, issued by the regulator, are mostly also due to deviations related to failures in the management of these SGSO management practices, as detailed in the ANP's Annual Safety Reports [2], which provide a comprehensive overview of the safety performance of offshore installations.

Thus, future work can address the correlation between the root causes of operational accidents and design deviations that were inadequately managed in the initial phase of the installations' life cycle. This approach aims to encourage operators to implement more effective actions in operational safety management during the design, construction and commissioning phase of the facilities.

References

[1] ANP. Resolução ANP n° 43/2007, Regulamento Técnico do Sistema de Gerenciamento de Segurança Operacional (SGSO) para Instalações Marítimas de Perfuração e Produção de Petróleo e Gás Natural. [Online]. Disponível em: http://www.anp.gov.br/?dw=603.

[2] Relatórios anuais de segurança operacional das atividades de exploração e produção de petróleo e gás natural no Brasil, 2010 a 2022. https://www.gov.br/anp/pt-br/assuntos/exploracao-e-producao-de-oleo-e-gas/seguranca-operacional/relatorios-anuais-de-seguranca-operacional. Accessed 31 January 2024.

[3] ISO 13702:2015, Petroleum and natural gas industries – Control and mitigation of fires and explosions on offshore production installations – Requirements and guidelines (2015);

[4] NFPA 15, Standard for Water Spray Fixed Systems for Fire Protection (2017);

[5] NFPA 12, Standard on Carbon Dioxide Extinguishing Systems (2008);

A Dynamic Process Safety Barriers Management: Challenges in the Governance of the Dynamic Monitoring Process at Petrobras' Business Unit

Keywords: Management of Change, Process Safety, Process Safety Culture, Dynamic Barrier Management, Reliability and Risk Analysis

Abstract

The Dynamic Barrier Management (DBM) allows for monitoring the integrity of preventive and mitigating process safety barriers at industrial plants.

This integrated and dynamic monitoring of information from several existing and mature systems and processes makes it possible to support the decisions of the plant's risk manager.

Since 2019, Petrobras has been carrying out actions to enable the implementation of Dynamic Management of Process Safety Barriers in its industrial facilities, as presented in previous paper references [3],[4], [5] e [6].

The Campos Basin Business Unit defined seven Exploration and Production Facilities (EPF) for the implementation of the process.

The challenge presented in this article is related to the implementation of a management process that monitors existing processes with the respective responsibilities and associated change management. Dynamic Barrier Management allows revisiting existing processes, contributing to a recurring improvement in the PDCA cycle of these primary processes and the DBM itself.

The current governance of the primary processes has not changed, but the governance of the DBM process is more clearly defined, allowing this new process to be included to support the industrial installation's risk manager.

1 Introduction

Dynamic process safety barriers management is a systematic monitoring of the integrity and reliability of preventive and mitigating barriers, defining responsibilities, and consolidating information across all levels of the corporation – operational, tactical, and strategic. This proactive process dynamically contributes to the prevention of possible accidental scenarios. [3]

Monitoring the status of process safety barriers in industrial facilities enables proactive masures to maintain their integrity and manage the risks of the facility.[3]

Since 2019, Petrobras has been implementing a Dynamic Process Safety Barrier Management System, systematic monitoring preventive and mitigating barriers status applying the bowtie methodology. [5]

Petrobras' Offshore Exploration and Production (E&P) Segment is organized into four business units (Búzios, Espirito Santo Basin, Santos Basin and Campos Basin), each with a set of Exploration and Production Facilities (EPF). Within the scope of this project, the Campos Business Unit (BU) has committed to the implementation of BDM in seven Exploration and Production Facilities (EPF) by 2025. This project, as presented in previous articles, presents a set of implementation phases in each industrial installation ranging from the availability of documentation to the assisted operation of DBM.

The post-implementation phase involves the development of a process to monitor the information consolidated in the management software that supports the DBM, identifying potential degradations and unavailability in the respective elements of the monitored processes.

DBM monitors processes already established at Petrobras each with a previously established governance. DBM is a new process that oversees these existing processes, giving risk managers the possibility of monitoring them in an integrated and dynamic way and take action to make any contingencies, if necessary.

The aim of this paper is to present the process of monitoring information obtained by DBM at the operational level with its consequences, highlighting tasks, responsibilities, and challenges in the implementation process.

Dynamic Barrier Management facilitates the reevaluation of existing processes, contributing to a recurring PDCA involving these primary processes and the DBM itself.

2 Development

2.1.Organizational structure

Petrobras' offshore E&P segment is governed by a structure consisting on 4 (four) business units (Santos Basin, Espirito Santo Basin, Búzios and Campos Basin), each with a set of Exploration and Production Facilities (EPF), as shown in Figure 1.

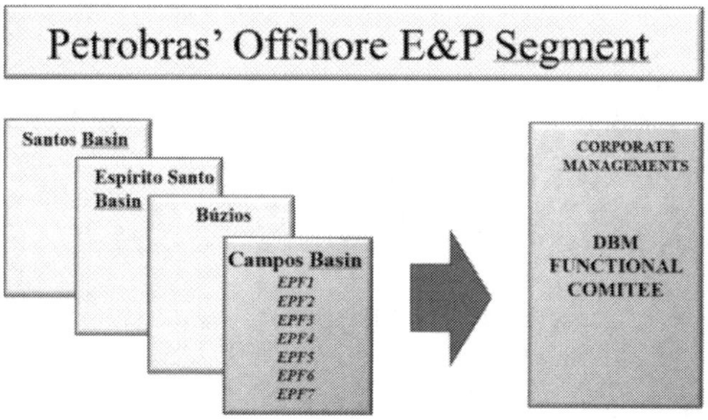

Figure 1 - Petrobras' Offshore E&P Segment

2.2. *Process diagram*

The process of monitoring the dynamic management of process safety barriers at Campos Basin Exploration and Production will be structured according to the process diagram- DBM, as shows in Figure 2.

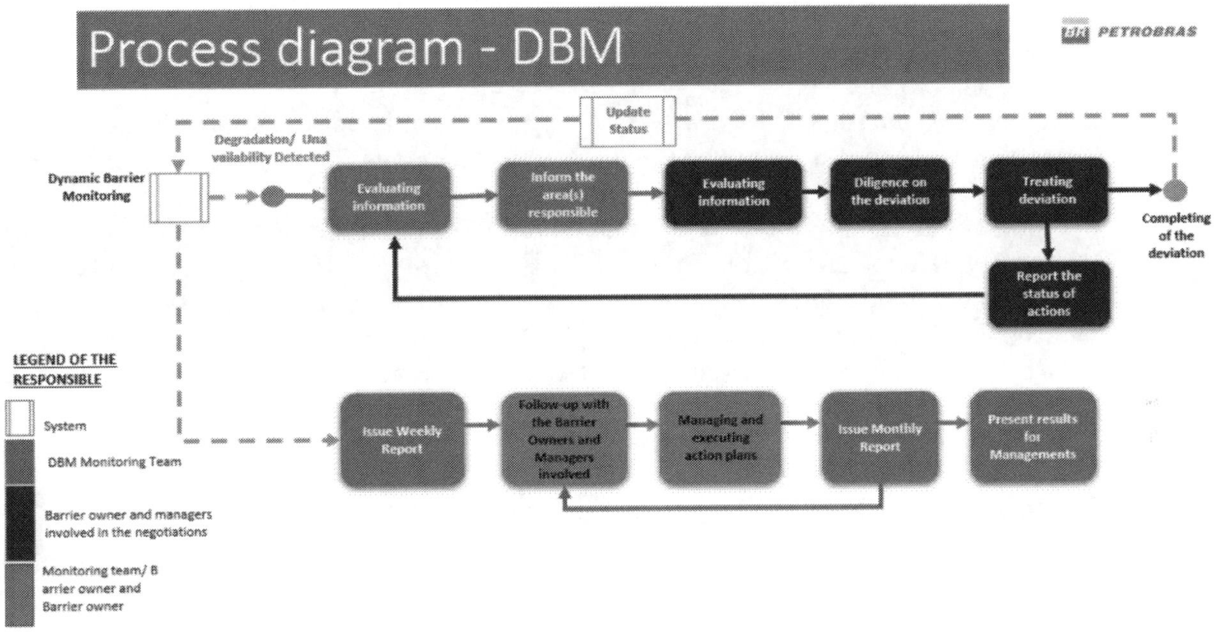

Figure 2– Process diagram –of the DBM at BU-Campos Basin.

Following is presented a schematic description of teams involved in DBM monitoring:

Light blue team → DBM monitoring team

- Activity: Conducts daily monitoring of the software tool that supports management, identifying deviations.

Dark Blue team → Barrier owners and managers involved.

- Activity: Addresses deviations upon request from the DBM Monitoring Team.

Green team → Collaboration between the Light blue team and Dark blue team.

2.2.1. Definitions Responsible persons and tasks:

DBM Monitoring Team

Evaluate the information: Access the data provided by the monitoring system to identify possible degradations or unavailability of safety functions. Update the the software with relevant information provided by the team responsible, as needed.
Notify responsible area(s): Identify which managements areas need to be informed and issue appropriate alerts.
Weekly Reporting: Issue a weekly report based on the information generated by the DBM software (Degraded Equipment and Common Elements Report), and send it to the relevant teams.
Monthly Reporting: Issue a monthly report based on the consolidated information from the weekly reports and validated by the teams involved. If necessary, conduct follow-up meetings with the DBM Monitoring Team, Barrier Owners and other stakeholders to provide feedback and support the process.
Present results to management: Present findings to the general manager and managers involved for deliberation. Upon approval by the General manager, disseminate results at operational, tactical and strategic levels (functional HSE committees).

DBM Monitoring Team, Barrier Owner and Management involved in the negotiations

Follow-up with the Barrier Owners and Managers: Conduct meetings to monitor the actions identified in the weekly report involving the DBM monitoring team, the Barrier Owners and relevant teams. Presennt a graph ilustrating the status evolution of safety elements totrack delegated actions thoughout the week.
Managing and executing action plans: Implement actions in the designated areas, managements, ordisciplines within the Operational Units. Monitor and report status weekly to the DBM monitoring team.

Barrier owners and managers involved

Evaluate information: Technically evaluate the information contained in the alert. Verify if any action has already been initiated or completed regarding the deviation.
Due diligence: Determine and plan the actions required to restore the integrity of the safety barriers in accordance with established standards.Define the teams responsible for carrying out the actions.
Execute planed actions: Carry out the planned action(s). Finalize the deviation and update its status accordingly.
Report status of actions: Provide and update on actions status and results achieved to the DBM monitoring team for updating records.

3 Results

3.1. Weekly Report

Every week, the owners of the barriers and the involved teams receives a status report on the barriers for critical analysis and progress monitoring of element degradation. This report is generated from the Degraded Equipment and Common Elements Report, which is updated using the DBM software, as samples in figure 3.

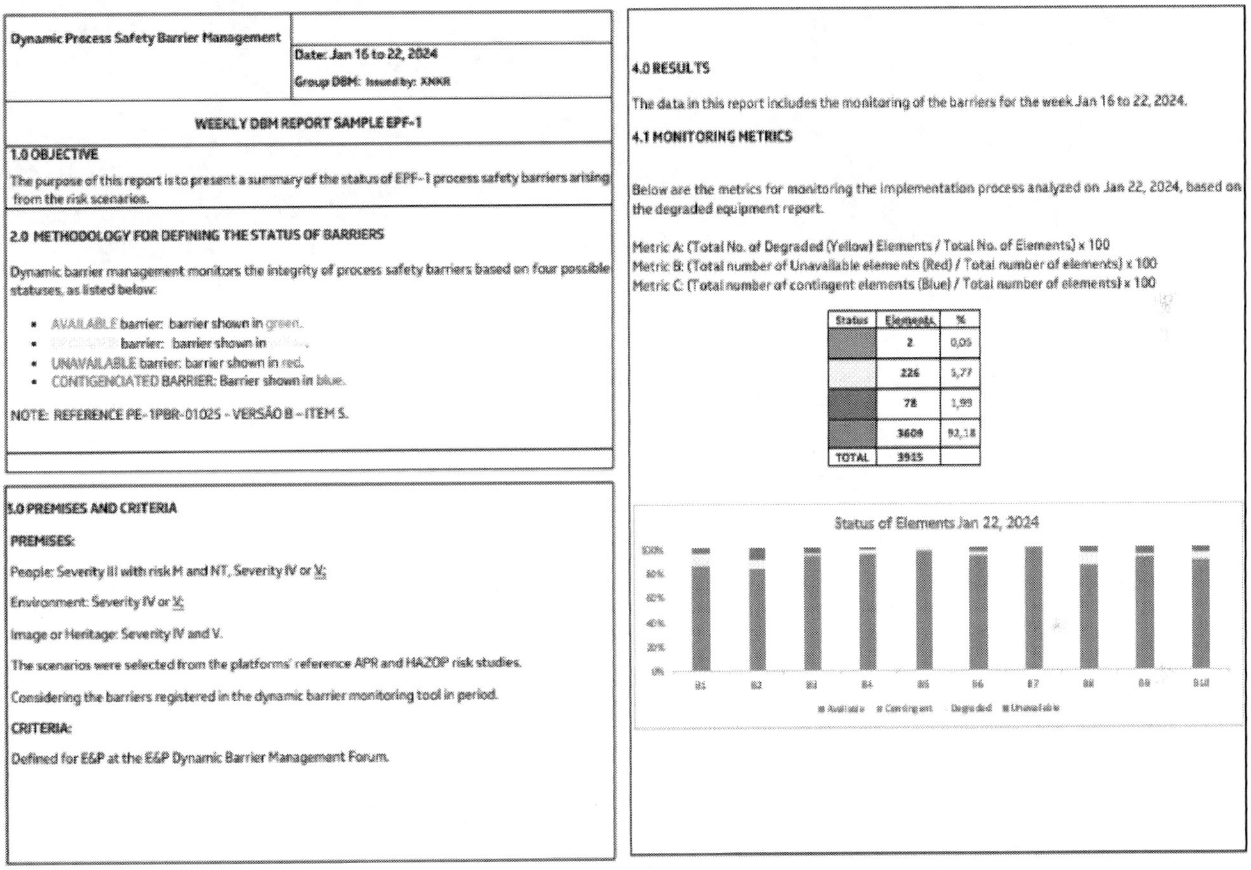

Dynamic Process Safety Barrier Management	
	Date: Jan 16 to 22, 2024 Group DBM: Issued by: XNKR

WEEKLY DBM REPORT SAMPLE EPF-1

1.0 OBJECTIVE

The purpose of this report is to present a summary of the status of EPF-1 process safety barriers arising from the risk scenarios.

2.0 METHODOLOGY FOR DEFINING THE STATUS OF BARRIERS

Dynamic barrier management monitors the integrity of process safety barriers based on four possible statuses, as listed below:

- AVAILABLE barrier: barrier shown in green.
- [illegible] barrier: barrier shown in [illegible].
- UNAVAILABLE barrier: barrier shown in red.
- CONTIGENCIATED BARRIER: Barrier shown in blue.

NOTE: REFERENCE PE-1PBR-01025 - VERSÃO B - ITEM 5.

3.0 PREMISES AND CRITERIA

PREMISES:

People: Severity III with risk M and NT, Severity IV or V;

Environment: Severity IV or V;

Image or Heritage: Severity IV and V.

The scenarios were selected from the platforms' reference APR and HAZOP risk studies.

Considering the barriers registered in the dynamic barrier monitoring tool in period.

CRITERIA:

Defined for E&P at the E&P Dynamic Barrier Management Forum.

4.0 RESULTS

The data in this report includes the monitoring of the barriers for the week Jan 16 to 22, 2024.

4.1 MONITORING METRICS

Below are the metrics for monitoring the implementation process analyzed on Jan 22, 2024, based on the degraded equipment report.

Metric A: (Total No. of Degraded (Yellow) Elements / Total No. of Elements) x 100
Metric B: (Total number of Unavailable elements (Red) / Total number of elements) x 100
Metric C: (Total number of contingent elements (Blue) / Total number of elements) x 100

Status	Elements	%
	2	0,05
	226	5,77
	78	1,99
	3609	92,18
TOTAL	3915	

Figure 3 - Weekly DBM Report sample EPF-1 - illustrative information

3.2.Barrier Status

For daily monitoring of the barriers and follow up with the involved teams, the DBM tool provides a color-coded view of the status of each one element and its availability on the barrier according to the degradation criteria, as shown in figure 4.

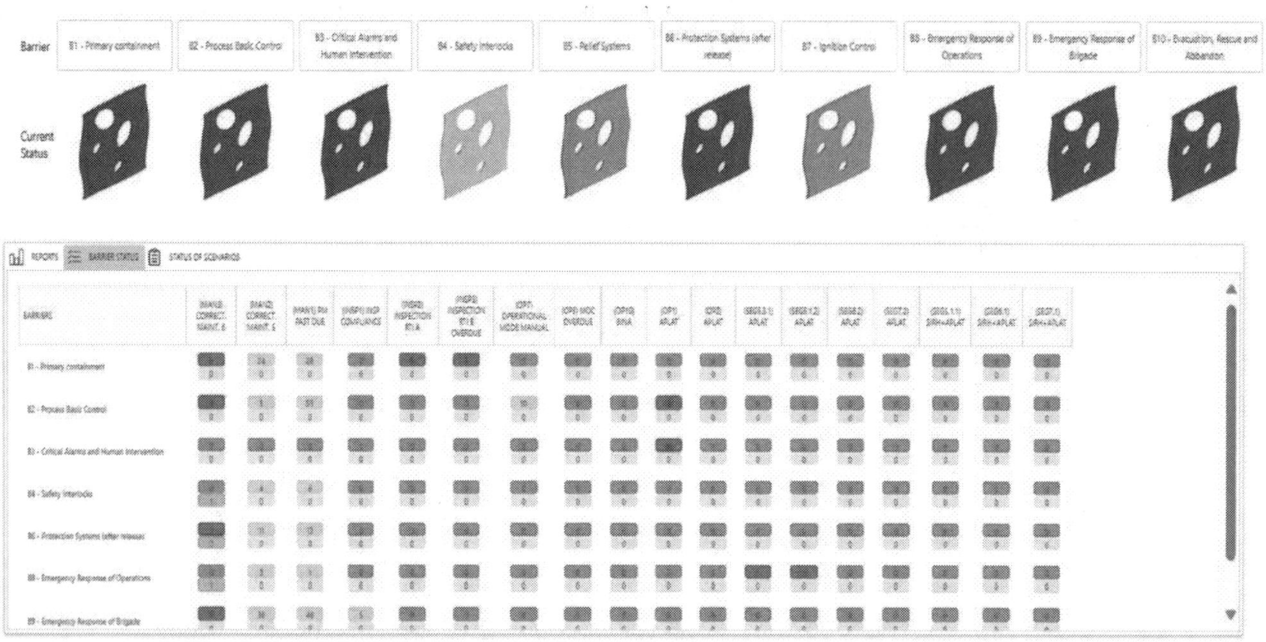

Figure 4: Barrier status of software DBM – illustrative information - https://petrobras.riskpoynt.com

3.3.Scenarios Status

The DBM tool has the option of displaying the barriers by scenarios originating from the bow-tie risk studies, with the display of colors according to their status, as shown in figure 5.

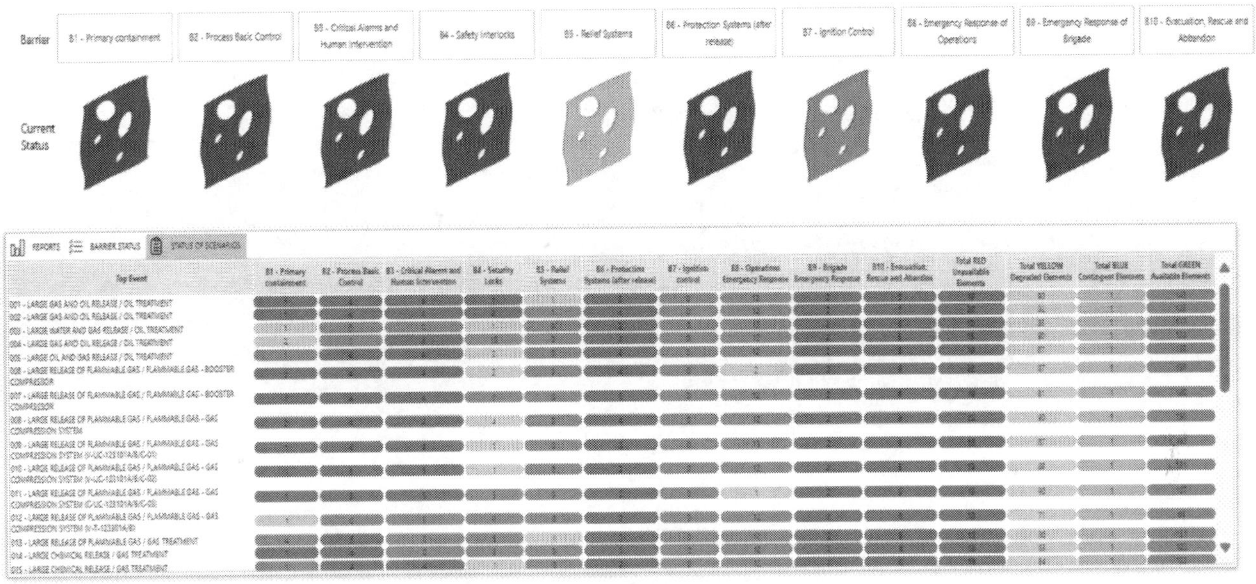

Figure 5 – Scenarios Status in the software DBM – illustrative information - https://petrobras.riskpoynt.com

3.4. Evolution Chart

During the follow up meetings with the owners of the barriers and the involved teams, the DBM monitoring team presents a weekly evolution chart illustration the status of unavailable, degraded and contingents' elements. This chart provides an overview of the actions that are taken during the meetings, allowing for effective tracking of progress and resolution efforts, as shown in figure 6.

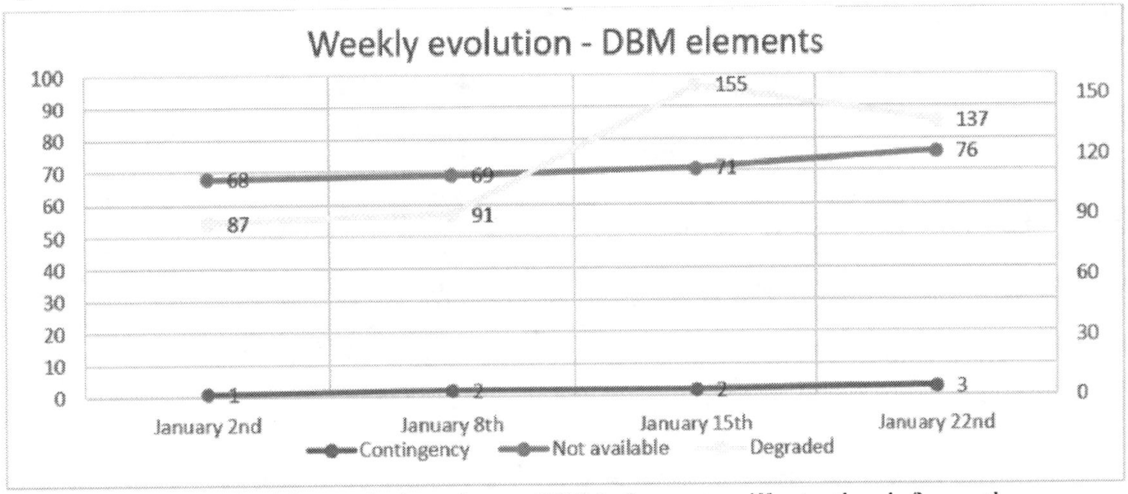

Figure 6: Weekly evolution chart - DBM elements - illustrative information

3.5. Monthly Report

On a monthly basis, this report will be presented to the managers involved, incorporating consolidated data from the weekly reports, showing the evolution of the elements of the process safety barriers. Once the results have been validated, the information will be disseminated to operational, tactical, and strategic levels, as shown in figure 7.

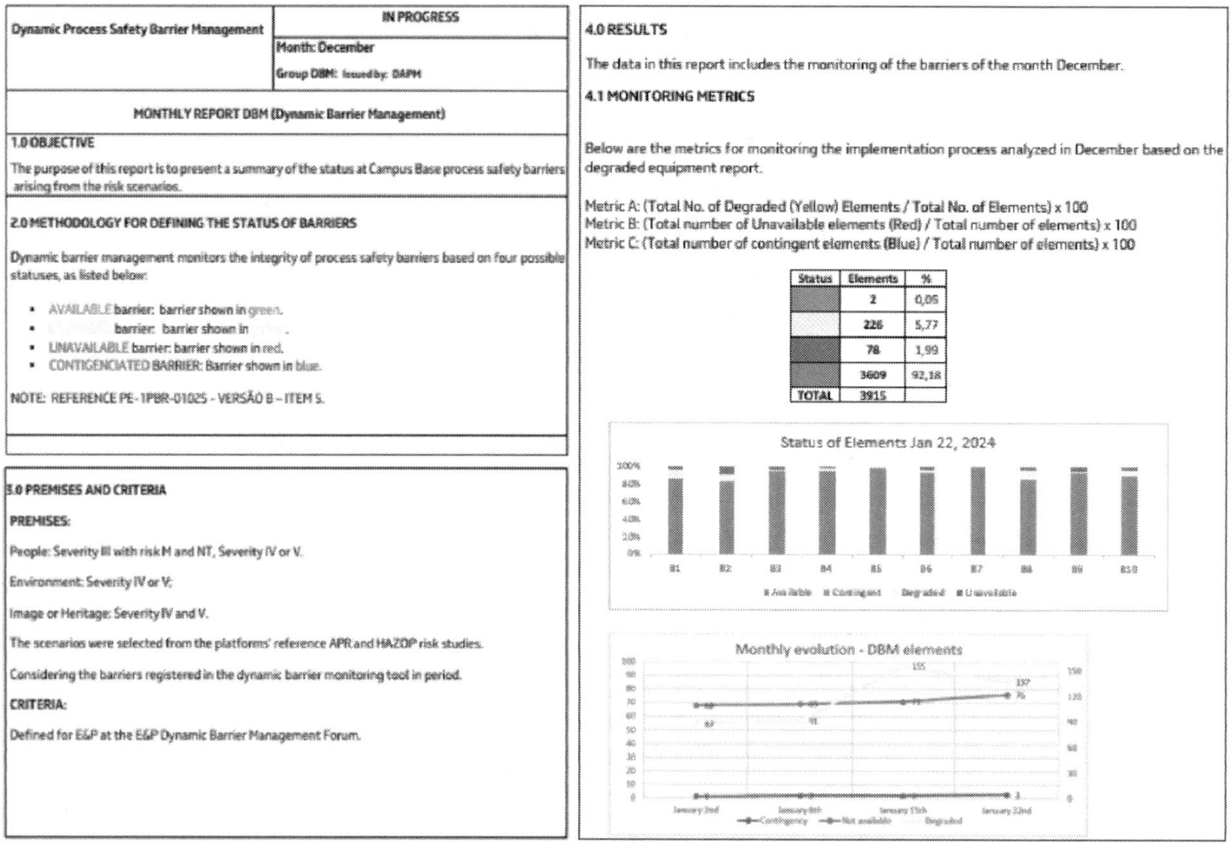

Dynamic Process Safety Barrier Management	IN PROGRESS
	Month: December
	Group DBM: Issued by: DAPM

MONTHLY REPORT DBM (Dynamic Barrier Management)

1.0 OBJECTIVE

The purpose of this report is to present a summary of the status at Campus Base process safety barriers arising from the risk scenarios.

2.0 METHODOLOGY FOR DEFINING THE STATUS OF BARRIERS

Dynamic barrier management monitors the integrity of process safety barriers based on four possible statuses, as listed below:

- AVAILABLE barrier: barrier shown in green.
- barrier: barrier shown in .
- UNAVAILABLE barrier: barrier shown in red.
- CONTIGENCIATED BARRIER: Barrier shown in blue.

NOTE: REFERENCE PE-1PBR-01025 - VERSÃO B – ITEM 5.

3.0 PREMISES AND CRITERIA

PREMISES:

People: Severity III with risk M and NT, Severity IV or V.

Environment: Severity IV or V;

Image or Heritage: Severity IV and V.

The scenarios were selected from the platforms' reference APR and HAZOP risk studies.

Considering the barriers registered in the dynamic barrier monitoring tool in period.

CRITERIA:

Defined for E&P at the E&P Dynamic Barrier Management Forum.

4.0 RESULTS

The data in this report includes the monitoring of the barriers of the month December.

4.1 MONITORING METRICS

Below are the metrics for monitoring the implementation process analyzed in December based on the degraded equipment report.

Metric A: (Total No. of Degraded (Yellow) Elements / Total No. of Elements) x 100
Metric B: (Total number of Unavailable elements (Red) / Total number of elements) x 100
Metric C: (Total number of contingent elements (Blue) / Total number of elements) x 100

Status	Elements	%
	2	0,05
	226	5,77
	78	1,99
	3609	92,18
TOTAL	3915	

Figure 6 - Monthly Report DBM (Dynamic Barrier Management) - illustrative information

4. Challenges

Barrier Dynamics Management is a relatively new process that monitors other existing processes in a company to support the risk manager's making decisions. Implementing the DBM process present several challenges, particularly in terms of Change Management. Some of the key challenges includes:

- **DBM Process Design** - Designing the DBM process requires an in-depth study of existing processes. It involves identifying relevant data related to each processes and their respective bases.

- **Responsibility Matrix -** Establishing a RACI (Responsible, Accountable, Consulted, and Informed) matrix for the DBM process requires the use of available human resources to new activities, in addition to their existing responsibilities. This decision could significantly impact the DBM-related tasks and responsibilities, requiring formal agreement with respective managers. Assessing the need for additional human resources to enable process implementation is essential.
- **Training** - The team involved in the DBM must undergo training in the software tool and the Bowtie methodology to effectively carry out their roles within the process.
- **Communication Plan** - A communication plan must be developed to keep all stakeholders informed.
- **Review of current processes** - Upon studying the existing processes, it may be necessary revisions due to inadequacies or to align them with the PDCA process. This situation can significantly impact on the DBM.
- **DBM PDCA** - Throughout the implementation and monitoring of the DBM process, opportunities for improvement may arise, such as new data, improvements in primary databases, and new processes that may require the revision of existing processes, degradation criteria, and/or development of new data integrations to increase the effectiveness of the DBM process.

5. Conclusion

The Dynamic Barrier Management enables an integrated and dynamic visualization of preventive and mitigating process safety barriers, along with their respective elements, systems, and degradation factor controls. Implementing DBM within a company involves multiple phases and should be viewed by managers as a process of continuous improvement, from the conception to the support.

The processes monitored by DBM are dynamic, as are the databases from which their information is extracted. Sustain DBM as a living and reliable process requires continuous evaluation of the processes it monitors. Any change to these processes, their databases or the quality and availability of data can significantly impact on DBM. Failing to address these changes may lead to a false perception of safety. Therefore, establishing DBM governance is crucial, including defining activities, responsibilities, and supporting tools, as well as a continuous critical analysis of the entire process.

This work has presented the challenges of implementing governance at the level of an Exploration and Production Facility (EPF) and its consequences for other facilities of this same Business Unit (BU).

Challenges related to establishing governance with the respective process designs and responsibility matrices at the tactical and strategic levels are in the construction phase for the entire company, involving the Exploration and Production, Refining, Gas and Thermoelectric Units, as well as the entire monitoring process at the corporate level. Construction of governance at tactical and strategic levels requires the evolution and maturation of DBM process at the operational level.

6 References

[1] CCPS. Guidelines for risk-based process. Center for Chemical Process Safety, American Institute of Chemical Engineers, New York, NY, 2007. Translation Petrobras/Recursos Humanos/Universidade Petrobras – 1 ed. Rio de Janeiro: Interciência. 2014; 872 p.

[2] CCPS. Bow Ties in Risk Management: A Concept Book for Process Safety. Center for Chemical Process Safety, American Institute of Chemical Engineers, New York, NY, and Energy Institute, London, UK, 2018.

[3] DANTAS, André Rodrigues. et al.: "Dynamic Management Implementation of Process Safety Barriers at Petrobras' FPSO Process Safety Barriers Management at Petrobras", 19Th GCPS – Texas – USA. 2023.

[4] DANTAS, André Rodrigues. et al.: **"Dynamic Process Safety Barriers Management at Petrobras",** 18 th Global Congress on Process Safety, San Antonio 2022.

[5] FIGUEIRA, Luciana Heil. et al.: **"Barrier Management, An Application for Events not related to loss of containment",** 17th Global Congress on Process Safety, Virtual Congress, Houston, 2021.

[6] FIGUEIRA, Luciana Heil. et al.: **"Process Safety at Offshore Aerodromes (Helideck) –New Application for Process Safety",** 19th AIChE Spring Meeting and15th Global Congress on Process Safety, in New Orleans, USA 2019.

[7] COSTA NETO, Felício P., RIBEIRO, Joao L. M., UGOLINO, Katia L. C. A., MINGRONE, Sérgio M., Safety Barriers Integrity Management System. **Chemical Engineering Transactions**, v. 36, p. 493-498, 2014.

[8] HUDSON, Patrick TW; GUCHELAAR, H.-J. Risk assessment in clinical pharmacy. **Pharmacy world and science**, v. 25, p. 98-103, 2003.

[9] BABU, J. Ramesh. Layer of protection analysis–an effective tool in PHA. **Cholamandalam MS Risk Services, Report,** 2007.

[10] KHAKZAD, Nima; KHAN, Faisal; AMYOTTE, Paul. Dynamic safety analysis of process systems by mapping bow-tie into Bayesian network. **Process Safety and Environmental Protection**, v. 91, n. 1-2, p. 46-53, 2013.

[11] LEWIS, Steve, SMITH, Kris, **"Lessons Learned from Real World Application of the Bowie Method".** 6th Global Congress on Process Safety, in San Antonio, 2010.

[12] OKSTAD, E. et al. "**Monitoring the risk picture by using QRA and barrier-based indicators. In: Safety, Reliability and Risk Analysis"**. CRC Press, 2013. p. 1739-1747.

[13] MOSTIA JR, W. B. L. "Got a risk reduction strategy?". **Journal of Loss Prevention in the Process Industries**, v. 22, n. 6, p. 778-782, 2009.

[14] CASSIDY, K. et al. The Cumulative Risk Assessment Barrier Model. In: **SPE Offshore Europe Oil and Gas Conference and Exhibition**. Society of Petroleum Engineers, 2011.

[15] OLIVEIRA, H.L.S et al. - Mixed application of CREAM & FTA to evaluate the non-technological barriers against the formation of explosive atmospheres in FPSO cargo tanks. In: 9th CCPS Latin American Conference on Process Safety 2022. Rio de Janeiro, 2022.

[16] DESTRI, A. C. et al. - Development of a Reliability and Maintenance Database System for Offshore Well Equipment in Brazil: MINERVA, Phase 1. In: Offshore Technology Conference Brasil. OTC, 2023. Rio de Janeiro, 2023.

[17] KAMPFFE, M. G. F. et al. - Estimating BOP Kill-Line Failure Probability Combining High level and Experts Opinion Informations. IEEE Latin America Transactions, v. 16, n. 8, p. 2199-2206, 2018.

[18] OLIVEIRA, H.L.S et al. - La Percepción de la Cultura de Seguridad de Procesos en la Industria de Petróleo y Gas Bajo la Visión de los Expertos en HSE. In: 8th CCPS Latin American Conference on Process Safety 2018. Buenos Aires, 2018.

[19] ALBUQUERQUE, T. M. C et al. - Diagnóstico de Aderência dos Elementos de Gestão de Segurança de Processo do E&P da Petrobras ao Sistema de Gestão de Segurança de Processo Baseada em Risco do CCPS. In: 2023 Spring Meeting and 19th Global Congress on Process Safety. Houston, March 12 - 16, 2023.

SPRING24
+20TH GCPS
A Joint AIChE and CCPS Meeting

March 24–28, 2024
New Orleans Ernest N. Morial
Convention Center
New Orleans, LA

SPRING24+20TH GCPS
A Joint AIChE and CCPS Meeting

March 24–28, 2024
New Orleans Ernest N. Morial
Convention Center,
New Orleans, LA

Process Safety Progress in China--2023

JIU Jiangbo, ZHAO Dongfeng, MENG Yifei
Mar.26,2024

SPRING24+20TH GCPS
A Joint AIChE and CCPS Meeting

March 24–28, 2024
New Orleans Hilton Riverside
New Orleans Convention Center

Content

- ☑ Two important working institutions related to process safety were established.
- ☑ One significant standard took effect and drove the improvement of multiple process safety efforts.
- ☑ Three important books related to process safety were published.
- ☑ Four influential and representative process safety conferences were held.

SPRING24+20TH GCPS
A Joint AIChE and CCPS Meeting

March 24–28, 2024
New Orleans Ernest N. Morial
Convention Center,
New Orleans, LA

Two important organizations dedicated to process safety

SPRING24+20TH GCPS
A Joint AIChE and CCPS Meeting

March 24–28, 2024
New Orleans Ernest N. Morial Convention Center,
New Orleans, LA

COSHA

China Occupational Safety and Health Association

It is the earliest established, largest, most influential, and prominent national and industrial social organization in the field of work safety, occupational health, and emergency management in China.

- **Establishment Date**: September 1983
- **Member Companies:** 26,000 (individuals)
- **Branches:** 63
- **Members of Expert Consultation Committee:** 250+
- **Expert Pool Members:** 2,000+
- **Professional Fields:** Nearly 20 fields including safety management, petrochemicals, fire prevention and explosion protection, dust and poison prevention, occupational health, community safety, etc.

SPRING24+20TH GCPS
A Joint AIChE and CCPS Meeting

March 24–28, 2024
New Orleans Ernest N. Morial Convention Center,
New Orleans, LA

COSHA has established two process safety-related working committees, and process safety-related work is carried out through these two committees.

★ Chemical Process Safety Management Working Committee

★ Chemical Safety Control Working Committee

Mission:

- Entroduce, digest, and absorb advanced concepts in chemical process safety management from both domestic and international sources
- Explore suitable methods for chemical process safety management that are tailored to China's national conditions and provide better technical support services for the government, industry, and enterprises in terms of chemical safety management.
- Promote the establishment of a standardized system for safety production with a focus on chemical process safety management within enterprises
- Enhance the scientific level of safety management in the chemical industry and facilitate the transformation towards proactive prevention in terms of security governance.

SPRING24+20TH GCPS
A Joint AIChE and CCPS Meeting

March 24–28, 2024
New Orleans Ernest N. Morial
Convention Center,
New Orleans, LA

COSHA has established two process safety-related working committees, which have added important driving forces to the improvement of process safety in China.

★ Chemical Process Safety Management Working Committee

★ Chemical Safety Control Working Committee

- Mr. Wang Haoshui, Vice Chairman of the China Occupational Safety and Health Association, has been elected as the Chairman of two working committees
- China University of Petroleum (East China) has been selected as the Secretariat for Chemical Process Safety Management Working Committee.
- These two working committees have almost encompassed most important organizations and top experts in process safety field within China.

SPRING24+20TH GCPS
A Joint AIChE and CCPS Meeting

March 24–28, 2024
New Orleans Ernest N. Morial
Convention Center,
New Orleans, LA

The official implementation of an important standard, which has driven multiple process safety enhancement efforts.

SPRING24+20TH GCPS
A Joint AIChE and CCPS Meeting

March 24–28, 2024
New Orleans Ernest N. Morial Convention Center,
New Orleans, LA

AQ

中华人民共和国安全生产行业标准

AQ/T 3034—2022

化工过程安全管理导则

Guidelines for chemical process safety management

AQ/T 3034-2010 12 elements

- based on OSHA 1910.119 Framework

AQ/T 3034-2022 20 elements

- based on CCPS RBPS Framework

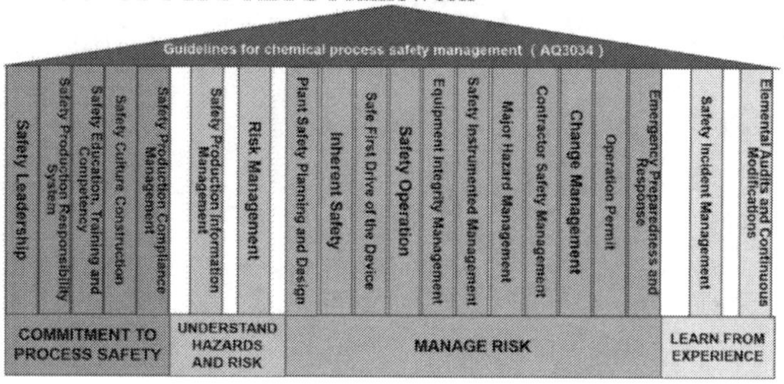

SPRING24+20TH GCPS
A Joint AIChE and CCPS Meeting

March 24–28, 2024
New Orleans Ernest N. Morial Convention Center,
New Orleans, LA

	CCPS RBPS	AQ 3034 RBPS
Commit to Process Safety		Safety Leadership
	Process Safety Culture	Safety Culture Construction
	Compliance with Standards	Safety Production Compliance Management
	Process Safety Competency	Safety Education, Training and Competency
	Workforce Involvement	Safety Production Responsibility System
	Stakeholders Outreach	----
Understand Hazards and Risk	Process Knowledge Management	Safety Production Information Management
	Hazard Identification and Risk Analysis	Risk Management
Manage Risk		Safety Planning and Design
		Major Hazard installation Management
		Inherent Safer
	Operating Procedures	Safety Operation
	Safe Work Practices	Work Permit
	Asset Integrity and Reliability	Asset Integrity Management
		Safety Instrument Management

SPRING24+20TH GCPS
A Joint AIChE and CCPS Meeting

March 24–28, 2024
New Orleans Ernest N. Morial Convention Center,
New Orleans, LA

	CCPS RBPS	AQ 3034 RBPS
Manage Risk	Asset Integrity and Reliability	Safety Instrument Management
	Contractor Management	Contractor Safety Management
	Training and Performance Assurance	Safety Education, Training and Competency
	Management of Change	Management of Change
	Operational Readiness	Safe Startup of New Facility
	Conduct Of Operations	Safety Production Responsibility System
	Emergency Management	Emergency Preparedness and Response
Learn from Experience	Incident Investigation	Incident Management
	Measurements and Metrics	Elemental Audits and Continuous Improvement
	Auditing	
	Management Review and Continuous Improvement	

SPRING24+20TH GCPS
A Joint AIChE and CCPS Meeting

March 24–28, 2024
New Orleans Ernest N. Morial Convention Center,
New Orleans, LA

I.

A comprehensive standard promotion campaign was carried out in China by various levels of government, major associations, institutions, enterprises, and social organizations

SPRING24+20TH GCPS
A Joint AIChE and CCPS Meeting

March 24–28, 2024
New Orleans Ernest N. Morial Convention Center, New Orleans, LA

II.
The pioneering implementation of enterprise safety management assessment in Shandong Province.

- In July 2023, the Shandong Provincial Safety Production Committee released the "Guidelines for Assessment Classification of Chemical and Hazardous Chemical Enterprises in Shandong Province (Trial)".
- This guideline is based on AQ3034 and provides assessment methods tailored to both large-scale enterprises and small-to-medium-sized enterprises in the chemical industry.
- The assessment mainly focuses on four aspects: completeness, effectiveness, execution, and execution results.

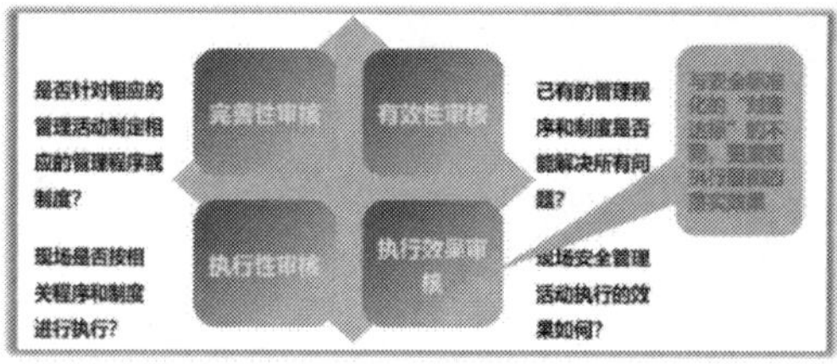

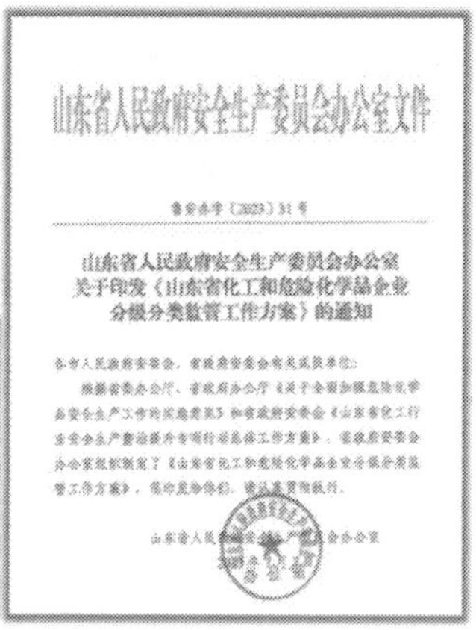
山东省人民政府安全生产委员会办公室文件

山东省人民政府安全生产委员会办公室
关于印发《山东省化工和危险化学品企业分级分类监管工作方案》的通知

SPRING24+20TH GCPS
A Joint AIChE and CCPS Meeting

March 24–28, 2024
New Orleans Ernest N. Morial Convention Center, New Orleans, LA

II.
The pioneering implementation of enterprise safety management assessment in Shandong Province.

Up to now:

★ Shandong Province Emergency Management Department selected 44 pilot enterprises to carry out process safety management level assessment and formulated standards and guidelines for process safety management assessment in Shandong Province.

★ Qingdao Emergency Bureau has already facilitated the completion of safety management assessments for 77 companies in Qingdao City.

- It establishes good practice in process safety management assessment
- It provides support for the construction and improvement of enterprise's process safety management system;
- The evaluation results from pilot enterprises demonstrate the systematic nature and superiority of AQ3034
- this assessment work is gradually expanding to multiple provinces and cities.

SPRING24+20TH GCPS
A Joint AIChE and CCPS Meeting

March 24–28, 2024
New Orleans Ernest N. Morial Convention Center,
New Orleans, LA

III.

The Ministry of Emergency Management is currently organizing the revision work for the safety production standardization rating system based on AQ 3034.

- Safety production standardization construction is a legal obligation for enterprises as defined in China's "Safety Production Law," and it serves as a key lever for the national promotion of enterprise safety production levels at a macro level.
- By incorporating the core content of AQ 3034 into the safety production standardization rating system, it facilitates the rapid dissemination of scientific RBPS in China.

Primary Elements of Standardized Production Safety:

- Objective Responsibilities
- Institutionalized Management
- Education and Training
- On-site Management
- Safety Risk Control and Hidden Danger Investigation and Governance
- Accident Management
- Continuous Improvement

SPRING24+20TH GCPS
A Joint AIChE and CCPS Meeting

March 24–28, 2024
New Orleans Ernest N. Morial Convention Center,
New Orleans, LA

The original safety standardization elements

- Objective responsibilities
- Institutionalized management
- Education and training
- On-site management
- Safety risk control and hidden hazard investigation and governance
- Emergency management
- Accident management
- Continuous improvement

The newly revised safety standardization elements

- Safety Leadership
- Safety Production Responsibility System
- Safety Production Information and Compliance Management
- Safety Education, Training, and Competency Building
- Risk Management and Dual Prevention Mechanism Construction
- Operational Safety
- Equipment Integrity Management
- Work Safety
- Contractor (Stakeholder) Management
- Chemical Safety, Major Hazard Sources, and High-Consequence Scenario Management
- Change Management
- Emergency Preparedness and Response
- Safety Incident (Accident) Investigation and Management
- Self-assessment and Improvement

SPRING24+20TH GCPS
A Joint AIChE and CCPS Meeting

March 24–28, 2024
New Orleans Ernest N. Morial Convention Center,
New Orleans, LA

The third development is the publication of three process safety books

SPRING24+20TH GCPS
A Joint AIChE and CCPS Meeting

March 24–28, 2024
New Orleans Ernest N. Morial Convention Center,
New Orleans, LA

Theory and Practice of Chemiccal Inherent Safety

- China's first specialized publication on inherent safety specifically tailored for the chemical industry.
- It systematically summarizes relevant theories on inherent safety in the chemical industry
- it presents specific case studies based on real practices from companies like Wanhua Chemicals to serve as practical examples for promoting inherent safety enhancements in chemical enterprises.

SPRING24+20TH GCPS
A Joint AIChE and CCPS Meeting

March 24–28, 2024
New Orleans Ernest N. Morial
Convention Center,
New Orleans, LA

Introduction to Process Safety for Undergraduates and Engineers (Chinese Version)

- It provides an overview textbook for undergraduate students majoring in chemical engineering and engineers in factories to understand process safety.
- It has received widespread praise from enterprises.
- It is a significant support for the widespread promotion of process safety in the entire industry.

SPRING24+20TH GCPS
A Joint AIChE and CCPS Meeting

March 24–28, 2024
New Orleans Ernest N. Morial
Convention Center,
New Orleans, LA

Dealing with Aging Process Facilities and Infrastructre (Chinese Version)

- The book describes a risk-based strategy in which plant leaders and supervisors can make more informed decisions on aging situations.
- The safe operation of aging equipment has long been a challenge for many Chinese companies.

SPRING24+20TH GCPS
A Joint AIChE and CCPS Meeting

March 24–28, 2024
New Orleans Ernest N. Morial Convention Center,
New Orleans, LA

Four significant conferences in the field of process safety

SPRING24+20TH GCPS
A Joint AIChE and CCPS Meeting

March 24–28, 2024
New Orleans Ernest N. Morial Convention Center,
New Orleans, LA

The 5th China International Symposium on Chemical Process Safety

- **Hosts:**
  - China Chemical Safety Association
  - International Cooperation Center of the Ministry of Emergency Management,
  - Chemical Registration Center of the Ministry of Emergency Management,
  - China Occupational Safety and Health Association,
  - China University of Petroleum (East China).
- **Date:** May 25th, 2023 - May 26th, 2023
- **Location:** Yantai, China
- **Theme:** New Technologies, Practices, and Solutions in Chemical Process Safety
- **Number of Keynote Reports:** 67
- **Number of Attendees:** 1200

The China International Symposium on Process Safety Management in the Chemical Industry

- **Theme:** PSM Best Practice Share
- **Organizers**: COSHA, UPC
- **Host:** COSHA CPSMWC
- **Date:** October 31, 2023
- **Location:** Shanghai
- **Number of Keynote Reports:** 21
- **Number of Attendees:** On-site 400+, Online: 30,000+

SPRING24+20TH GCPS
A Joint AIChE and CCPS Meeting
March 24–28, 2024
New Orleans Ernest N. Morial Convention Center,
New Orleans, LA

Seminar on Chemical Safety Control to Support Safe Production

- **Theme:** Continuously promote the transformation of hazardous chemical safety risk management towards automation, informatization, digitization, networking, and intelligence, aiming to enhance the intrinsic safety level of the chemical industry.

- **Host:** COSHA
- **Date:** June 28, 2023
- **Location:** Shanghai
- **Number of Keynote Reports:** 17
- **Attendees:** 180 technical experts from 40 industry enterprises

SPRING24+20TH GCPS
A Joint AIChE and CCPS Meeting

March 24-28, 2024
New Orleans Ernest N. Morial Convention Center,
New Orleans, LA

Process Safety Entrepreneur Salon

- **Theme:** Formation and Best Practices of Enterprise Safety Leadership
- **Organizers:** COSHA CPSMWC, UPC
- **Date:** March 16-17, 2024
- **Location:** Qingdao, Shandong Province
- **Participants:** 20+ chemical enterprise leaders from Qingdao and 26+ from other cities in Shandong Province

SPRING24+20TH GCPS
A Joint AIChE and CCPS Meeting

March 24-28, 2024
New Orleans Ernest N. Morial Convention Center,
New Orleans, LA

Thanks for your attention

SPRING24+20TH GCPS
A Joint AIChE and CCPS Meeting

March 24–28, 2024
New Orleans Ernest N. Morial
Convention Center,
New Orleans, LA

THE ROLE OF COMPANY MANAGEMENT IN ESTABLISHING A STRONG PROCESS SAFETY PROGRAM AND CULTURE

Scott Berger
President, Scott Berger and Associates, LLC
ScottBergerLLC @gmail.com
+1.609.462.5057

SPRING24+20TH GCPS
A Joint AIChE and CCPS Meeting

March 24–28, 2024
New Orleans Ernest N. Morial
Convention Center,
New Orleans, LA

What do you think?

What is our gap in knowledge of how to design, operate, and maintain safe chemical plants and petroleum operations?

What don't we know how to do?

SPRING24+20TH GCPS
A Joint AIChE and CCPS Meeting

March 24–28, 2024
New Orleans Ernest N. Morial Convention Center,
New Orleans, LA

I believe:

- In a "World without process safety Incidents"
- We can eliminate process safety incidents in less than 10 years
- We have all the knowledge we need to do this
- If we can't commit to managing our risks we should get out of the business
- Process Safety is everyone's job - from the Boardroom to the Frontline
- Process safety must be executed professionally and integrated into the business
- We must learn from all incidents and near-misses, and implement what we learn
- We must continuously improve our process safety management systems
- There is a Business Case for Process Safety.

SPRING24+20TH GCPS
A Joint AIChE and CCPS Meeting

March 24–28, 2024
New Orleans Ernest N. Morial Convention Center,
New Orleans, LA

Useful reference

- Authored for CCPS by Kenan Stevick and Scott Berger
- Written for executives and leaders at all levels

SPRING24+20TH GCPS
A Joint AIChE and CCPS Meeting

March 24–28, 2024
New Orleans Ernest N. Morial Convention Center, New Orleans, LA

Process Safety Leadership at a Glance

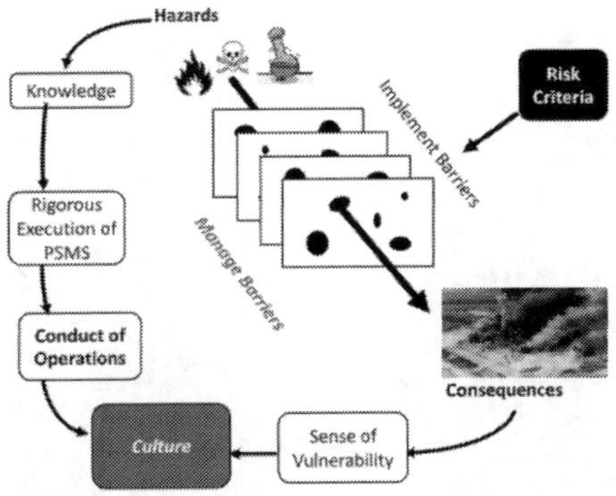

Leaders drive proactive culture & continuous improvement

SPRING24+20TH GCPS
A Joint AIChE and CCPS Meeting

March 24–28, 2024
New Orleans Ernest N. Morial Convention Center, New Orleans, LA

Hazards, Consequences, and Barriers

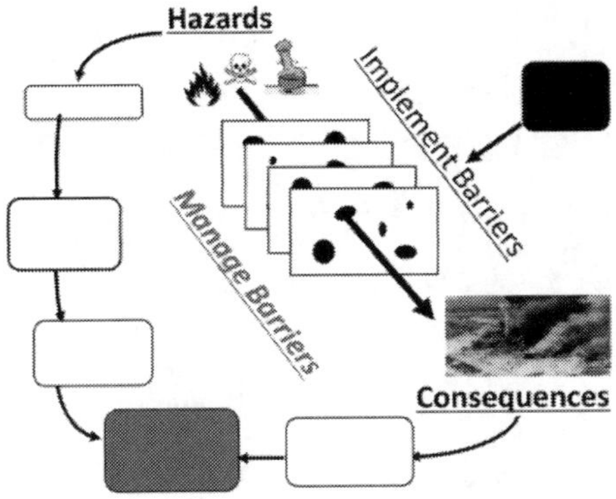

Leaders at all levels must know:

- The process safety hazards under their organization's control
- The potential consequences if these hazards are not managed professionally
- The barriers that exist to prevent these consequences
  - Senior leaders: Barriers = programs
  - Operational leaders: Barriers = actual safeguards
- That these barriers were implemented correctly and are functioning reliably

Risk Criteria

- Risk ≈ probability x consequence
- The risk to which processes must be designed, operated, and maintained
- Usually expressed as a risk matrix
- Most companies call this "PHA risk matrix"
- Understates the importance of risk criteria
- Should be "Board and senior leadership's risk matrix"
  - Risk that shareholders / owners and leadership are willing to accept

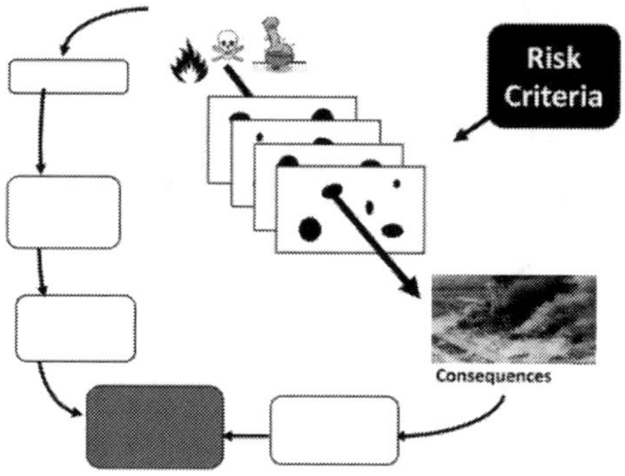

SPRING24+20TH GCPS
A Joint AIChE and CCPS Meeting
March 24–28, 2024
New Orleans Ernest N. Morial Convention Center,
New Orleans, LA

Example risk criteria

Example Risk Matrix

Consequence Frequency (Yr^{-1})	Level 1	Level 2	Level 3	Level 4	Level 5
> E-2					
9.9E-3 to E-3					
9.9E-4 to E-4					
9.9E-5 to E-5					
9.9E-6 to E-6					
< E-6					

Defines:

- Unacceptable range: shut down or do not start
- Target risk: a long-range target deemed generally acceptable
- One or more "ALARP" zones, where risk must be reduced "As Low As Reasonably Practical"
  - Cost and timing
  - Acceptance at the appropriate leadership level

Sense of Vulnerability

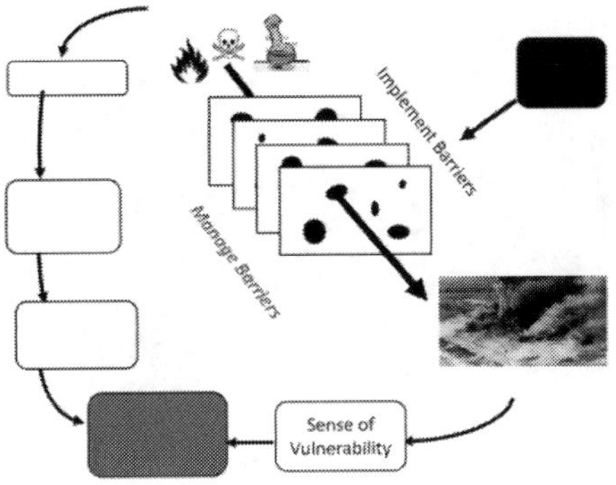

- Drive out and avoid complacency
- Potential consequences need to be felt as well as known
- Leaders must:
  - Act and make process safety decisions, always keeping in mind the potential process safety consequences
  - Drive that sense of vulnerability through the organization, so everyone does the same

SPRING24+20[TH]GCPS
A Joint AIChE and CCPS Meeting
March 24–28, 2024
New Orleans Ernest N. Morial Convention Center,
New Orleans, LA

Sense of Vulnerability:

How would you feel if…?

Knowledge, execution, and professionalism

Leaders must ensure:

- Everyone in the organization is **competent** to carry out their process safety responsibilities:
  - With knowledge, ability, and skills needed
- Every part of the Process Safety Management System has been designed and is being **executed rigorously** as intended
- Their activities, and activities at all levels of the organization are being **conducted** professionally and reliably, no short-cuts

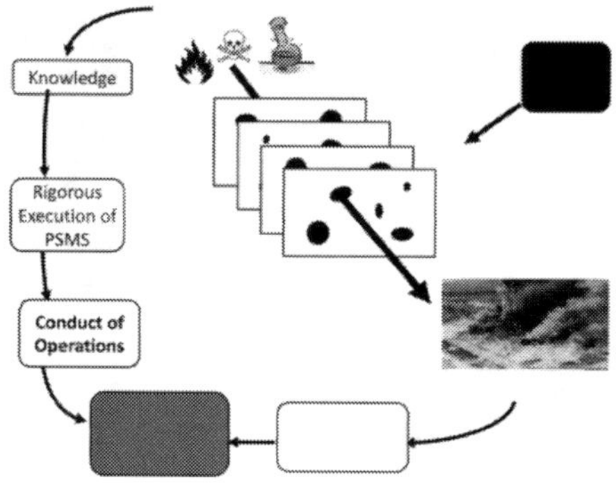

SPRING24+20TH GCPS
A Joint AIChE and CCPS Meeting
March 24–28, 2024
New Orleans Ernest N. Morial
Convention Center,
New Orleans, LA

The primary role of any leader is to drive culture

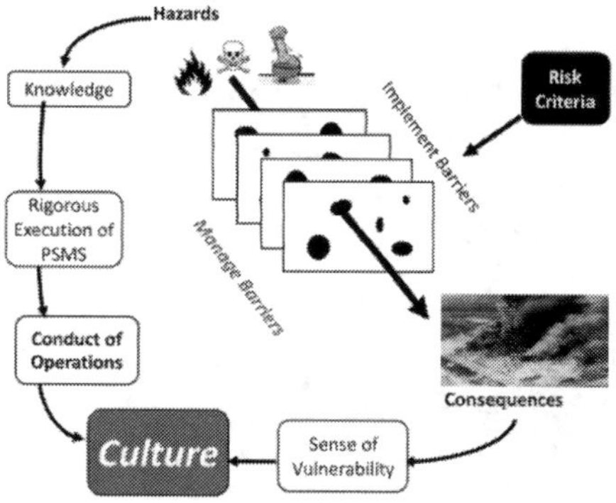

- The process safety culture of the organization is what leaders expect, measure, and reinforce through their own behaviors
- Risk criteria, to make clear what is unacceptable and what is the target
- Aggressive improvement goals
- Avoiding complacency
- Hiring and training competent personnel
- Executing the management system rigorously and professionally based on data

SPRING24+20TH GCPS
A Joint AIChE and CCPS Meeting

March 24–28, 2024
New Orleans Ernest N. Morial
Convention Center,
New Orleans, LA

But I'm only a process safety expert, what can I do?

- Communicate these concepts
- Help build the sense of vulnerability
- Promote PHA risk criteria to Leadership risk criteria
- Make a business case
- Support your position with data and professional analysis
- Demonstrate your personal commitment
- Be respectful, but firm
- Keep process safety visible

SPRING24+20TH GCPS
A Joint AIChE and CCPS Meeting

March 24–28, 2024
New Orleans Ernest N. Morial
Convention Center,
New Orleans, LA

Thank you!

Scott Berger
President, Scott Berger and Associates, LLC
ScottBergerLLC@gmail.com
+1.609.462.5057

Estrategias Para La Optimización Del Proceso De Revisión De Seguridad Previa a La Puesta En Marcha

Luis Andrés Landa
YPF S.A.
Complejo Industrial Luján de Cuyo
Luján de Cuyo – Mendoza - Argentina
luis.landa@ypf.com

Augusto Racca
YPF S.A.
Complejo Industrial Luján de Cuyo
Luján de Cuyo – Mendoza - Argentina
augusto.racca@ypf.com

Florencia Soledad Cesar Tita
YPF S.A.
Complejo Industrial Luján de Cuyo
Luján de Cuyo – Mendoza - Argentina
florencia.s.cesartita@ypf.com

Keywords: PSM, PSSR, Preparación para la operación, Revisión previa a la puesta en marcha, RPPM

Abstract

La experiencia ha demostrado que la frecuencia de accidentes es más alta durante transiciones del proceso, tales como la puesta en marcha. Esto lo podemos explicitar mostrando datos estadísticos de incidentes de seguridad de procesos categorizados por modo de operación, por unidad de tiempo. Si además consideramos las conclusiones de los investigadores de accidentes, donde a menudo identifican como un aporte significativo a la aparición de accidentes industriales mayores a una nula o pobre preparación para la puesta en marcha de las instalaciones, se concluye la importancia de realizar una correcta revisión previa a la puesta en marcha (PSSR, de sus siglas en ingles), para confirmar que todos los elementos apropiados de la gestión de la seguridad de los procesos se han abordado satisfactoriamente y que la instalación es segura para arranque.

El presente trabajo expondrá cómo YPF S.A. ha incorporado el elemento PSSR dentro de su PSM, sus antecedentes, su proceso actual (marco normativo, desarrollo del flujo de trabajo, sus hitos y lecciones aprendidas), y próximos desafíos.

1 Introducción

El proceso de Revisión de Seguridad Previa a la Puesta en Marcha se encuentra definido dentro del marco del sistema de Gestión de Seguridad de Procesos Basado en Riesgos desarrollado por el CCPS bajo el elemento Preparación de las Operaciones en el pilar Administrar Riesgos, y tiene como objeto principal asegurar la puesta en marcha segura de los procesos.

Si bien dicho elemento aborda todos los distintos tipos, condiciones y características de paradas de planta, las cuales serán mencionadas en el presente trabajo, su desarrollo estará enfocado a puestas en marcha de plantas luego de grandes intervenciones como paros programados por mantenimiento, modificaciones mayores, plantas nuevas e intervenciones mayores.

De esta forma, se describirá en el presente artículo cómo evolucionó la forma de verificar las condiciones de seguridad de las instalaciones antes de la puesta en marcha, en el Complejo Industrial Luján de Cuyo (CILC), perteneciente a YPF S.A., a partir de los últimos lineamientos corporativos y recomendaciones de organismos externos y los beneficios obtenidos.

2 Proceso de Preparación para la Operación

El proceso de preparación para la operación tiene una gran relevancia si uno mira la situación de vulnerabilidad en que se encuentra el proceso, en especial cuando fue modificado. Esto puede verse reflejado en datos estadísticos de la compañía, los cuales están en sintonía con los recopilados por fuentes externas [5]. Es por ello la necesidad de verificar y asegurar que estén dadas las condiciones de seguridad de la instalación antes de proceder con el ingreso de material al proceso.

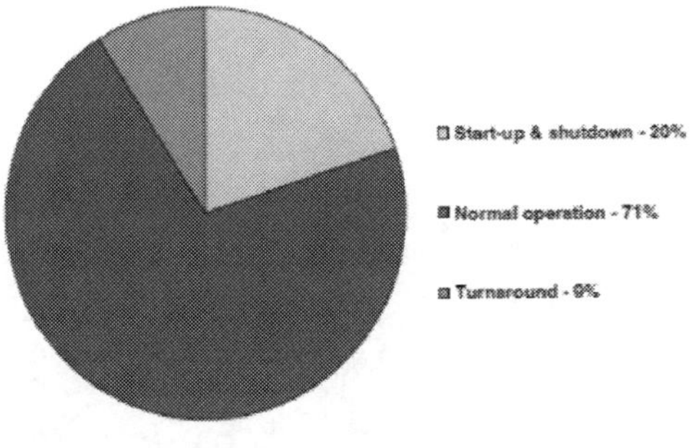

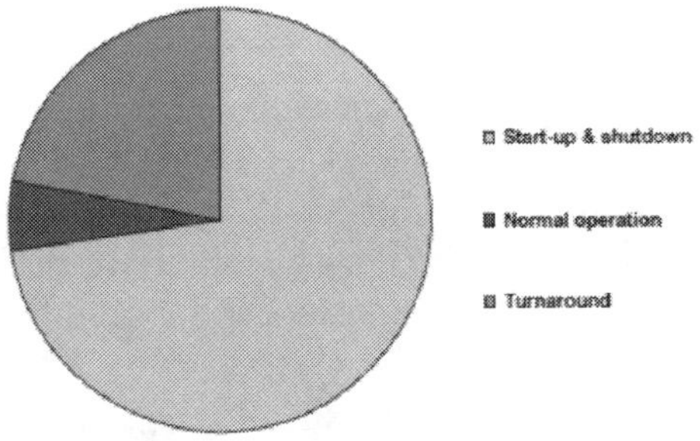

Figuras 1 y 2. % de eventos registrados por modo de operación y normalizado por tiempo

Lo que el proceso pretende asegurar es

- Confirmar que la instalación se encuentra lista para ser puesta en marcha de manera segura
- O una lista de las acciones que se deben tomar para que el proceso esté listo.

2.1 Antecedentes

Históricamente el proceso de preparación para la operación se realizaba en el CILC bajo un proceso poco integrado, no sistémico, con debilidades en la trazabilidad y formalidad. Su mención y desarrollo en el Sistema de Gestión de Excelencia Operativa de la compañía no tenía entidad propia sino que se encontraba referenciada indirectamente en otros elementos como *Identificación Operativa de Peligros / Aspectos y Evaluación de riesgos /Impactos*.

Dada la importancia del proceso, en el año 2019 se comienza a trabajar en un procedimiento específico de revisión para las plantas de Topping IV y Vacío integrado por Producción y Seguridad de Procesos, consolidándose una lista de chequeo base, la cual se utilizó como piloto durante los paros programados de dicho año, generándose una serie de lecciones aprendidas.

A partir de la publicación del procedimiento corporativo en el 2021, se comienza a trabajar en un proceso formal el cual se pone a prueba en los paros programados del 2022, de cuyo aprendizaje permitió la elaboración de un procedimiento específico del CILC en el 2023 y su aplicación en los paros programados e intervenciones de dicho año.

En el marco de la revisión del sistema de gestión de seguridad de procesos (PSM) impulsado por la compañía, se identifica la "preparación para la operación" como un elemento con entidad propia dentro del PSM.

2.2 Subprocesos de RPPM

Dado que el elemento Preparación para la Operación abarca los distintos tipos de paro, el proceso de revisión debe contemplar todas las necesidades según características de cada caso, como el tipo de la instalación y magnitud de modificaciones implementadas (planta nueva o modificación de existente), la naturaleza del proceso y sus peligros, el nivel de complejidad del proceso de puesta en marcha, el tiempo del paro (o estado fuera de servicio), entre otros.

En función de estas características, se requerirá

- un proceso de RPPM más extenso y exhaustivo para PEM más complejas (plantas nuevas, paros programados de mantenimiento, etc.) el cual necesita ser llevado adelante por equipos más numerosos y multidisciplinarios,

- procesos de RPPM de alcance acotado para intervenciones de nodos u operaciones unitarias,

- y procesos de revisión más simples para PEM pequeñas (entregas de equipos o accesorios por parte de mantenimiento, etc.), donde el equipo revisor puede estar acotado a 2 o 3 personas según tipo de actividad realizada.

De esta forma se definieron tres subprocesos, atendiendo a esta preclasificación.

2.3 Hitos del proceso RPPM

Sin tener la intención de desarrollar la metodología, se describen los hitos que permitieron lograr la mejora del proceso de RPPM en el CILC, fruto de recomendaciones y aprendizajes sobre el mismo.

2.3.1 Preparación previa

Surgida la necesidad de iniciar el proceso, una de las primeras actividades es la conformación del equipo revisor. Para ello la primera asignación es la del Líder de RPPM, quien definirá el equipo revisor, el alcance de la RPPM y desarrollará el programa general de la actividad.

Es importante que la definición del alcance se comience a tratar con la suficiente antelación como para asegurar que en caso de surgir actividades adicionales sean incluidas a tiempo en el alcance de paro. El alcance debe quedar claramente reflejado en la lista de verificación a utilizar.

Respecto a la conformación del equipo revisor, dependerá de las características del paro y puesta en marcha, como se indicó anteriormente, pero en grandes intervenciones es esperable la participación de al menos personal de Operaciones, Mantenimiento, Seguridad de Procesos, Ingeniería, a quienes se sumarán otras especialidades.

Básicamente la estructura queda:

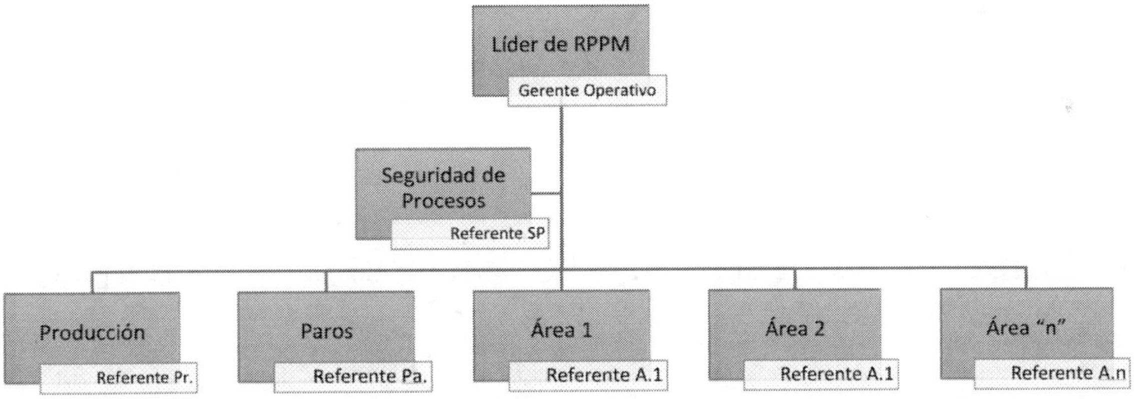

Siendo el líder de RPPM pertenece al área de operaciones, sector responsable de recibir y operar los activos que se deben poner en funcionamiento, es la responsable del aseguramiento de la aplicación de este proceso. Es quien debe definir los miembros del equipo revisor, el programa general (estrategia de reuniones y visitas a campo), la gestión documental y la aprobación de la puesta en marcha (o no) en función de la documentación recibida por el equipo revisor.

El referente de seguridad de Procesos asiste al líder en la conformación de la lista de chequeo y en el aseguramiento del conocimiento del proceso de los integrantes del equipo. Tiene la responsabilidad de mantener actualizada la planilla y definir la tipología de "condicionante" de aquellos ítems que por sus implicancias un desvío es capaz de generar un incidente de seguridad de procesos o afectar de manera directa un elemento crítico de seguridad de procesos. También es responsable de realizar las auditorías requeridas por el procedimiento corporativo y llevar indicadores del proceso.

Ambas figuras tienen la responsabilidad de aceptar o rechazar acciones de mitigación propuestas por el equipo revisor para cubrir posibles desvíos sobre ítems condicionantes (condición "y")

Cada integrante del equipo revisor tiene la responsabilidad de realizar la actividad asignada, registrando las conclusiones, dando el visto bueno o comunicando el pendiente o desvío, la acción correctiva, su resolución y/o acciones de mitigación y/o acciones con responsable y fecha comprometida de finalización para su aceptación.

2.3.2 Reunión de lanzamiento

El líder de la RPPM debe convocar a una reunión de lanzamiento del proceso con el objeto de afianzar el propósito de la actividad y coordinar actividades.

Se presenta la lista de verificación, revisando los puntos incluidos para asegurar su correcto entendimiento. Esto implica repasar el alcance de cada línea, la documentación de respaldo que se espera recopilar y su repositorio, las implicancias de posibles desvíos o hallazgos.

Además, se indica al equipo cuales ítems fueron categorizados como "condicionantes" de puesta en marcha. Esta condición implica que su no realización o el hallazgo de desvíos no permite la aprobación de la puesta en marcha de la instalación.

En la reunión se consolida la planificación de hitos, reuniones y visitas a campo. Es esperable que la reunión de lanzamiento incluya una visita a campo, para identificar allí aspectos contemplados en la revisión.

2.3.3 Desarrollo de la revisión de seguridad

En esta etapa el equipo revisor trabaja sobre la lista de chequeo a medida que avanza el paro y la preparación para la PEM.

Se realizan reuniones y visitas a campo periódicas para el seguimiento de la revisión, presentando avances, potenciales amenazas y hallazgos que pudieran haberse identificado para su tratamiento a tiempo.

En el caso de cierre de actividades críticas o tratamiento de hallazgos condicionantes se coordinan visitas a campo o sala de control para la confirmación visual.

Dentro del programa se define una fecha como reunión multidisciplinaria, la cual se establece en un momento avanzado del paro pero con un margen tiempo para dar tratamiento a pendientes y hallazgos que puedan amenazar la puesta en marcha de la planta.

En dicha reunión se repasan todos los ítems, definiendo las acciones que deben ser implementadas antes de la PEM (denominadas acciones Tipo A) y acciones que pueden darse cumplimiento luego de la PEM (denominadas acciones Tipo B).

Tanto las acciones de tipo A como las B deben tener un responsable y fecha de finalización, asegurando que las "A" estén en el período entre esta reunión y la fecha programada de PEM.

A partir de esta reunión, las reuniones de seguimiento son diarias y se pone foco en el tratamiento de acciones tipo A. En estas reuniones diarias la asistencia es obligatoria para quienes están involucrados en las acciones y en la programación del paro, dado que el no tratamiento de alguna de ellas puede implicar la postergación de la fecha programada para la PEM.

2.3.4 Aprobación de la puesta en marcha

Una vez que se ha asegurado la revisión de todos los puntos, cerradas todas las acciones Tipo A y concluido que están dadas las condiciones de seguridad para proceder a la PEM, el equipo revisor pone a disposición el informe elaborado y la documentación soporte al Líder de la RPPM, quien formalmente dará aprobación correspondiente.

2.3.5 Cierre del proceso

El proceso se cierra con el archivo documental. El informe se adjunta en la herramienta corporativa (Módulo *Informe Seguridad y Medioambiente*) y se cargan las acciones tipo B con los responsables y fechas comprometidas previamente.

Finalizado el proceso de RPPM, y pasado el período de revisión post puesta en marcha, se coordina un proceso de auditoría para verificar que se ha implementado el proceso de acuerdo con la normativa de la compañía.

Los KPI definidos para el elemento "Preparación para la Operación" de nuestro PSM son medidos y reportados para asegurar la madurez del proceso.

Luego de cada proceso de revisión, auditoría y análisis de indicadores, se realiza una actividad para identificar lecciones aprendidas y oportunidades de mejora para retroalimentar el procedimiento específico y en particular la lista de chequeo.

3 Desafíos y Mejoras derivadas de paros programados 2023

En el 2023 se desafió el nuevo procedimiento específico con los paros programados del año, los cuales se logró implementar en tiempo y forma.

Las dos intervenciones más grandes corresponden a el paro de ciertas unidades del área de Upgreading (pantas Isomax, Hidrógeno I y Gascon I) y al paro del área de Refino (plantas Topping III y Coque I).

3.1 Dimensionamiento de las actividades realizadas

A continuación se listan referencias para dimensionar la actividad de ambas intervenciones:

Paro programado de instalaciones del área Upgrading

- Inducción del proceso

  2 sesiones previas al inicio del proceso de RPPM para nivelar conocimientos, roles y responsabilidades de participantes directos e indirectos (3 horas).

- Reunión previa

  1 reunión con Líder de RPPM, jefe de planta y referente de seguridad de procesos - Alcance y presentación de ítems condicionantes (2 horas)

- Reunión de lanzamiento

  1 reunión de lanzamiento con el equipo RPPM (3 horas)

- Reuniones y consultas de seguimiento durante el paro

  9 reuniones formales (presenciales y remotas), sin considerar reuniones/consultas diarias con referentes (1,5 horas promedio por reunión)

- Reunión multidisciplinaria en gabinete y visita a campo previa a la PEM

  1 Reunión multidisciplinaria final presencial, estado de RPPM (4 horas – 19 personas)

  Visita en campo con el equipo RPPM para la verificación de pendientes y cierres comentados.

- Reuniones de seguimiento de líneas condicionantes y acciones tipo A

  Se identificaron 26 Acciones Tipo "A"

  7 reuniones presenciales con áreas involucradas para el cierre de condicionantes (1 hora promedio cada reunión)

- Recopilación de documentación soporte

- Carga de informe y acciones tipo B en SIGEO

  Se identificaron y registraron 34 Acciones Tipo "B"

- Auditoría del proceso, coordinación de actividades con Jefe de Calidad.

Paro programado de instalaciones del área Refinación

- Inducción del proceso

  1 sesión previa al inicio del proceso de RPPM para nivelar conocimientos, roles y responsabilidades de participantes directos e indirectos (1,5 horas).

- Reunión previa

  3 reuniones con Líder de RPPM, jefe de planta y referente de seguridad de procesos - Alcance y presentación de ítems condicionantes (2 horas total)

- Reunión de lanzamiento

  1 reunión de lanzamiento con el equipo RPPM (3 horas)

- Reuniones y consultas de seguimiento durante el paro

  5 reuniones formales (presenciales y remotas), sin considerar reuniones/consultas diarias con referentes (1,5 horas promedio por reunión)

- Reunión multidisciplinaria en gabinete y visita a campo previa a la PEM

  1 Reunión multidisciplinaria final presencial, estado de RPPM (3 horas – 15 personas)

  Visita en campo con el equipo RPPM para la verificación de pendientes y cierres comentados.

- Reuniones de seguimiento de líneas condicionantes y acciones tipo A

  Se identificaron 25 Acciones Tipo "A"

  5 reuniones presenciales con áreas involucradas para el cierre de condicionantes (1 hora promedio cada reunión)

- Recopilación de documentación soporte

- Carga de informe y acciones tipo B en SIGEO

  Se identificaron y registraron 21 Acciones Tipo "B"

- Auditoría del proceso, coordinación de actividades con Jefe de Calidad.

3.2 Aspectos positivos identificados en el proceso 2023

- Liderazgo y valoración del proceso por parte de la Gerencia, el Equipo RPPM y sus Colaboradores.

  Clave para poder realizar el proceso en tiempo y forma, producto de actividades derivadas del programa de cultura y formación.

  Fuerte impulso desde áreas corporativas de la compañía y liderazgo del CILC.

  Se definió como líder de RPPM al gerente operativo. El referente de cada área / especialidad corresponde a una jefatura, independientemente del equipo de trabajo que lo respalda. La necesidad es agrupar actividades para evitar un equipo RPPM demasiado grande y que dificulte la gestión y comunicación, y tener capacidad de toma de decisión.

- Realización de revisión final de todo el equipo 1 semana antes de la PEM (sábado) y recorrida en campo.

  Derivado de aprendizajes de paros anteriores, fue clave la definición como hito del proceso una reunión general para la analizar el estado de cada ítem de manera multidisciplinaria, definición de acciones tipo A y B, posibles acciones de mitigación requeridas hasta el cierre de pendientes, asignación de responsables y fechas de implementación.

- Reuniones diarias de seguimiento de acciones tipo "A". Seguimiento de acciones tipo "B" en la herramienta informática corporativa con validadores cruzados.

  El seguimiento diario con las áreas intervinientes permitió darle agilidad al tratamiento, pudiendo llegar al 100% de acciones tipo A cerradas e implementadas todas las acciones mitigantes antes de la fecha programada de PEM

  Con el objeto de robustecer la gestión de acciones tipo B, se asignaron aprobadores de cierre de acciones de manera de lograr una verificación cruzada.

- Implementación de recomendación de reaseguradores respecto a clasificación de "condicionantes" de la lista de chequeo

  Una observación recibida en el 2022 fue la manera en que se clasificaba cada ítem según su condición de "Condicionante" o no de PEM. El mismo era realizado principalmente bajo el criterio del área de Producción, junto con seguridad de proceso. Esto se interpretaba como un potencial conflicto de interés, siendo que el Líder que aprueba la PEM pertenece a esa área, por lo que durante el proceso de revisión de la lista de chequeo a fines del 2022 se realizó una nueva clasificación, realizada por personal del departamento de seguridad de procesos, basado en las implicancias de cada ítem (peligro y riesgo asociado)

- Repositorio de documentación soporte en One Drive, disponible para todo el equipo.

  Esto le dio no solamente documentación de respaldo al líder, sino que trazabilidad y transparencia al compartir con el equipo los protocolos de ensayos, sus resultados, acciones correctivas, etc.

  La documentación queda disponible en un repositorio compartido para consulta posterior (lección aprendida derivada de la investigación de un evento ocurrido años posteriores al paro, momento en que ocurre el desvío)

- Incorporación de las acciones derivadas en tablero de control de Acciones del CILC – Power BI

  Al margen de la formalidad lograda al cargar las acciones en la herramienta corporativa, se incorporaron las acciones tipo B en el tablero de comando bajo la tipología RPPM. Estas son reportadas periódicamente y presentadas en comité mensual para seguimiento de pendientes y vencimientos.

- Colaboración de especialistas de otros negocios

  Permitieron reforzar el equipo, realizar verificaciones cruzadas con una mirada "no contaminada" evitando tunelizaciones del proceso.

- Proceso de análisis FODA del Paro coordinada por consultora (Iniciativa Paros)

3.3 Aspectos de mejora identificados en el proceso 2023

- Incorporación de hitos al programa de paros.

  Comienzo de la definición del alcance de RPPM con las reuniones de macro alcance del paro (año anterior) / KOM con el equipo con más anticipación (previo al paro)

  Es importante iniciar la definición del alcance de RPPM mientras se define el alcance del paro, para asegurar la inclusión actividades requeridas para una correcta revisión de seguridad y evitar cambios de alcance del paro.

  Incluir en el programa del paro los hitos asociados al proceso de RPPM, de manera de incorporar la necesidad como parte de la actividad del paro y hacer

- Secuenciar ítems (en particular los condicionantes) según etapas de la PEM

  El proceso se había definido para avanzar en una modalidad tipo flujo pistón, esto generó algunas dificultades en el seguimiento, porque el avance de las actividades de paro, la preparación para la puesta en marcha y las actividades propias de puesta en marcha están definidas con cierta secuencia.

  Es por ello que se definieron una serie de mejoras al proceso para subdividir ciertas actividades, como por ejemplo, separar una revisión simplificada para la habilitación de servicios (momento que no es considerado PEM de la planta, pero es requerido para realizar actividades posteriores que están sujetas a ítems del proceso de revisión)

- Procedimiento Específico / Formalizar proceso de aceptación sobre tratamiento de desvíos en ítems condicionantes

  Incluir en el procedimiento específico el tratamiento de desvíos derivados de ítems condicionantes, para acotar subjetividades y darle tratamiento sistémico y formal. Esto incluye la propuesta de acciones mitigantes hasta corregir el desvío, su aceptación por el equipo, por el departamento de seguridad de procesos y por el líder de RPPM.

3.4 Debilidades y amenazas identificadas en el proceso 2023

Si bien el resultado general de la PEM fue positivo, se identificaron algunas debilidades que derivaron en eventos menores (cuasi-accidentes) que en condiciones distintas pudieron escalar, y básicamente se trata de actividades que, producto de hallazgos no esperados, quedaron fuera del alcance original de la revisión, lo que derivó en una línea de trabajo de mejora del proceso que permita la identificación del hallazgo como necesidad de cambio de alcance, para su tratamiento inmediato.

4 Conclusiones

La revisión de seguridad previa a la puesta en marcha es un proceso crítico dentro del PSM, el cual tiene como objetivo evitar la ocurrencia de incidentes de seguridad de procesos en el momento en que la instalación se encuentra en el momento más vulnerable, los estados transientes (en este caso la puesta en marcha)

Su éxito depende en gran medida del liderazgo y cultura madura en seguridad de la compañía, por lo que es recomendable evaluar este aspecto antes de procedimentar e implementar el proceso.

El equipo revisor debe ser multidisciplinario, tratando de acotar el número para que el proceso y la comunicación sea viable (8 integrantes es recomendable, pero dependerá de las características de la PEM a preparar), pero siempre asegurando que los integrantes tengan las competencias cubiertas para la actividad.

Otro factor clave es definir claramente y con suficiente antelación el alcance de la revisión, para asegurar la verificación de todas las condiciones de seguridad requeridas, poner foco en lo que se quiere revisar, la documentación soporte esperada y su correcto reporte, de manera que los resultados lleguen en tiempo y forma.

Las listas de verificación deben ser específicas y no perderse en requisitos ambiguos (si es necesario referenciar a documentos o planillas particulares), debe ser flexible para adaptarse a las distintas necesidades (tipo de instalación, características del proceso, complejidad de puesta en marcha, magnitud de la intervención, MoC o proyectos involucrados, etc.), siendo consistente el esfuerzo a los peligros y riesgos asociados. La lista no puede ser estática, debe estar continuamente alimentada por distintos aprendizajes derivados de incidentes, auditorías, nuevas prácticas recomendadas, etc.

La planilla de verificación contemplar requisitos estándares, que normalmente se repiten entre una instalación y otra, pero deben incluirse ítems específicos o particulares para cada planta.

Mantener viva la gestión, medir la performance y la eficiencia, la calidad del proceso, el cumplimiento de los requisitos normativos, la correcta gestión de acciones y el mantenimiento de mitigaciones hasta tanto se cierren las acciones, como así también medir la efectividad del proceso (asegurando un correcto registro e identificando causas asociadas) contabilizando y analizando eventos durante la intervención para entender qué parte del proceso falló.

Si bien es prematuro sacar conclusiones a partir de datos estadísticos sobre las mediciones de performance del proceso (la cantidad de mediciones no son suficientes para elaborar reportes estadísticos) sí puede afirmarse que la evolución a un proceso sistémico, auditable, transparente y bajo un programa sujeto a reportes de seguimiento y auditorías ha mejorado el proceso de revisión de seguridad previa a la puesta en marcha, aportando una visión multidisciplinaria y más profunda de las revisiones, mejorando la gestión del conocimiento, del cambio y de incidentes dado que no se han registrado eventos de seguridad de procesos (considerando el alcance definido en el proceso), generando un contexto de círculo virtuoso para la mejora continua del proceso de RPPM.

Alcanzar la madurez mínima requerida para que el proceso alcance el objetivo propuesto requirió desarrollo y trabajo previo en distintos aspectos, que implicó involucrar varios elementos del PSM, además del elemento *Preparación para la Operación*.

5 *Referencias y Fuentes de Información*

[1] Center for Chemical Process Safety. *Guidelines for Performing Effective Pre-Startup Safety Review (PSSR)*. New York, John Wiley & Sons; 2007.

[2] Center for Chemical Process Safety. *Guidelines for Risk Based Process Safety*. New York, John Wiley & Sons; 2007.

[3] Marsh *Pre-Startup Safety Review*. Risk Engineering Position Paper – 06 (2011). https://www.marsh.com/content/dam/marsh/Documents/PDF/UK-en/Engineering%20Position%20Paper%20Pre-Start-Up%20Safety%20Review-03-2016.pdf

[4] Procedimiento Corporativo YPF S.A. *PR__-0001932. Gestión de Riesgos y Cambios en Activos Industriales*; 2021.

[4] Procedimiento Corporativo YPF S.A. *ES__-00100333. Revisión de Seguridad Previa a la Puesta en Marcha*; 2021.

[5] BP. *Hazards of Oil Refining Distillation Units*, 1st Edition, 2008

Arquetipos de liderazgo y sentido de vulnerabilidad

De Aprendiz a Sabio, el camino organizacional para construir una Cultura de Seguridad de Procesos basadas en la Disciplina Operacional, a través del liderazgo.

Martín Enrique Fernández
Presidente, Whycomm S.A.
Chacabuco 271 piso 14 (1069). CABA. Argentina.
martin.fernandez@whycommsystem.com

Keywords: Sentido de Vulnerabilidad, Liderazgo, Disciplina Operacional

Abstract

Diversos autores hablan de la velocidad con la que ocurren los cambios en el presente. Que lo más probable es que las desiciones permanentes, las formas de gestión estáticas y las culturas sedimentadas por años, no vuelvan a ocurrir. Pareciera ser que el cambio será, finalmente, el nuevo paradigma de gestión.

En un mundo que se vuelve, cada día, un rompecabezas más difícil de armar; un rompecabezas vivo al que le crecen nuevas piezas y nuevos bordes; que no nos deja casi nunca tomar distancia para apreciar una imagen clara y completa de lo que ocurre en el día a día (y por ende, en nuestra Gestión de Riesgos), pensar en un estilo de liderazgo propio de la época, con características delimitadas y específicas de esta tercera década del siglo XXI, parece más una búsqueda teórica que una realidad posible. Porque definir un estilo implicaría encontrar algún tipo de nuevo estatus quo en las formas de liderar; de nuevas ideas que formen parte de nuestra vida cotidiana sin cambios permanentes. De una visión sobre el liderazgo que llegue para quedarse, para ser "lo normal". Y todo parece indicar que eso no volverá a ocurrir.

En la búsqueda de un contraste a este presente de cambios continuos y vertiginosos es que recurrimos a los Arquetipos, esas formas permanentes del comportamiento humano, que nos pemiten distanciarnos de la velocidad cotidiana, y contribuyen al armado general de ese rompecabezas que nos toca enfrentar cada día.

En el presente trabajo abordamos, desde una perspectiva filosófica y adaptada de la visión de los arquetipos del psiquiatra y ensayista suizo Carl Gustav Jung, los diferentes estadios de madurez del liderazgo a medida que avanzan en su carrera organizacional, y cómo en ese recorrido asimila la dimensión cultural (CCPS, 2018) "Mantener presente el Sentido de Vulnerabilidad".

1. INTRODUCCIÓN

1.1. La velocidad del cambio y el liderazgo para la gestión de riesgos

El concepto de liderazgo es dificil de abordar en toda su complejidad. Muy por el contrario, las nociones que hacen al entendimiento del liderazgo deben ser analizadas con calma, para evitar el pozo profundo de la lectura angustiante y simplificadora, de ideas que pretenden abarcarlo todo. Liderar es un arte noble, pero para comprender parte de su complejidad necesitamos aislar conclusiones, desarmarlas, para luego re conectarlas en un todo más coherente. Son tantas las facetas del comportamiento humano implicadas en la naturaleza de liderar que es poco probable que puedan ser abordadas e interconectadas, todas juntas, con la profundidad y rigurosidad adecuada. Sin embargo, si cerramos el foco de la mira, y nos centramos en las claves del liderazgo para la seguridad de procesos y la transformación de la cultura para gestionar riesgos, podemos separar lo esencial de lo accesorio.

¿Qué consideramos esencial para este trabajo? La forma en la que los diferentes segmentos de liderazgo van construyendo su sentido de vulnerabilidad y sus nociones sobre Disciplina Operacional. Esas dos temáticas fueron definidas como centrales en esta mirada filosófica sobre la Seguridad de Procesos y su forma de liderarla.

Durante esa cápsula temporal extraña, ese presente eterno y confuso que fue la pandemia, pensar en el futuro generaba incertidumbre. Si bien ese nuevo mañana siempre parecía lejano, estaba claro que algo iba a ser diferente; si bien esperábamos (y deseábamos) que el mundo volviera a ser normal otra vez, sobre todo, sabíamos, iba a ser nuevo.

La noción abstracta de "nueva normalidad", además de la incertidumbre por lo nuevo, traía adjunta una falsa sensación de tranquilidad, de que todo iba a volver a ser diferente pero normal. Que la rutina iba a volver a ser parte de nuestras vidas. La idea de "Nueva Normalidad" más que una categoría realista fue una etiqueta con la que quisimos visualizar un futuro estático o un mundo que -por lo menos por un rato- no se moviera. Y si vale el dramatismo, fue un grito desesperado por entender ese presente caótico, inabarcable, "ancho y ajeno", que nos tocó vivir a todos juntos de un momento a otro. Para combatir la angustia producida por ese empujón planetario llamado Pandemia, que nos sacó de nuestra zona de confort, que nos obligó a reformular nuestra vida cotidiana, es que encontramos ese relato tranquilizaro: el futuro sería incierto, pero otra vez normal.

Finalmente, la nueva normalidad llegó y la estamos viviendo.

Eso que fue un concepto inquietante, ahora, visto con cierta distancia, lo podemos resignificar. Porque ya sabemos que no es una, sino que serán varias, que el ritmo del cambio al acelerarse más y más genera infinidad de nuevos escenarios interconectados. Esta nueva normalidad no es la única, no será la última, ni siquiera será nueva. Ya no hace falta el empujón global para vernos expulsados de nuestra zona de confort. El presente y sus límites difusos es cada vez más difícil liderar con habilidades del viejo mundo, de aquel mundo que ya contaba con la tecnología para ser híbrido, pero no con la mentalidad (ni individual ni organizacional). La pandemia dejó un mundo híper acelerado, en el que lo permanente (la llamada "nueva normalidad"), será el cambio. Y esto también les ocurrirá a las formas del liderazgo para la gestión de riesgos.

¿Cuál es la nueva normalidad de la gestión de riesgos? ¿Qué desafíos enfrentamos?

El legado de la pandemia deja dos finales superpuestos: el fin de los estilos de liderazgo tradicionales y el fin de la noción de "Zero Accidents".

El fin de los estilos de liderazgo significa que está llegando el fin de una época en la que tomar una actitud proactiva de liderazgo era una opción que podía tomarse o no, ya que aún lideraban, principalmente, quienes contaban con la jerarquía formal. La nueva normalidad nos colocó a todos del mismo lado: del lado de los que lideran. Del otro lado ya no queda nadie. El liderazgo dejó de ser una opción y pasó a ser un atributo de toda posición organizacional. Por eso es importante separar liderazgo de jerarquía. Las jerarquías seguirán existiendo. Pero para liderar la gestión de riesgos, será el entendimiento, la proactividdad y la capacidad de colaborar con otros lo que prime más allá de la jerarquía. El liderazgo entendido a través de estilos es una forma del pasado, de cuando el mundo se podía dividir en etapas estáticas y permanentes.

Esta idea de que "todos somos líderes" no es una sentencia idealista o una declaración de intenciones, sino el estatus quo que irrumpió con la complejidad. Quien no sea capaz de liderar interacciones cotidianas, su propia agenda y la de sus pares y sus líderes, estará en desventaja.

Hay demasiadas pistas que nos sugieren que nada de lo que estamos viviendo va a quedarse quieto tanto tiempo como para ser normal. Ese mundo estático, en que exístian líderes a los cuales seguir desde un rol pasivo, llegó a su fin.

1.2. Del "Zero Accidents" a la gestión de barreras, la nueva normalidad del liderazgo

La transición que está ocurriendo globalmente hace más de una década, finalmente, parece desembocar en un cambio de era: la visión Zero Accidents está quedando definitivamente atrás, a medida que se instala otra visión superadora: el foco en la gestión.

Zero Accidents, como utopía simplificadora, ha dejado de ser el discurso dominante de la gestión de riesgos. Zero Accidents -que sirvió como una fórmula potente para instalar temáticas resistidas- ha perdido su efectividad discursiva. En términos de Escalera de Cultura (IOGP, 2010), Zero Accidents corresponde a los estadíos patológico o reactivo, ya que conserva -aunque sea en parte- el foco en las personas. El Zero Accidents conectaba el liderazgo a la jerarquía, y no a la proactividad y la efectividad para liderar procesos. Zero Accidents pasó a ser entendida como una mirada que focaliza sus esfuerzos en las personas y no en los sistemas.

Además de mutar conceptualmente de "seguridad" a "gestión de riesgos e impactos", el cambio de mindset es de 180 grados: del modo cumplimiento, con foco en el individuo, al modo compromiso, con foco en los sistemas. De la seguridad personal a la gestión de eventos mayores. Del zero incidents a la gestión de barreras. De avolir los riesgos a manejar los riesgos. Del "modo cumplimineto" al "modo compromiso". From avoid to managing. El cambio de puede sintetizarse como un largo viaje que fue del Zero a la gestión.

1.3. Los "embrujos del lenguaje", de liderazgos disciplinadores a la disciplina operacional

Ludwig Wittgenstein fue una luminaria intelectual del siglo XX. Un lingüista y matemático vienés que "luchaba contra los embrujos del lenguaje", esas formas confusas de la comunicación humana que atentan contra el entendimiento de las personas. A lo largo de su obra, en sus investigaciones sobre filosofía del lenguaje y su intento por acercar el mundo erudito a un saber popular, acuñó una definición que se volvió célebre: "el significado de las palabras es su uso en el lenguaje". A través de una idea simple, expresada con belleza, desnuda que existen diferentes trampas semánticas, trampas del lenguaje, es decir, palabras con múltiples significados. Y propone entonces pensar el significado a través del "uso en el lenguaje". Esto significa que las palabras tienen un significado dominante, un sentido común instalado, un uso cotidiano que va a prevalecer sobre su semántica definida por la Real Academia del lenguaje.

"El significado de una palabra es su uso en el lenguaje" expresa que según el contexto en el que la usemos, según la época, según la cultura, una palabra puede significar una cosa o, incluso, lo opuesto. En esta línea, nos preguntamos preguntarnos cuál fue el significado cotidiano del término disciplina en la industria de Oil and Gas, a lo largo del tiempo, es decir, en diferentes contextos. Nos interesa visualizar el recorrido y los desplazamientos en su significado durante la segunda mitad del siglo XXI y hasta el primer cuarto del siglo XXI.

Siguiendo a Wittgenstein podríamos decir que la transformación que sufrió el significado de la palabra Disciplina fue un "embrujo del lenguaje", y que además, cambió su significado "a partir de su uso en el lenguaje".

1.4. De liderazgos disciplinadores a la disciplina operacional; de la obediencia al entendimiento

¿Cuál es el sentido común instalado o uso cotidiano de la palabra disciplina? Es decir, si pensamos en disciplina, ¿en qué pensamos?

Probablemente, en obediencia.

Porque el significado dominante de la palabra disciplina ha estado tradicionalmente vinculado al carácter individual y no a los sistemas organizacionales. La tradición cultural de la palabra disciplina en el siglo XX está vinculada a la formación militar y a la educación basada en la repetición sin cuestionamientos. También a las artes marciales o a los esfuerzos extremos. La disciplina, entre otras cosas, premiaba el esfuerzo repetitivo y muchas veces irreflexivo.

La reflexión no está asociado a la idea de disciplina. El universo de significados vinculados a la disciplina fuer rígido, sin lugar al cuestionamiento. ¿Quiénes son las personas disciplinadas? Aquellas que obedecen, que cumplen con lo esperado y definido por una autoridad incuestionable, qué respeta la norma. En términos de liderazgo, la tradición disciplinadora, fue expresada a través de liderazgos verticales, autoritarios o paternalistas, que generaban obediencia a través del temor.

La industria de Oil and Gas del primer cuarto del siglo XXI toma la idea de Disciplina Operacional para expresar un nuevo universo de significados, ya vinculados a la comprensión y el entendimiento. Complejiza esa idea inicial de obediencia sin cuestionamientos. Surge la idea de "conciencia" conectada a la idea de entendimiento. A modo de ejemplo, la definición de dos de operadoras importantes de América Latina, como ejemplo de esta tendencia:

"Entender y realizar cada tarea de la forma correcta, siempre."

Definición de Disciplina Operacional utilizada por YPF S.A.

"Hacer la tarea correcta, de la manera correcta. Todo el tiempo. Con consciencia y consistencia."

Definición de Disciplina Operacional utilizada por Pluspetrol.

En ambos casos el entendimiento aparece como eje rector de la Disciplina. En el primero, presentado en forma literal como la primera palabra de la definición. En el segundo, a través de la idea de consciencia. Cumplir los procedimientos, pero entendiéndolos.

Existe una distinción conceptual que el experto James Reason aborda en el libro "El error humano" y es la de Contexto semántico. Significa que palabras o nociones quedan fuertemente asociadas unas a otras porque operan dentro de campos de significados similares. Esto ocurre cuando “una palabra está cerca de otra en términos de significado y tienden a usarse en forma intercambiable”.

Un ejemplo de esto ocurre con los términos “cultura” y “cultura organizacional” y “cultura de seguridad”. Son conceptos que están “cerca” unos de otros, y tieneden a utilizarse en forma aleatoria. Lo mismo ocurre con la palabra Disciplina y el término disciplina operacional: comparten “Contexto Semántico”. Y eso hace que el desplazamiento del significado se ponga en marcha.

De un liderazgo disciplinador a la disciplina operacional puede entenderse como el pasaje de la obediencia al entendimiento, pero usando un mismo campo de significados: el de la disciplina, lo que genera confusión. Las ideas de entendimiento y obediencias están tomando la misma palabra de referencia: disciplina.

¿Y entonces cómo se resuelven estos “embrujos del lenguaje”?

Justamente, generando un lenguaje propio y adecuado para la gestión. El lenguaje de la Disciplina Operacional, signado entre otras dimensiones (como la noción de aprendizaje organizacional y la de Combatir la normalización de desvíos), por el sentido de vulnerabilidad

Para desarmar estas trampas semánticas y contribuir a un mayor entendimiento de cómo opera el sentido de vulnerabilidad a lo largo de la vida organizacional de los individuos, utilizaremos las ideas vinculadas a los Arquetipos.

2. AQUETIPOS DE LIDERAZGO Y SENTIDO DE VULNERABILIDAD

2.1. El concepto base: ¿qué es un arquetipo?

Un arquetipo es un modelo, una referencia reconocible.

Los arquetipos representan patrones de comportamiento, formas universales de conducta que se referencian en una imagen genérica; mapas de la psiquis colectiva que trascienden al comportamiento individual. Los arquetipos -y en esto se encuentra parte de la quintaesencia de su significado- nos representan a todos, y a la vez representan esa imagen reconocible por todos, ya que se manifiestan de diversas formas en las personas y en las historias que se narran, y con las cuales, como individuos, podemos identificar los diferentes ciclos vitales de nuestra existencia. Cada momento de nuestra vida posee un reflejo arquetípico. Y como toda idea universal, el arquetipo nos ayuda a dar sentido a la realidad que nos rodea, en suma, a ordenar el mundo.

Para contribuir a la comprensión de la noción de arquetipo, es conveniente dejar la esfera del pensamiento conceptual y abstracto, y desplazarnos hacia las miradas pragmáticas y basadas en ejemplos (lo que en la antropología buscó desde inicios del siglo XX a través del trabajo de campo impulsado por uno de sus creadores, Bronislaw Malinowski). Pensar con ejemplos favorece un acercamiento directo y sintetizador de nociones complejas como la de arquetipo, introducida por el psicólogo suizo Carl Gustav Jung como una parte fundamental de su teoría del inconsciente colectivo. Pensar con ejemplos, permite "hacer visible el pensamiento" (f). Con el fin en mente de ejemplificar para comprender, abordaremos un breve semblante de uno de los Arquetipos más extensamente conocido en la cultura occidental: el del Rebelde, interpretado por James Dean.

2.2. El Rebelde, un ejemplo universal para profundizar la noción de arquetipo

Fueron muchas y muy bien combinadas entre sí las razones por las cuales James Dean se convirtió en un arquetipo de rebeldía para occidente. La primera y la más conocida fue que el actor estadounidense, nacido en 1931, en el Medio Oeste de Estados Unidos, portagonizó dos obras emblemáticas con la que da vigor al arquetipo. La popular "Rebelde sin causa" y la menos conocida "Al este del Edén", ambas películas que reflejan la angustia de la juventud estadounidense de la década de 1950.

El mundo, desde entonces, cuenta con una nueva imagen del Rebelde global en la que reflejarse. El arquetipo está en su clímax. Una narrativa que acompañará a la cultura popular durante décadas ha sido forjada por Hollywood. El carácter de los personajes que interpretó en estas dos películas reflejan la imagen y la energía combativa de la juventud: luchar contra las normas, contra la autoridad establecida, contra el estatus quo. En "Rebelde sin causa", el Jim Stark encarnado por Dean, desafía toda expectativa social y enfrenta a problemas familiares y conflictos internos.

Por otra parte, su vida fuera de la pantalla también ayudó a la construcción del arquetipo del Rebelde. Mas allá de sus personajes James Dean personificaba la rebeldía en su esfera personal. Con un estilo de vida poco convencional, con su belleza hegemónica, con su actitud desafiante, con su amor por las carreras y la velocidad, en suma, con cada detalle, alimentaba la imagen de "chico malo", que, como ya expresamos, serviría de reflejo durante décadas para gran parte de la cultura occidental.

Su muerte fue prematura y trágica. En un accidente vehicular en una carretera californiana, con solo 24 años, conduciendo su Porsche blanco, modificado exclusivamente (conocido como "Little bastard" o "pequeño bastardo"), colisionó de frente contra otro automóvil e inició un sinfín de interpretaciones sobre su vida y su muerte. La fatalidad fue declarada en el lugar del accidente y desde ese momento comenzó su nuevo viaje: la inmortalización como un símbolo de juventud y rebeldía, favoreciendo que una vez más, arquetípicamente, que juventud y rebeldía estrecharan su unión universal y perdurable a través de la historia humana. Murió como vivió, metafóricamente, en un choque frontal contra la sociedad.

Una breve digresión para aclarar los límites del un arquetipo, tomando como ejemplo esta última idea en la que se asocia juventud a rebeldía. Esta asociación no implica que, por ejemplo, no se puedan tener atributos de rebelía a una edad más avanzada. Lo que señala el arquetipo es que no se puede ser arquetípicamente reblede a los 55 años. La imagen universal de la rebeldía está asociada a la energía vital y desafiante de la juventud temprana, que aún tiene un mundo por descubrir y desafiar. La rebeldía de la mitad de la vida sigue siendo rebeldía, pero nunca es arquetípica.

2.3. ¿Se puede establecer un vínculo entre Arquetipos y Sentido de Vulnerabilidad?

Jung, además de fundar la psicología analítica, acuñó el concepto de Inconsciente Colectivo "un espacio mental en el que se une la experiencia heredada por la humanidad, transmitida de generación en generación, que posee características comúnes (arquetípicas) a todos los seres humanos". El arquetipo contiene las experiencias y el conocimiento compartido por la especie humana.

En su obra "Arquetipos e inconsciente colectivo", publicada inicialmente en 1959 y con un gran impacto en la psicología, la antropología y otras disciplinas, abordó una serie de arquetipos que listaremos solo modo de referencia, y que buscan presentar formas universales del carácter con los que se puede clasificar a los seres humanos:

- El héroe.
- El sabio.
- El amante.
- El cuidador.
- El explorador.
- El rebelde.
- El creador.
- El mago.
- El gobernante.
- El bufón.
- El saboteador.
- La madre.
- El anciano sabio.
- La sombra.
- El niño divino.
- El anima y el animus.

Es importante remarcar que Jung no proporcionó una lista exhaustiva y fija de arquetipos en sus obras, sino que discute una variedad de caracteres que pueden manifestarse de diferentes maneras y variar según las culturas y contextos individuales. Los arquetipos son conceptos abstractos que pueden manifestarse de diferentes maneras en la vida real. Las personas raramente nos ajustamos por completo a un único arquetipo. Es un patrón semi permanente, que se caracteriza por trascender el tiempo, y en muchas ocasiones, la distancia. Expresa Jung: "los arquetipos están en el inconsciente y en el acervo cultural, en el conocimiento cultural de la humanidad, y por lo tanto en nuestros sueños".

En medio de la complejidad y el cambio permanente introducido en la primera parte del paper, el ejercicio que proponemos a continuación es una reflexión desde lo opuesto al cambio y la inmediatez, desde lo que -justamente- no cambia con velocidad, desde lo que no está sujeto a la tiranía de un tiempo acelerado. En suma, desde lo arquetípico.

El ejercicio que nos proponemos realizar en este trabajo es definir una serie de arquetipos organizacionales, es decir, de formas y "personajes" que se repiten en el mundo del trabajo (tomando prestadas las palabras de Jung diríamos en el "Inconsciente Colectivo" de las organizaciones de la industria de la energía), para luego establecer un vínculo entre cada uno de ellos y el sentido de vulnerabilidad.

Las personas vamos pasando por diferentes arquetipos a medida que avanzamos en nuestra vida profesional, encarnando diferentes roles de liderazgo.

Los seis arquetipos que vamos a analizar, que se presentan como un camino a través de los diferentes momentos de la vida humana, son los siguientes:

- Aprendiz
- Rebelde
- Guerreo
- Gobernante
- Mago
- Sabio

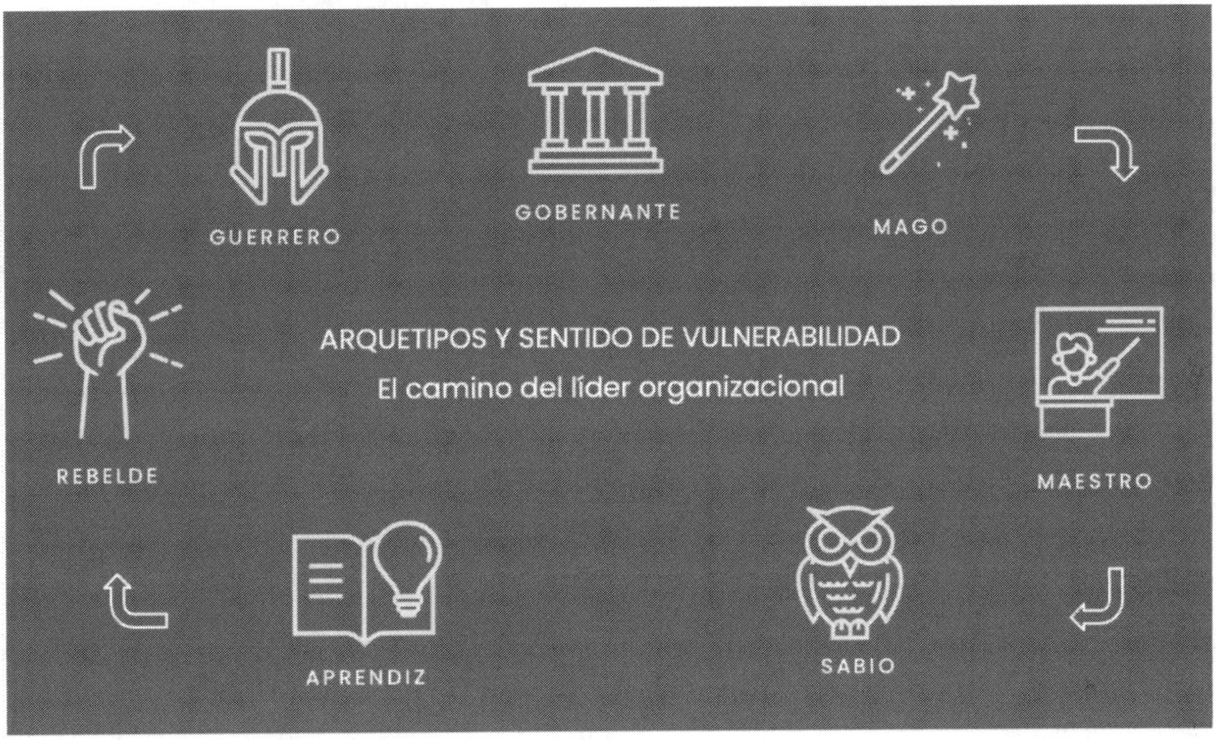

Figura 1.

2.4. Primer arquetipo organizacional: Aprendiz

El primero de los arquetipos en la vida organizacional definidos en este viaje del liderazgo, es el Aprendiz, que es la forma forma universal en la que se inicia un camino de aprendizaje. ¿Quién es un Aprendiz? Alguien que está en su etapa formativa y tiene la necesidad de incorporar conocimiento y experiencia. Un rol que por definición debe aprender.

Si bien Jung no dedicó un trabajo específico a este arquetipo, muchas de sus ideas sobre el proceso de "individuación" y del desarrollo de la personalidad son relevantes para comprenderlo. Integraremos también las ideas de Robert Moore y Douglas Gillette, en su libro "Rey, Guerrero, Mago, Amante: Arquetipos de la madurez masculina" en el que presentan al Aprendiz como una de las cuatro estructuras arquetípicas fundamentales en la psique masculina. Este estadio inicial es una base filosófica y estructural de la habilidad para luchar y proteger que encontraremos en el Guerrero, la sabiduría y el liderazgo que encontraremos en el Gobernante, y el dominio de habilidades prácticas y espirituales del Mago. Pero no nos adelantemos.

Durante la Edad Media, el Aprendiz se manifestaba en una variedad de contextos, desde el aprendizaje de artes hasta la educación académica formal, el servicio en la corte o la iglesia, los viajes de aprendizaje y la formación monástica. Todas formas de en la que los jóvenes buscaban adquirir conocimientos, habilidades y experiencia práctica en diferentes áreas de la vida medieval. Pero sin dudas, fue el aprendizaje artesanal y gremial el que ubicamos como arquetípico. En sociedad medieval, el aprendizaje de un oficio era fundamental para el desarrollo de habilidades y la adquisición de conocimientos prácticos. Los jóvenes interesados en convertirse en artesanos o comerciantes se convertían en aprendices y trabajaban bajo la tutela de un maestro artesano. Así, un Aprendiz de panadero debía estar al servicio de un maestro panadero, que en muchas ocasiones también era su padre o un familiar. En el nivel más descriptivo de la experiencia, un panadero experto enseñaba a otro a hacer el pan.

En nuestra industria (Oil & Gas) el Aprendiz es el nuevo ingresante, que puede iniciarse como joven profesional y, en muchos casos, llegan preparados por las universidades pero en busca de "Maestros" de lo práctico, que le enseñen los secretos de operaciones reales.

¿En qué nivel de desarrollo se encuentra el Sentido de Vulnerabilidad de ese arquetipo del Aprendiz? ¿Cómo se manifiesta?

Sin dudas, el sentido de vulnerabilidad de un Aprendiz es bajo.

Para comprender esto podemos recordar la forma en la que percibimos el riesgo, que si bien no es tema principal de este trabajo, complementa el entendimiento y justifica la hipótesis sobre por qué el arquetipo del Aprendiz posee un bajo sentido de vulnerabilidad.

2.5.5.1 "No vemos el riesgo por los ojos, sino que lo percibimos a través del cerebro"

La frase "el riesgo se percibe a través del cerebro" enfatiza que la percepción del riesgo no ocurre exclusivamente a través de un sentido específico como la vista, sino que es un proceso complejo que implica la interpretación de información sensorial por parte del cerebro.

Inicialmente, son los estímulos visuales los que desempeñar un papel importante en la percepción del riesgo (por ejemplo, al identificar un peligro evidente como un precipicio), aunque la percepción del riesgo no se limita solo a lo que se ve. También se basa en la información recopilada a través de otros sentidos como el oído, el tacto, el olfato y el gusto.

Así y todo, es una percepción limitada, basada en un sentido común precariamente elaborado. Mantener presente el sentido de vulnerabilidad requiere rutinas de pensamiento más complejas, que favorezcan su desarrollo (entrenamiento, análisis de casos, momentos de seguridad, investigaciones, otras). Una vez que los estímulos sensoriales son recibidos, se lleva acabo un proceso de interpretación cognitiva. Esto implica analizar y evaluar la información sensorial en función de experiencias pasadas, conocimientos previos, valores personales y expectativas. La percepción del riesgo varía de persona a persona y es influenciada por una variedad de factores, que incluyen experiencias previas, conocimientos, cultura, educación y contexto social. Y si de algo carece el Aprendiz, es de experiencias significativas en ámbitos operativo que alimenten su percepción. En el estadío de Aprendiz, el procesamiento del riesgo en las operaciones es mayormente sensorial, u obedece a un sentido común construido en base a experiencia ordinaria y no a competencias desarrolladas profesionalmente.

La noción de vulnerabilidad de un proceso productivo y sus implicancias es incipiente. No conocer las operaciones, por añadidura, implica no conocer los riesgos. Sus ideas preconcebidas sobre lo que es riesgoso y lo que no, son defincientes, y esto hace que los entonos operativos complejos (un frente de trabajo con tareas simultaneas, por ejemplo) no sean seguros para el Aprendiz. Su sentido de vulnerabilidad no ha sido desarrollado organizacionalmente. Y mientras esto no ocurra, su percepción del riesgo será una evaluación subjetiva, construida con elementos y cálculos rudimentarios sobre sobre la probabilidad y las consecuencias de un evento adverso o peligroso. En la vida organizacional, quien se encuentra transitando el estadio del Aprendiz, por definición, posee un sentido de vulnerablidad bajo.

2.5.5.2 Los primeros cien días en la gestión de riesgos

La idea más popularizada relacionada con "los 100 días para algo" es probablemente el concepto de "los primeros 100 días de un presidente". Este término se refiere al período inicial en el cargo, durante el cual se espera que establezca su agenda, implemente políticas clave y establezca el tono para su administración. La noción de los primeros 100 días se remonta a la presidencia de Franklin D. Roosevelt en 1933, cuando prometió una serie de acciones para abordar la Gran Depresión. Desde entonces, se ha convertido en un punto de referencia común para evaluar el éxito y la eficacia de un presidente en sus primeros meses en el cargo.

Tomando la noción de "los primeros 100 días" para el ámbito laboral es un período crítico durante el cual se espera que se adapte al nuevo entorno, comprenda sus responsabilidades, establezca relaciones con colegas y comiencen a demostrar su valor. Un libro relevante que adapta este concepto al mundo laboral es "The First 90 Days: Proven Strategies for Getting Up to Speed Faster and Smarter" (Los primeros 90 días: Estrategias probadas para ponerse al día más rápido y de manera más inteligente). El autor, Michael D. Watkins, ofrece consejos prácticos y estrategias para ayudar a los profesionales a navegar con éxito los desafíos comunes que enfrentan durante la transición a un nuevo trabajo o rol dentro de una organización:

> *"Los primeros 90 días en una nueva carrera pueden estar llenos de desafíos y riesgos potenciales. Desde la presión de adaptarse rápidamente a un nuevo entorno hasta la posibilidad de cometer errores debido a la falta de experiencia, es crucial mantenerse alerta y consciente de los posibles peligros. Mantén una actitud de precaución, busca orientación cuando sea necesario y no tengas miedo de admitir tus limitaciones. Recuerda que priorizar la seguridad y el aprendizaje durante este período te ayudará a construir una base sólida para tu éxito futuro."*

Un concepto clave para percibir el riesgo es el entendimiento. Por eso los primeros 100 días de un nuevo ingresante a un frente de trabajo, que es una etapa crítica donde ocurren eventos no deseados, suelen ser críticos. La falta de experiencia se traduce, linealmente, en falta de entendimiento. En esos primeros cien días los Aprendices tienen un contacto inicial con la noción de vulnerabilidad.

Algunas característica, que por añadidura, se experimentan en el arquetipo son la curiosidad y la sed de conocimiento. Absorber el mundo que lo rodea, acompañado de cierta humildad y disposición a aprender, ya que reconoce que hay un mundo "ancho y ajeno" por descubrir y está dispuesto a admitir su propia ignorancia. Está abierto a recibir orientación y retroalimentación de otros. Esto cambiará drásticamente en el siguiente estadio cuando el fuerte impulso interno para crecer y abandone la actitud de humildad, que como Aprendiz mantiene ante un mundo nuevo, que desconoce completamente.

A medida que el ser humano madura, el arquetipo del Aprendiz no desaparece, sino que evoluciona, integrándose con los otros arquetipos para formar una identidad o personalidad

¿Cuál es el próximo arquetipo en este viaje organizacional por el sentido de vulnerabilidad?

¿Por qué una persona deja de ser Aprendiz?, ¿qué ocurre en ese primer proceso de transformación organizacional?, ¿cómo es esa metamorfosis hacia el siguiente estadio de evolución?, ¿cuándo ese joven profesional deja de ser Aprendiz? Cuando ya incorporó algún tipo de conocimiento y siente "el llamado interno" de la rebeldía. La adolescencia del sentido de vulnerabilidad ha comenzado.

2.5. Segundo arquetipo organizacional: Rebelde

El arquetipo del Rebelde es el segundo estadio en este recorrido, y comienza cuando ya ha madurado en él alguna forma del conocimiento, sea teórica o práctica. Así como el Aprendiz tomo cierta posición de humildad ante la falta de saber, el Rebelde asume una posición de desafío ante la necesidad de poner en valor el mínimo de conocimiento incorporado. Desea contrastar sus nuevas experiencias, necesita probarse ante el mundo.

En términos del sentido de vulnerabilidad, es una etapa crítica. Los primeros cien días han pasado (o un año, según el perfil) el Aprendiz dejó de percibirse Aprendiz y, arquetípicamente, comienza la etapa de rebeldía. Rebeldía y gestión de riesgos son enemigos por antonomasía. Ya vimos en la primera parte de este paper que el significado de disciplina operacional fue desplazándose desde la obediencia al entendimiento. Al Rebelde no le gusta la disciplina, principalmente, porque no entiende el mundo bajo esa noción. El Rebelde se caracteriza por desafiar la autoridad establecida, ya sea en forma de figuras de autoridad, normas o convenciones culturales. No acepta ciegamente las reglas impuestas, sino que busca cuestionarlas para crear cambios. El Rebelde se niega a conformarse con las expectativas o normas impuestas por su entorno, sino que necesita seguir su propio camino, incluso si eso significa ser percibido como diferente o excéntrico.

Mucho conocimiento sobre algo nos hace humildes, como ya veremos en el estadio del sabio ("solo sé que no sé nada"), pero un poco de conocimiento puede generar rebeldía. Entre otras cosas, la imagen de James Dean y su "Rebelde sin causa", se sostienen en que "no tiene nada que perder", en que todavía no construyó nada, y por lo tanto no tiene nada que perder. Es una de las características de la adolescencia. Es Rebelde tiene "precios muy bajos que pagar" ante su rebeldía.

¿Cómo es el Sentido de Vulnerabilidad del arquetipo Rebelde? Así como el sentido de vulnerabilidad del arquetipo Aprendiz fue definido como bajo por su falta de conocimiento y experiencia, a este nuevo estadio, en el que se mantiene la falta de conocimiento, pero se añade una actitud temerosa, desafiante y cerrada, la problemática del bajo sentido de vulnerabilidad de agudiza. Para el Aprendiz su actitud abierta al aprendizaje y humilde, funcionaba de paliativo ante su bajo sentido de vulnerabilidad. No percibía el riesgo porque "su cerebro no lo veía", en cambio, el Rebelde, no solo no lo ve, sino que no desea verlo. Este es, quizás, es el más peligroso de todos los estadios. No solo hay falta de vulnerabilidad sino que hay exceso de confianza en relación al propio saber, sumado a una actitud inadecuada. ¿Cómo puede ser el sentido de vulnerabilidad de un arquetipo cuya característica central es la resistencia? Necesariamente bajo.

El dolor del nacimiento, de la integración social el Rebelde hay un gran dolor inicial que tiene que ver con la construcción de la propia personalidad que se da en la juventud

Desde una perspectiva psicológica, la rebeldía también es una de las caras del miedo a lo desconocido o a la incertidumbre. Ante un mundo que se complejiza y empieza a exigir cada vez más de las personas, reaccionamos con la falsa acción de la rebeldía. Rebeldía e inacción suelen estar acompañadas. La queja como sustituto de la acción. Es un comportamiento más bien estéril o pasivo.

Continuando el recorrido hacia la madurez para gestionar riesgos, en un inicio, el estadio de Aprendiz dejo paso a una breve etapa de rebeldía (la rebeldía siempre es un tránsito breve) y llega el momento de iniciar la acción. Se inicia un nuevo ciclo, el primer estadio de este ciclo vital que aporta valor a la vida organizacional. Llega el momento de dar pelea, a través del arquetipo de la acción explícita: el Guerrero.

2.6. Tercer arquetipo organizacional: Guerrero

Es estadio arquetípico del guerrero es el último dentro del ciclo de la juventud o el primero de la madurez. El guerrero ya incorporó herramientas de gestión y cierta experiencia, y quiere ponerlas en juego. Quiere probarse en batalla. La transición ocurre cuando se supera el estadio a la queja, cuando en su interior siente la llamada (f) y se produce un proceso interno de transformación de una posición pasiva hacia una activa. La lucha, la confrontación, es el ámbito natural del guerreo. En este estadio las personas que comienzan a forjar carrera profesional, de forma desordenada y viceral, como ocurren casi todas las peleas iniciales. Energía e ímpetu y cierta cierta inocencia que irá perdiendo en cada batalla son algunos de sus atributos.

El sentido de Vulnerabilidad ha sido definido como "la percepción consciente y permanente de estar en una situación riesgosa que podría causarle daños a las personas, los activos y el ambiente" (f). Todo lo que nos mantiene atentos favorece el sentido de vulnerabilidad; todo lo que nos vuelve complacientes alimenta la narrativa de invulnerabilidad. A diferencia del Rebelde, el guerrero lucha por mantener alto el Sentido de Vulnerabilidad en sus ámbitos de influencia, porque empieza a entender. El período de dar batalla es un período largo. Por eso es importante que tenga un sentido por el cual luchar, para que puedan hacerlo por ellas mismas (f). El guerrero necesita ser liderado. Tiene mucha energía y está forjando su experiencia, y experimentando cómo aporta valor a un colectivo que lo supera. Pero necesita ser liderado. Está trabajando para algo más que él mismo, característica que no se realizaba en los dos primeros arquetipos. Tanto Aprendiz como Rebelde están demasiado concentrados en cerrar sus propias brechas de entendemiento como para pensar en las ajenas.

Los guerreros son personas decididas y perseverantes que se mantienen firmes en la búsqueda de sus metas, incluso cuando enfrentan obstáculos o contratiempos. Su determinación les permite superar las dificultades y seguir adelante a pesar de las adversidades.

El guerrero desatiende a la política organizacional (ese será terreno del Gobernante), en parte, porque no la controla y en parte porque no la comprende.

Es una etapa de múltiples desafíos y de gran despligue, aunque ese despligue no necesariamente se traduce en impacto de alta calidad. Suelen ser los primero 15 años de carrera. En esta etapa se incorpora un atributo que debería acompañar a las personas por el resto de sus carreras: la resiliencia. Los guerreros son resilientes y pueden recuperarse rápidamente de los contratiempos y las derrotas. No se desaniman fácilmente por los fracasos y están dispuestos a aprender de sus errores y seguir adelante con determinación renovada.

En esta etapa existe una apropiación inicial del Sentido de Vulnerabilidad. Es el joven Aprendiz que ha madurado y quiere (y necesita) probar su valor. Encontrar su lugar profesional. El Aprendiz tiene bajo sentido de vulnerabilidad porque no percibe el riesgo, el Rebelde también posee un bajo sentido de vulnerabilidad ya que a su inexperiencia agrega un componente temerario. El estadio del guerrero es de un despertar a un Sentido de Vulnerabilidad más permanente y generado por si mismo.

El guerrero se va volviendo experimentado en unos pocos años. La batalla aceleara los procesos de aprendizaje. Aprende a dar golpes y a recibirlos. También le da un sentido verdadero y profundo al término disciplina. Los guerreros suelen tener una fuerte disciplina personal, lo que les permite mantenerse enfocados en sus objetivos y trabajar para alcanzarlos. Están dispuestos a hacer sacrificios y a seguir una rutina rigurosa para lograr el éxito. Es la última etapa de la juventud, con la energía a flor de piel. En términos organizacionales, los Guerreros están dispuestos a enfrentarse a desafíos difíciles y a tomar riesgos calculados en nombre de sus objetivos y valores. Quienes salen airosos y finalizan la etapa de Guerreros con éxito, encuentran un nuevo títulos en sus tarjetas personales: Gerentes. La graduación del guerrero llega con su primera gerencia y cambio de arquetipo. De ahora en adelante dejará de ser un Guerrero para ser un Gobernante.

2.7. Cuarto arquetipo organizacional: Gobernante

El Gobernante, en el inicio de su recorrido por el Arquetipo, puede ser descripto como un guerrero institucionalizado. Es el arquetipo que gobierna la vida organizacional. A diferencia del guerrero, el Gobernante ya está dispuesto a negociar. Entiende que diferentes actores poseen diferentes intereses, y por lo tanto parte de su trabajo es conocido como "política" organizacional. Administrar conflictos, conciliar intereses, negociar posiciones; en suma, liderar.

El Gobernante es la clave para una gestión conjunta entre sectores y con otros actores externos. En términos arquetípicos es quien está en la edad de gobernar su propia tribu y su familia. Ya no posee la energía de los arquetipos iniciales, pero si un temperamento que ha sido fraguado con la experiencia. Posee un manejo más sutil de las situaciones. A diferencia del Rebelde, el Gobernante comienza a tener mucho por perder. Es tiempo de defender lo conseguido. Ya se cumplió la primera mitad de su carrera, y queda de cara a su un salto cualitativo definitivo.

¿Cómo es el Sentido de Vulnerabilidad en este arquetipo? Ya vimos que el Aprendiz poseía un bajo sentido de vulnerabilidad porque "no veía el riesgo", que en el Rebelde el sentido de vulnerabilidad no solo es bajo sino que a la falta de experiencia se suma una actitud inadecuada, muchas veces temeraria; también analizamos que en el Guerrero se produce el despertar a la vulnerabilidad y a los desafíos incipientes y esfuerzos por incrementarlo. Para el Gobernante se produce un punto de inflexión: surge la complejidad en toda su magnitud. La visión global y las responsabilidades de su posición le muestra los diferentes intereses de los actores que debe gestionar. En el Gobernante debe consolidar el Sentido de Vulnerabilidad no solo propio, sino de la organización. El Gobernante se debate entre la consolidación y la negociación. Es una de las etapas en la que más valor se aporta a una organización y su gestión de riesgos, o en la que más daño se puede realizar. El conocimiento sobre riesgos puede ser utilizado o negociado. Culturas orientadas al barril generan gobernantes con escaso sentido de vulnerabilidad.

El Gobernante, necesariamente -arquetípicamente- es quien alcanza una posición de liderazgo. Esto no significa que será un buen Gobernante, o un Gobernante eficiente. Puede tratarse de un líder errante carente de visión, o, por el contrario, expresar el más alto valor del liderazgo: el de liderar con el ejemplo. Gobernar implica pagar costos. Los liderazgos emocionalmente maduros son los que soportan los embates de organizaciones, y en términos de Gestión de Riesgos, quienes son capaces de definir los riesgos tolerables y asegurar que la porción de la organización que está bajo su mando obre en consecuencia.

El camino que un arquetipo de Gobernante debe recorrer para convertirse en un buen líder puede ser complejo y variado, pero generalmente incluye el autoconocimiento, con una comprensión profunda de sus fortalezas y debilidades, sus valores y su capacidad de reflexión, que le permita desarrollar un sentido claro de identidad y propósito. En términos de Simon Sinek (f), un buen Gobernante debe entender su "Why", y el de su organización.

El tránsito por este arquetipo contempla la fase final de su formación. Puede encontrarse ante la necesidad de cerrar brechas en áreas técnicas y las llamadas competencias "blandas" (f). El Gobernante ya ha adquirido la suficiente experiencia operativa pero ahora debe adquirir experiencia práctica en asuntos de gobierno y liderazgo. Un buen Gobernante debe desarrollar una serie de habilidades de liderazgo, como la comunicación efectiva, la toma de decisiones, la resolución de problemas, la negociación y la gestión de conflictos. En términos de gestión de riesgos, entender y actuar en función de riesgos es lo que asegurará su efectividad (f). En ocasiones busca la mentoría de líderes experimentados o participar en programas de desarrollo de su propio liderazgo.

Es en este estadio donde el desarrollo de la noción compleja de sentido de vulnerabilidad toma un papel decisivo en la construcción de liderazgos fuertes y culturas que gestionen en función de riesgos.

En una gran cantidad de ocasiones, "los riesgos que generan fatalidades" o dañan los activos y el ambiente versus los riesgos que se "perciben", son diferentes. Así, las acciones definidas y las barreras colocadas suelen estar en función de los riesgos percibidos, y no necesariamente de los riesgos reales. El Gobernante es quien debe asegurar el entendimiento adecuado

Las personas desarrollan su sentido de vulnerabilidad de acuerdo a las experiencias vividas, pero también al impacto de los sistemas en los que se encuentran inmersos. Si la normalización de desvíos es alta, es probable que el sentido de vulnerabilidad sea bajo, lo que indica una menor conciencia de los riesgos y una mayor tolerancia a los desvíos (f). Por el contrario, si la normalización de desvíos es baja, es probable que el sentido de vulnerabilidad sea alto, lo que indica una mayor conciencia de los riesgos, una menor tolerancia a los desvíos y un mayor sentido de disciplina operacional. La tarea del Gobernante es asegurar una cultura efectiva para la gestión de riesgos e impactos. Como quedó expresado anteriormente, cambiar el foco en el individuo por la gestión de sistemas y barreras.

2.8. Quinto arquetipo organizacional: Mago

El arquetipo del Mago es una figura mística que representa el conocimiento profundo, el poder espiritual y la capacidad de transformación. Aparece en mitologías, leyendas y obras de ficción de todas las culturas. Sin dudas, es uno de los arquetipos más atractivos. Es una figura poderosa y fascinante en la imaginación humana; un símbolo de conocimiento profundo.

El Mago posee un entendimiento del mundo que va más allá de lo superficial y tiene acceso a verdades universales. Su poder está conectado con fuerzas espirituales y energías más allá de la comprensión humana ordinaria. Puede canalizar estas energías para manifestar cambios en el mundo físico o para ayudar en la búsqueda del autoconocimiento y la iluminación espiritual. Puede alterar situaciones, sanar heridas emocionales o físicas, y ayudar a otros a crecer y evolucionar en su camino. A menudo es un ser enigmático y misterioso cuyos poderes y motivaciones no siempre son completamente entendidos por los demás. Este misterio agrega una capa de intriga y fascinación al arquetipo.

El Mago tiene la capacidad de transformar la realidad a través de su conocimiento y poder. Y es en este punto donde establecemos la conexión organizacional: el lugar del Mago, organizacionalmente, es el lugar del CEO.

¿Qué es lo que define a un Mago? Ni más ni menos que la posibilidad de hacer magia. Sin embargo, todos sabemos que en este mundo de tres dimensiones la magia no existe. Lo que ocurre con los "magos organizacionales" es que no se conocen sus trucos (f). El Mago es un Gobernante que conoce los trucos de la Alta Política. Por su lugar en el organigrama están lejos de las tareas cotidianas y corrientes de una organización.

El término "energía en el trabajo" es utilizado en procesos de selección de personal y es un aspecto fundamental en la decisión final. Específicamente, se refiere a la vitalidad, motivación y disposición de un individuo para desempeñarse eficazmente en su puesto laboral. Esta "energía" puede manifestarse de diversas formas, que van desde el entusiasmo por las tareas asignadas, la capacidad para mantenerse concentrado y productivo durante la jornada laboral, hasta la habilidad para colaborar de manera efectiva con diferentes equipos de trabajo. En esta línea, los arquetipos nos ayudan a trazar interpretaciones lineales.

¿Qué deberíamos esperar de un Mago? Una energía en el trabajo que refleje su nivel de experiencia, competencia y liderazgo. Se espera la capacidad de tener una visión a largo plazo, a la vez que sea altamente eficaz, con un extraordinario manejo del tiempo y los recursos, para alcanzar resultados de alta calidad de manera consistente. También se espera que el Mago sea flexible para enfrentar situaciones nuevas y desconocidas, con confianza y consistencia, y encontrar soluciones creativas a los desafíos que surjan. Por último, es deseable que demuestre un alto nivel de integridad y ética. Esto incluye mantener altos estándares de calidad en sus decisiones y transparente en sus interacciones, además de actuar de manera justa y ética en todas las situaciones.

El Mago es el estadio de la energía más sutil.

Breve digresión. En la primera de las tres películas de la saga de "El Señor de los Anillos", llamada "La Comunidad del Anillo", se puede apreciar el arquetipo del Mago haciendo gala de todos sus poderes. Gandalf el Gris, el Mago de la saga, lidera a su grupo a una encrucijada en la que se bifurcaba el camino, en el paraje conocido como las Minas de Moria.

En un momento de indecisión, Gandalf menciona que no puede recordar cuál de los dos caminos es correcto tomar. Y entonces se produce la escena curiosa y "mágica": recurre a su nariz. Gandalf se pone a oler el aire para intentar discernir cuál es la mejor opción. Finalmente decide seguir el camino que considera más seguro, como se lo indica su olfato.

La escena es una metáfora apropiada para el tipo de reflexión filosófica que propone este paper, ya que los magos organizacionales (las posiciones más altas de la jerarquía organizacional) muchas veces toman decisiones "siguiendo a su nariz". Aunque esto no representa ningún poder mágico, sino la suma de experiencia concentrada y la posibilidad de realizarlo si tener que dar explicaciones inmediatas. Cargarán en soledad con el peso de la decisión.

Pero en términos de lo que nos interesa en este trabajo, ante este tipo de encrucijadas que se experimentan en altos niveles organizacionales, ¿cómo funciona el sentido de vulnerabilidad?

Este tipo de situaciones, tienden a generar un sentido de vulnerabilidad cada vez más alto, cuando se cuenta con los equipos adecuados, capaces de gestionar la información necesaria para la toma de decisiones. El alto nivel de las organizaciones recibe todo el peso del accountability ante los eventos mayores. A modo de comentario de color, cerramos este arquetipo diciendo que en la ficción podemos encontrar "personajes" similares a las personas reales que encontramos en la realidad: existen los magos "blancos" y los magos "negros". Con diferente sensibilidad para tomar decisiones en función de riesgos y alocar recursos, y diferente tolerancia al riesgo.

2.9. Sexto arquetipo organizacional: Maestro

Acercándonos al cierre del círculo, el ante último arquetipo de este viaje por el liderazgo y su vínculo con el Sentido de Vulnerabilidad, es la figura del Maestro, símbolo de la transmisión del conocimiento.

El arquetipo del Maestro es un símbolo de sabiduría. A través de la enseñanza y el ejemplo, ayuda a otros a encontrar su camino en el mundo y a descubrir su verdadero potencial. El Maestro actúa como mentor y guía para aquellos que buscan aprender y crecer. Proporciona orientación, apoyo y dirección a sus alumnos, ayudándoles a desarrollar sus habilidades y a alcanzar su máximo potencial e inspirándolos a convertirse en versiones más completas y realizadas de sí mismos.

A pesar de su sabiduría y su papel fundamental como guía y mentor, el arquetipo del Maestro en la vida organizacional presenta debilidades o desafíos. A veces, puede volverse demasiado rígido en sus creencias y enfoques, lo que puede dificultar su capacidad para adaptarse a nuevas ideas o perspectivas. Si un maestro se vuelve complaciente o deja de buscar continuamente el aprendizaje y el crecimiento personal, puede quedarse atrás en cuanto a nuevos conocimientos y prácticas educativas.

A su vez, si el Arquetipo a recorrido el camino organizacional con riqueza de experiencia, está listo para transmitir las tres preguntas clave para desarrollar rutinas de pensamiento que impulsen el Sentido de Vulnerabilidad de los líderes y sus equipos ("¿qué puede salir mal?, ¿qué tan mal puede salir? y ¿con qué frecuencia puede ocurrir?") (f) en los diferentes segmentos jerárquicos de una organización.
El maestro es el que entiende el poder de muchos (f), el poder de la cultura. El sentido de vulnerabilidad está en un punto alto. Porque está el rol de transmitirlo y a la vez reforzarlo en el interior. Un Maestro es alguien que recorrió el camino, recuerda cómo lo hizo y lo puede narrar, acorde a la necesidad de quien escucha. Todavía tiene la energía y la misión de enseñar. Y comienza el interés por el legado. El Maestro es un arquetipo que las organizaciones tienden a valorar poco, y que rara vez encuentra un lugar preponderante. La llama del Maestro se apaga

lentamente en un mundo organizacional cada vez más existente, pero suelen encontrar un lugar apropiado en organizaciones del tercer sector o generadoras de conocimiento.

Es el ante último arquetipo de este recorrido, pero el último que se encuentra en las organizaciones, que no tienen lugar para el Sabio.

2.10. Séptimo arquetipo organizacional: Sabio

El sabio es quien completa el ciclo y cierra el círculo del conocimiento. Es una figura poderosa en y representa la búsqueda de la comprensión profunda y el conocimiento trascendental. Este arquetipo, al igual que el Mago, aparece en mitologías y leyendas de las más diversas tradiciones espirituales alrededor del globo.

La popular frase "Solo sé que no sé nada" atribuida a Sócrates refleja la idea de que el conocimiento verdadero comienza con el reconocimiento de nuestros propios límites y de nuestra propia ignorancia. Sugiere que la búsqueda constante de la verdad y el conocimiento requiere un profundo conocimiento y comprensión del mundo, así como una capacidad para ver más allá de lo superficial y entender las verdades universales, pero a la vez reconocer que siempre hay más por aprender. Estar dispuesto a escuchar y a considerar diferentes perspectivas.

El arquetipo del Sabio tiene como una de sus virtudes la paciencia. Acepta el ritmo natural de la vida y comprende que incorporar conocimiento profundo sobre las cosas requiere tiempo de calidad para la reflexión. El sabio tiene una mente clara y enfocada, capaz de discernir la verdad de la ilusión y encontrar soluciones innovadoras a los problemas complejos. Pero posee menor energía vital.

Es posible pensar su Sentido de Vulnerabilidad dentro de la lógica propia del arquetipo, pero por fuera del mundo organizacional, con un profundo respecto por los ciclos de vida, los procesos naturales y la paciencia y tolerancia ante el error ajeno.

3. CONCLUSIONES: EL RIESGO NO RESPETA JERARQUÍAS

Antropomorfizar es atribuir características humanas a seres no humanos, objetos o conceptos abstractos. Esta atribución puede incluir emociones, motivaciones, comportamientos o características físicas propias de los humanos a entidades que no las poseen en realidad. Por ejemplo, cuando se dice que un animal "sonríe" o que una máquina "piensa", se está antropomorfizando, ya que se están aplicando cualidades humanas a entidades que no son humanas. En esta línea fue que Mike Mullane, astronauta experto de la NASA y divulgador del concepto Combatir la Normalización de Desvíos, acuñó la expresión "el riesgo no tiene memoria". (f)

Si continuamos antropomorfizando a la noción de riesgo podemos decir que: el riesgo odia a las personas que no cumplen los procedimientos, a los que hacen las cosas a medias, a los que romantizan la improvisación, a los incapaces de cuestionar sus propias ideas, a los que no aprenden de la experiencia y a los que normalizan desvíos. Con los improvisados directamente se ensaña, los detesta, los persigue, les hace la vida imposible. Pero, además, el riesgo odia a todas las personas que lo molestan, que lo invocan, que lo despiertan, al no realizar la tarea correcta, de la forma correcta.

Por el contrario, el riesgo se aburre con las personas que planifican y cumplen con su palabra, que aprenden de la experiencia, que reflexionan a partir de evidencia, que involucran en forma temprana a los acores relevantes, en suma, se aburre con las personas que gestiona los riesgos de forma eficiente. Y entonces los deja en paz, los evita, trata de no cruzárselos en ningún lado.

Así como el recurso de la antroporfización puede ser útil en contextos comunicacionales como un paper, ya que contribuye a hacer determinados conceptos más comprensible o para generar empatía hacia las ideas presentadas, a través de la narrativa de los arquetipos y su impacto en la psique humana, con sus aspectos positivos y sus desafíos pendientes, se busca favorecer una comprensión de la complejidad y la profundidad del liderazgo para la gestión de riesgos.

La intención de este ensayo filosófico, presentado en formato escrito por primera vez en este congreso, es estimular el pensamiento creativo con foco en el liderazgo para la gestión de riesgos e impactos, desde la perspectiva de la psicología positiva de Jung. Se buscó tomar distancia sobre las definiciones elaboradas por las ciencias duras, y visitar otro sector de los trabajos académicos, los elaborados por las ciencias interpretativas.

El futuro es un futuro en el que cada vez se controlan menos variables, en el que imprevisibilidad parece ser la norma, y donde la noción de liderazgo se está redefiniendo desde la práctica y a medida que surge cada nueva tecnología.

El desarrollo del Sentido de Vulnerabilidad es una tarea inherente de los nuevos requerimientos de época para liderar la Gestión de Riesgos, y un desafío actual para llevar a las organizaciones hacia un futuro sostenible. Así como en palabras de Mullane el riesgo no tienen memoria, creemos que tampoco respeta jerarquías, con lo que una visión arquetípica puede ayudarnos en el camino hacia un entendimiento profundo.

Referencias

Any literature directly referenced within the text, or in any tables or figures, must be notated using Arabic superscript numerals. The reference number must correspond to the numbered citation provided in the references section. Citations must be listed in the order in which they appear in the article. References should be styled in accordance with the reference style of the AMA (American Medical Association). For more information about this reference style, please see the AMA Manual of Style.

ANÁLISIS DE ESCENARIO NATECH: DERRAMES DE PETRÓLEO DESENCADENADOS POR EL IMPACTO DE FENÓMENOS HIDROLÓGICOS EXTREMOS EN TANQUES DE ALMACENAMIENTO

JULIO ALFONSO DEL VASTO URUEÑA[1]
Ing. en recursos hídricos y gestión ambiental. M.Sc (c)
SACS Consultores
Bogotá D.C., Colombia
Julio.delvasto@sacsconsultores.com

CAROLINA LÓPEZ PÉREZ
Ing. Química Esp.
SACS Consultores
Bogotá D.C., Colombia
director.consultoria@sacsconsultores.com

Keywords: Inundación, riesgo tecnológico, tanque de almacenamiento de hidrocarburo, derrame de petróleo, NaTech, simulación computacional de derrame.

Abstract

RESUMEN

La creciente industrialización trae consigo un incremento en el almacenamiento y manejo de sustancias peligrosas y la ocupación de nuevas zonas para el emplazamiento de instalaciones, cuya localización es seleccionada, a menudo, sin tener en cuenta la exposición a amenazas naturales, configurando de esta forma un escenario de riesgo tecnológico, en el que como consecuencia de la manifestación de un evento natural sobre la infraestructura industrial, pueden liberarse accidentalmente agentes nocivos para la salud pública y el ambiente, eventos que se conocen bajo el término de NaTech (accidentes tecnológicos desencadenados por eventos de origen natural - Natural disasters triggering TECHnological accidents), (Cruz et al., 2004; Showalter & Myers, 1994).

El diseño y operación de equipos, procesos, estructuras e instalaciones petroleras, sigue un conjunto de normas y estándares internacionales de la industria, que en términos de seguridad, son propuestos para soportar las fuerzas operacionales o internas del proceso (temperatura, presión, caudal, flujos, etc.), sin considerar la acción de fuerzas externas (Cozzani et al., 2009), como por ejemplo, aquellas ejercidas por la velocidad (esfuerzos

[1] *Aspirante a título de maestría en ingeniería civil, énfasis en recursos hidráulicos y medio ambiente. Escuela Colombiana de Ingeniería Julio Garavito. Director de tesis: Germán Ricardo Santos Granados P.hD. Bogotá D.C.*

cortantes) y la profundidad de la lámina de agua, durante la manifestación de eventos de inundación.

Es así como el presente documento tiene como objetivo analizar la fuerza ejercida por un evento hidrológico extremo sobre una instalación industrial, específicamente sobre un tanque vertical de almacenamiento de petróleo; establecer el modo o tipo en que falla la vasija y las consecuencias desencadenadas por la pérdida de contención (LOC) del hidrocarburo, empleando técnicas de modelamiento bidimensional para analizar la inundación, el comportamiento del derrame y el incendio de piscina generado por el derrame.

ABSTRACT

The increasing industrialization brings with it an increase in the handling of hazardous substances (toxic and ecotoxic) and the occupation of new areas for the placement of facilities, whose location is often chosen without considering exposure to natural threats, thus creating a technological risk scenario in which, as a consequence of the occurrence of a natural event on the infrastructure, harmful agents can be accidentally released into public health and the environment, events known as NaTech (technological accidents triggered by natural events - Natural disasters triggering TECHnological accidents).

The design and operation of equipment, processes, structures, and oil facilities follow a set of international industry standards and norms that, in terms of safety, are designed to withstand operational or internal process forces (temperature, pressure, flow rate, flows, etc.), without considering the action of external forces (Cozzani et al., 2009), such as those exerted by velocity (shear forces) and water depth during flooding events.

Thus, the present work aims to analyze the force exerted by an extreme hydrological event on an industrial installation, specifically on a vertical petroleum storage tank; the mode or type of vessel failure and the consequences resulting from the loss of hydrocarbon containment (LOC), using two-dimensional modeling techniques to analyze the flooding, spill behavior, and pool fire generated by the spill.

1 Generalidades del riesgo NaTech

La creciente industrialización trae consigo un incremento en el almacenamiento y manejo de sustancias peligrosas y la ocupación de nuevas zonas para el emplazamiento de instalaciones, cuya localización es seleccionada, a menudo, sin tener en cuenta la exposición a amenazas naturales, configurando de esta forma un escenario de riesgo tecnológico, en el que como consecuencia de la manifestación de un evento natural sobre la infraestructura industrial, pueden liberarse accidentalmente agentes nocivos para la salud pública y el ambiente, eventos que se conocen bajo el término de NaTech (accidentes tecnológicos desencadenados por eventos de origen natural - Natural disasters triggering TECHnological accidents), (Cruz et al., 2004; Showalter & Myers, 1994).

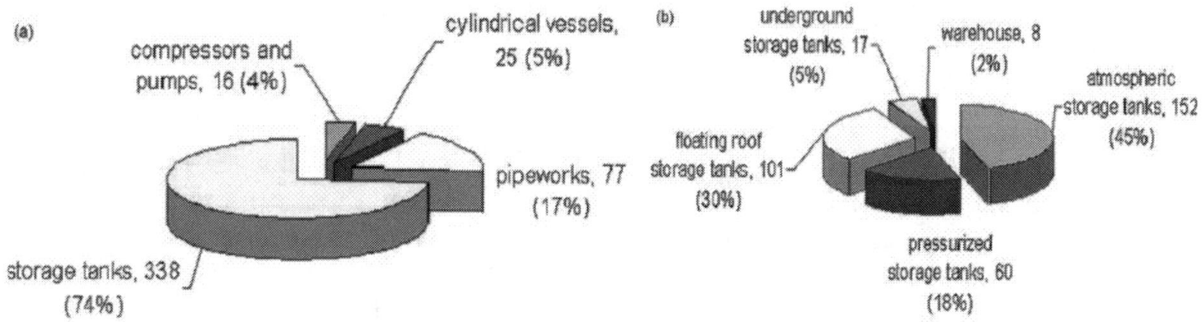

Imagen 1. *Categorías de equipos principalmente implicados en los accidentes provocados por inundaciones: (a) categorías generales; (b) detalle de tanques de almacenamiento. Fuente: V. Cozzani et al., 2010*

Aspectos como la industrialización, el crecimiento poblacional, la densificación de comunidades y el cambio climático, han provocado que la frecuencia y la severidad de los eventos NaTech vayan en aumento (Krausmann et al., 2017) con las graves consecuencias que traen consigo, como las conocidas en los sucesos catastróficos ocurridos en Kocaeli – Turquía (1999), Tohoku – Japón (2011), Ventanilla – Perú (2022) y Cantón Quijos – Ecuador (2023), entre otros, demostrando, en términos generales, la fragilidad de los procesos, equipos e instalaciones frente a fuerzas externas (Di Maio, F et al., 2022) y de otro lado, legislaciones de adaptación al cambio climático con vacíos técnicos en la normatividad de gestión del riesgo alrededor de los eventos NaTech, que dificultan la institucionalidad y la gobernanza (Pilone, E. et al., 2022).

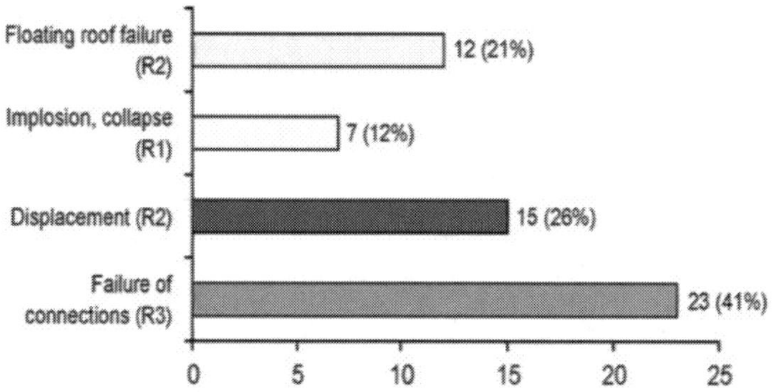

Imagen 2. *Datos disponibles sobre los tipos de daños estructurales experimentados por los equipos de proceso durante las inundaciones y la categoría de liberación asociada. Resultados obtenidos del análisis de 57 sucesos. Fuente: V. Cozzani et al., 2010*

La fase y peligrosidad de la sustancia implicada en el LOC en un evento NaTech, depende del proceso, equipo e instalación afectada por el fenómeno natural, y la severidad del suceso en términos de cantidad liberada, dispersión y receptores afectados, depende del tipo de fenómeno natural que desencadena la pérdida de contención accidental, siendo las inundaciones, tormentas eléctricas y sismos (Krausmann et al., 2008), los principales iniciadores de este tipo de incidentes. En ese orden de ideas, la infraestructura petrolera, como industria de alto riesgo (Yanting & Liyun, 2011), cuya localización o emplazamiento en el contexto mundial, cubre vastas zonas del territorio, está expuesta a diferentes amenazas naturales, siendo los tanques de almacenamiento de hidrocarburos, los equipos más afectados (V. Cozzani et al., 2010) y con mayor criticidad por los altos inventarios de petróleo que potencialmente serían derramados en un evento de este tipo (Imagen 1).

Entre los principales daños o afectaciones que sufren este tipo de equipos debido a las inundaciones, según Campedel, 2009, se encuentran daños estructurales (desplazamientos, impactos con objetos flotantes, colapso) y fallas eléctricas en la sensórica que automatiza los procesos asociados (imagen 2), mientras que en las consecuencias más graves están aquellas relacionadas con la dispersión y reacción química de la sustancia peligrosa en agua (Imagen 3).

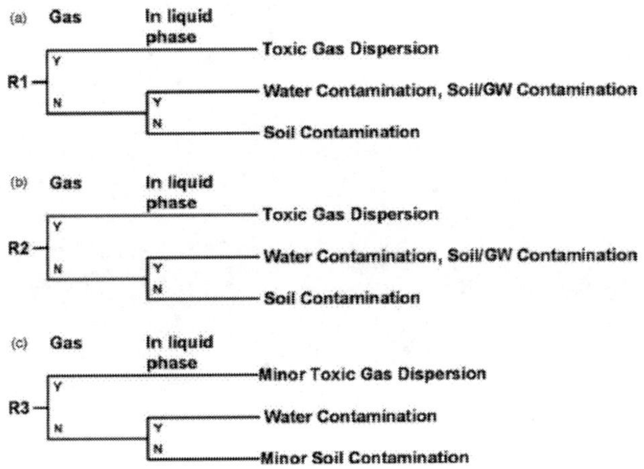

Imagen 3. *Árboles de eventos posteriores a la liberación de sustancias tóxicas en accidentes NaTech provocados por inundaciones. (a) Categoría de vertido R1; (b) Categoría de vertido R2; (c) Categoría de vertido R3. Fuente: V. Cozzani et al., 2010*

De acuerdo con la U. C. Lovain (2023), desde la década de 1970, el incremento en la ocurrencia de eventos naturales en el mundo se ha acentuado considerablemente, registrando aumentos de más del 500% con respecto al año 2022 *(imagen no. 4)*, mientras que en el caso específico de los eventos hidrológicos extremos *(imagen no. 5)*, el incremento llega a más del 400% con respecto al mismo año, cifras que guardan una estrecha correlación con la ocurrencia de eventos NaTech, según Ricci et. al, 2022 *(imagen no. 6)*.

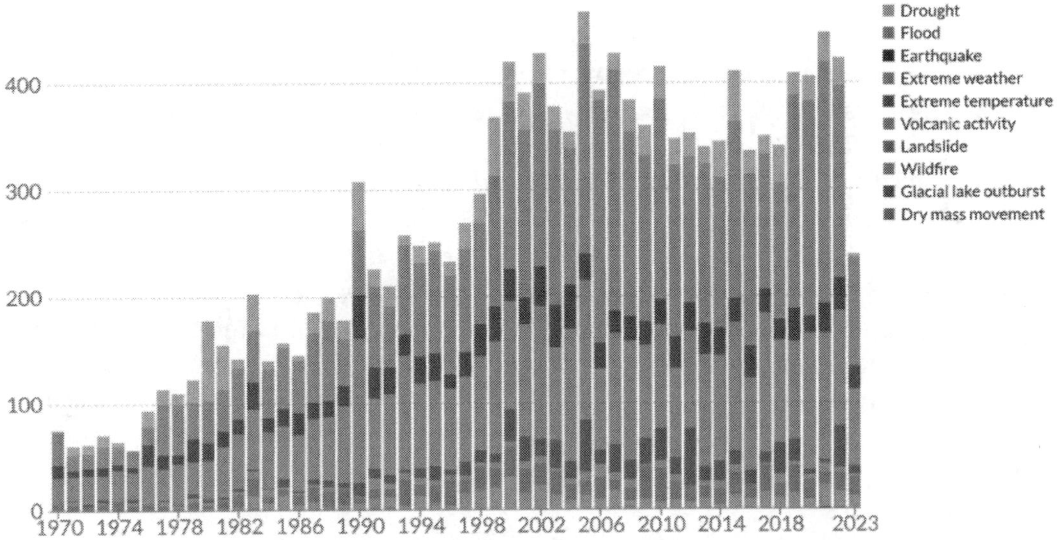

Imagen 4. *– Reporte global de desastres naturales por tipo entre 1970 y 2023.*
Fuente: EM-DAT, CRED / UCLouvain (2023). https://ourworldindata.org/natural-disasters

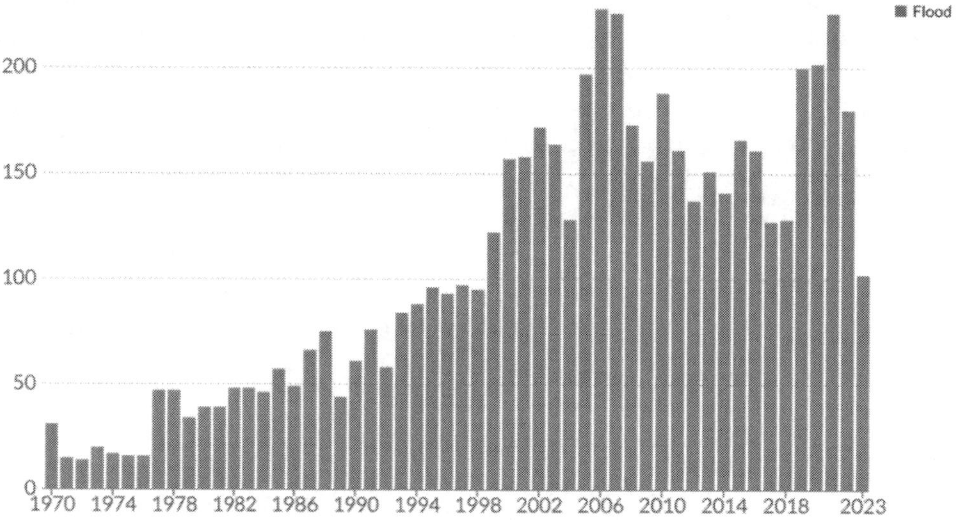

Imagen 5. *– Reporte global de desastres naturales por inundaciones entre 1970 y 2023. Fuente: EM-DAT, CRED / UCLouvain (2023).* https://ourworldindata.org/natural-disasters

Por su parte, la dinámica petrolera en el ámbito mundial ha estado marcada por la ocupación de diferentes zonas de la geografía mediante la concesión y explotación de bloques o campos de producción, algunos de estos ubicados en áreas susceptibles a inundaciones, de allí que, diferentes facilidades operativas de hidrocarburo (pozos, líneas de producción, oleoductos, estaciones, refinerías, etc.), se encuentren bajo alguna condición o nivel de riesgo respecto a este tipo de fenómenos naturales.

Por ejemplo, en Colombia, un país particularmente expuesto a los incrementos en la precipitación por la ocurrencia de fenómenos climáticos de meso y macro escala, como en el caso de La Niña, que se traducen en la manifestación de eventos hidrológicos extremos como las inundaciones (IDEAM, 2011), sumado a la vocación petrolífera a lo largo y ancho del territorio, se crean las condiciones para la ocurrencia de eventos tipo NaTech, en los que, entre otros sucesos consecuentes, existe el alto riesgo de incidentes relacionados con los derrames de hidrocarburos en ambientes acuáticos y terrestres, cuya particularidad es la ausencia de un marco metodológico que permita analizar predictivamente el comportamiento de la sustancia liberada al ambiente y sus efectos sobre los receptores o elementos vulnerables *(imagen no. 7).*

Particularmente, el diseño y operación de equipos, procesos, estructuras e instalaciones petroleras, sigue un conjunto de normas y estándares internacionales de la industria, que en términos de seguridad, son propuestos para soportar las fuerzas operacionales o internas del proceso (temperatura, presión, caudal, flujos, etc.), sin considerar la acción de fuerzas externas (Cozzani et al., 2009), como por ejemplo, aquellas ejercidas por la velocidad (esfuerzos cortantes) y la profundidad de la lámina de agua, durante la manifestación de eventos de inundación.

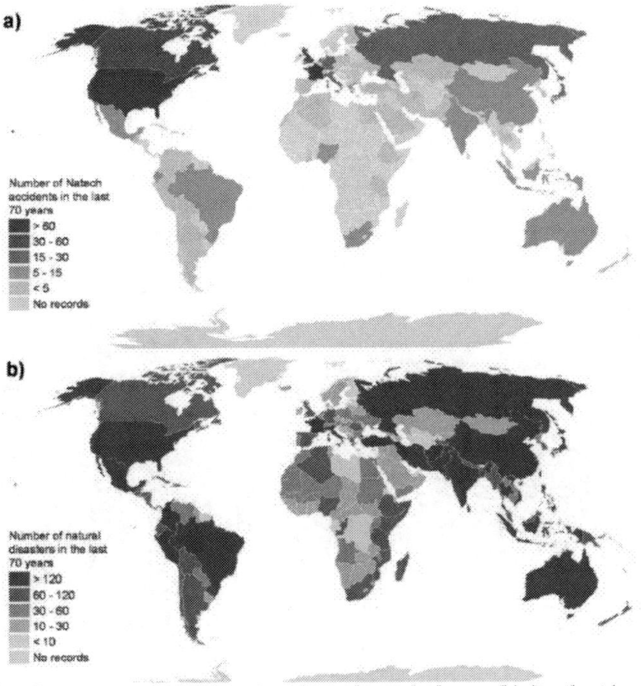

Imagen 6. *Distribución geográfica de eventos NaTech (a) registrados en bases de datos y (b) distribución geográfica de desastres naturales registrados por el Centre of Research on the Epidemiology of Disasters (CRED,2020) en los últimos 70 años.*

En este sentido, la ausencia de estas barreras de seguridad en los estándares de diseño, construcción y operación de equipos y procesos, constituyen una condición de debilitamiento de las instalaciones frente a amenazas naturales y por ende, abre la posibilidad a la ocurrencia de un evento tipo NaTech, cuyo efecto dominó puede conllevar a la pérdida de contención (LOC por sus siglas en inglés) de sustancias altamente peligrosas.

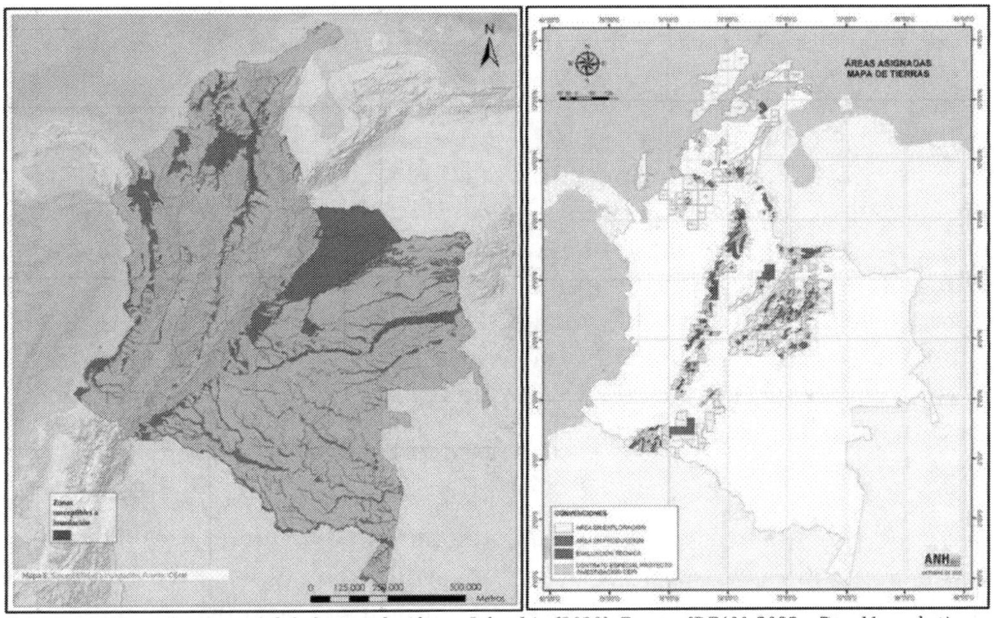

Imagen No. 7. – Izq. *Mapa de susceptibilidad a inundación en Colombia (2023). Fuente: IDEAM, 2023 –* ***Der.*** *Mapa de tierras de áreas asignada para la exploración y explotación de hidrocarburos. Fuente: ANH, 2023*

Misuri & Cozzani (2021), proponen un marco de evaluación del riesgo para eventos NaTech en el que no solo se considera un impacto directo del fenómeno natural sobre el equipo o el proceso (imagen 8, a), sino que, dada la complejidad de estos eventos en relación con el nivel de daño, extensión de las afectaciones y evolución o desarrollo multiamenaza de dicho fenómeno, el marco de evaluación puede llegar a ser panorámicamente más amplio, de forma tal que se incluyan rutas indirectas de afectación (imagen 8, b), las cuales se caracterizan por el daño o deterioro de los sistemas de sustentación y/o respaldo que permiten un adecuado funcionamiento de los equipos o procesos, ocasionando la pérdida de contención de la sustancia peligrosa.

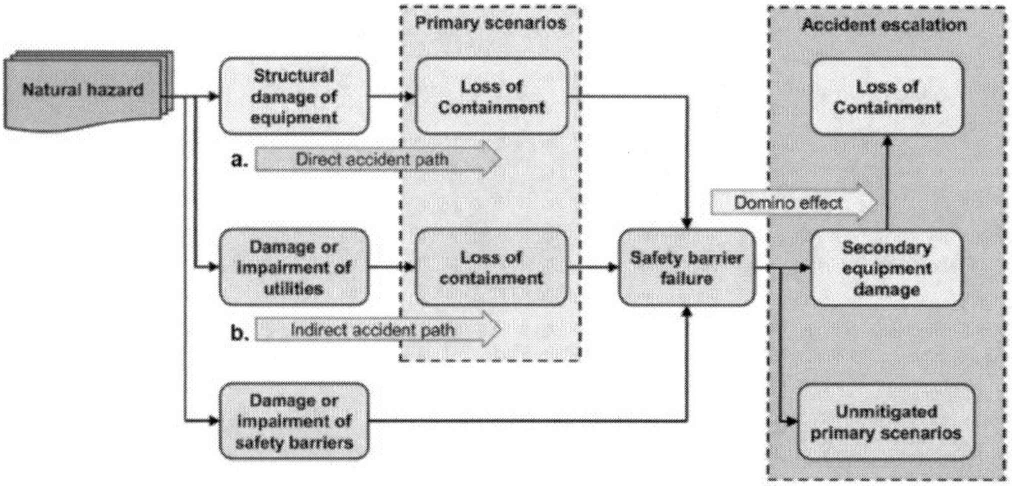

Imagen No. 8. *– Evaluación de sucesos Natech basado en las lecciones aprendidas de sucesos anteriores relativos al fallo de sistemas auxiliares y barreras de seguridad. Fuente: Cozzani & Misuri. 2021. A paradigm shift in the assessment of Natech scenarios in chemicaland process facilities*

2 Análisis de escenario NaTech: Derrames de hidrocarburos en tanques de almacenamiento, desencadenados por inundaciones

Investigaciones realizadas por Antonioni et al., 2009, Antonioni et al., 2015, respecto a la evaluación cuantitativa del riesgo (ECR) por derrames de petróleo en tanques de almacenamiento debido a inundaciones, proponen el análisis a partir de la caracterización de la amenaza natural y de la vulnerabilidad de los tanques, este último en función de modelos de fragilidad que determinan la probabilidad de daño en el equipo, mientras que para el fenómeno amenazante plantean el análisis del evento hidrológico, a partir de las fuerzas externas ejercidas por la profundidad de la lámina de agua y por la velocidad de la corriente, asociadas a un determinado periodo de retorno (Imagen 8), semejante a lo evidenciado por V. Cozzani et al., 2010 y V. Cozzani et al., 2014, en cuanto a la relación hallada entre la profundidad de la inundación y la ocurrencia de eventos NaTech (Imagen 9, derecha).

En ese orden de ideas, la selección del evento hidrológico, como punto de partida del análisis, dependerá de los criterios establecidos en la evaluación en cuanto a la magnitud y probabilidad de la inundación de interés, en consideración a que la ocurrencia de estos sucesos está determinada por un periodo de retorno (t_r) en años cuya frecuencia se estima mediante (Antonioni et al., 2015):

$$f = \frac{1}{t_r} \quad (1)$$

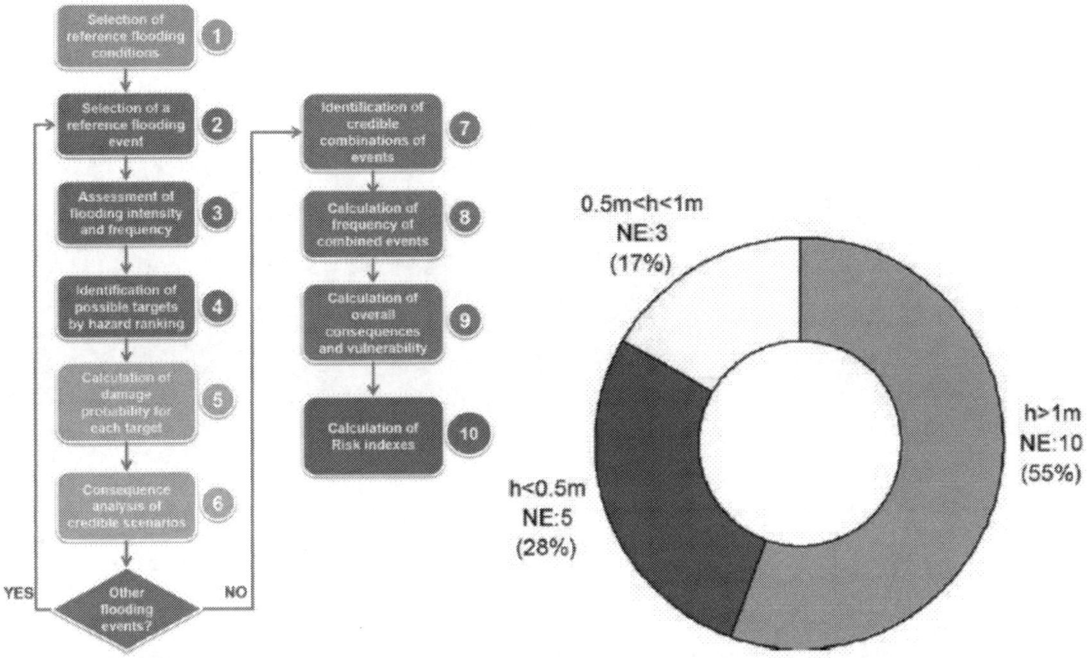

Imagen No. 9. – Izq*. Metodología para la evaluación cuantitativa del riesgo de los escenarios NaTech desencadenados por eventos hidrológicos extremos. Fuente: Antonioni et al., 2015 –* ***Der****. Clasificación de 18 eventos NaTech provocados por inundaciones en función de la altura máxima del agua (h: altura máxima del agua, m; NE: número de eventos). Fuente: V. Cozzani et al., 2010*

A partir de allí, empleando métodos de simulación hidráulica, se obtienen profundidades (h_W) y velocidades (v_W) de la mancha de inundación que afecta el equipo en análisis (tanque de almacenamiento), teniendo en cuenta que mediante estas dos variables se simplifican los efectos hidráulicos asociados a las fuerzas externas ejercidas por la inmersión lenta, oleaje de alta profundidad, oleaje de velocidad moderada, oleaje de alta velocidad y finalmente oleaje de alta velocidad con profundidad limitada (V. Cozzani et al., 2014).

Posteriormente es necesario determinar el tipo y el nivel de daño ocasionado en el equipo expuesto, debido a la inundación. Para ello diversos autores han propuesto modelos mecánicos de falla que permiten estimar con cierta aproximación la vulnerabilidad del equipo, entre ellos Salzano et al., 2003, G. Antonioni et al., 2009, G. Landucci et al., 2012 y G. Antonioni et al., 2015, cuyas contribuciones se representan mediante curvas de fragilidad (imagen 10 e imagen 11), modelos mecánicos de falla bajo carga de inundación ((2) a (8)) y Tabla 1.

Determinado el modelo de vulnerabilidad del equipo frente al evento hidrológico extremo, se establece el modo de falla que genera la consecuencia de interés, para lo cual V. Cozzani et al., 2010, proponen a partir del árbol de eventos (Imagen 3), tres (3) modos de falla en el tanque a raíz de las fuerzas externas ejercidas por la inundación: *R1* - Fuga instantánea de todo el tanque; *R2* – Fuga continúa de todo el contenido del tanque en un determinado periodo, debido a la rotura catastrófica de la conexión de mayor diámetro o por roturas en su estructura. *R3* – Fugas o descargas menores por roturas parciales en las conexiones del tanque o por roturas catastróficas de conexiones de menor diámetro, mientras que Antonioni et al., 2015, plantean una falla catastrófica que desencadena una fuga instantánea de todo el recipiente.

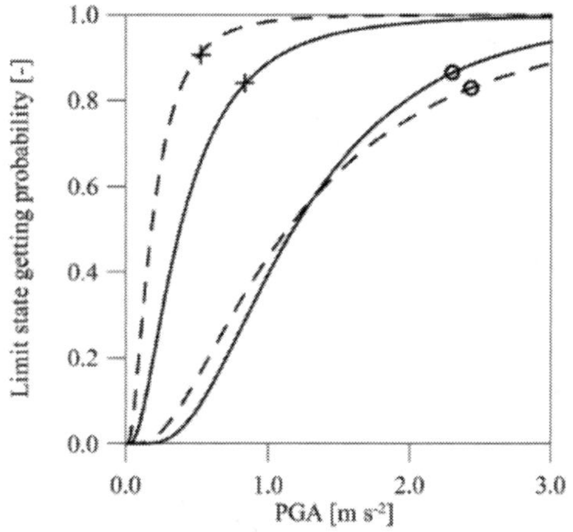

***Imagen No. 10.** – Curvas de fragilidad calculadas para tanques atmosféricos de acero afectados por sismos. Fuente: Salzano et al., 2003*

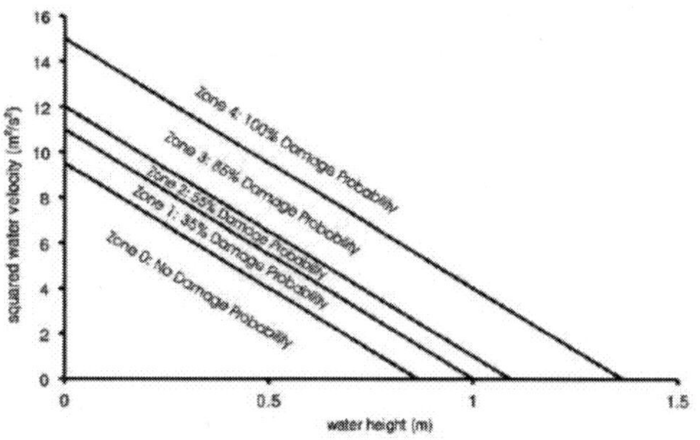

***Imagen No. 11.** – Estimación de la probabilidad de daño del equipo con respecto a la altura máxima del agua (h) y del cuadrado de la velocidad máxima del agua (V²). Fuente: Antonioni et al., 2009*

$$P_w = P_{ws} + P_{wd} \quad \textbf{(2)}$$

$$P_{ws} = \rho_w \, g \, h_w \quad \textbf{(3)}$$

$$P_{wd} = \frac{1}{2}\rho_w k_w v_w^2 \quad \textbf{(4)}$$

$$P_f = \rho_f g H \phi \quad \textbf{(5)}$$

$$P_{net} = P_{ws} + P_{wd} - P_f \quad \textbf{(6)}$$

$$P_{net} \geq P_{cr} \quad \textbf{(7)}$$

$$P_{ws} + P_{wd} - P_f \geq P_{cr} \Rightarrow P_{ws} + P_{wd} \geq P_{cr} + P_f \Rightarrow P_w \geq P_{cr} + P_f \quad \textbf{(8)}$$

Donde:

- P_w: Carga ejercida por la inundación.
- P_{ws}: Carga hidrostática ejercida por la lámina del agua.
- P_{wd}: Carga asociada a la energía cinética del agua.
- g: Constante de gravedad.
- h_w: Profundidad de la lámina de agua.
- ρ_w: Densidad del agua.
- k_w: Coeficiente hidrodinámico.
- V_w: velocidad del agua.
- P_f: Presión interna del tanque.
- ρ_f: Densidad del fluido contenido en el tanque.
- H: Altura del tanque.
- Φ: Nivel de llenado del tanque.
- P_{net}: Presión neta sobre el tanque.
- P_{cr}: Presión crítica

Ecuaciones del modelo mecánico de falla propuesto por Salzano et al., 2003. - Fuente: Salzano et al., 2003

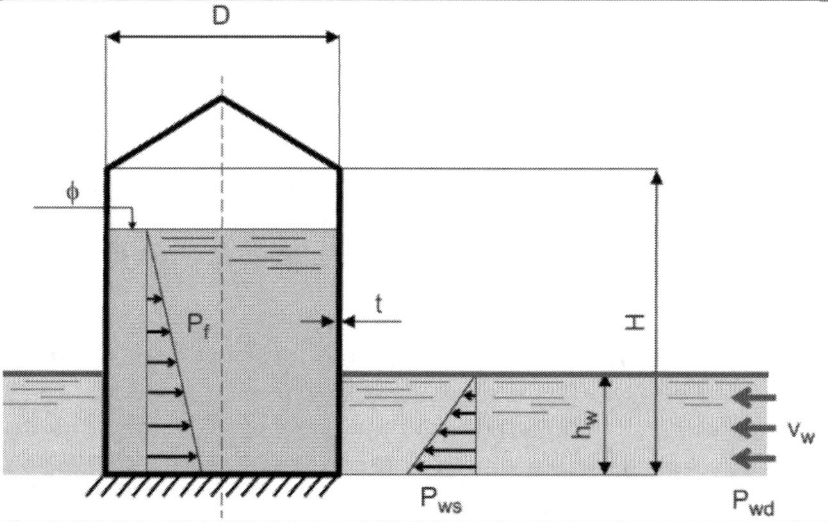

***Imagen 12**. Esquematización de las fuerzas que actúan sobre tanques atmosféricos verticales inundados.*

Fuente: Salzano et al., 2003

Item	Definition	Value/Equation
Vulnerability model equations		
CFL_v	Critical Filling Level for vertical atmospheric vessels	$CFL_v = \left(\frac{\rho_w k_w}{2} v_w^2 + \rho_w g h_w - P_{cr}\right) / \rho_f g H$
P_{cr}	Vessel critical pressure (Pa) evaluated with the proposed simplified correlation	$P_{cr} = J_1 C + J_2$ in which $J_1 = -0.199$ $J_2 = 6950$
Ψ	Vessel vulnerability due to flooding	$\Psi = (CFL - \varphi_{min}) / (\varphi_{max} - \varphi_{min})$
Input parameters		
C	Vessel capacity	Small capacity $C < 5000$ m^3 Medium capacity 5000–10000 m^3 Large capacity > 10000 m^3
v_w	Flood water speed [a]	0–3.5 m/s
h_w	Flood water depth [a]	0–4 m
ρ_w	Flood water density	1100 kg/m^3
ρ_f	Stored liquid density	650-1300 kg/m^3
k_w	Hydrodynamic coefficient	1.8
H	Vessel height	Small capacity 3.6–18 m Medium capacity 3.6–16.2 m Large capacity 3.6-7.2 m
g	Gravity acceleration	9.81 m/s^2
ϕ_{min}	Minimum operative filling level	0.01
ϕ_{max}	Maximum operative filling level	0.75

Tabla 1. *Modelo de vulnerabilidad y parámetros de entrada para tanques cilíndricos atmosféricos involucrados en eventos de inundación. CFL: Nivel de llenado crítico. Fuente: Antonioni et al., 2015*

Por su parte Rojas, S., 2005, Guerrero, Y., 2007 y Zapata C. et al.,2002, han evaluado diversas aplicaciones computacionales para analizar predictivamente el comportamiento de derrames de hidrocarburo en ambientes acuáticos, principalmente en aguas continentales en Colombia, a fin de dimensionar las consecuencias potenciales, empleando diferentes software de modelado de tipo Lagrangiano que convergen en el cálculo de los procesos de desgaste o intemperización del derrame y de los efectos de la hidrodinámica en el transporte del derrame, siendo comunes la estimación de la advección, trayectoria, destino y las pérdidas por evaporación, dispersión mecánica, adherencia a orillas o frontera terrestre y la disolución, en modelos con diferentes capacidades y tipos de licencia.

Imagen No. 13. *– Marea negra formada después de los daños por inundación en los tanques de almacenamiento en Coffeyville, EE. UU., en 2007. Fuente: Necci, Amos & Wood, Maureen & Krausmann, Elisabeth & Girgin, Serkan. (2020). Natech risk management Common Inspections Criteria.*

Las formulaciones matemáticas más empleadas en el análisis del comportamiento de derrames en ambientes acuáticos se muestran en la siguiente tabla:

<table>
<tr><th>Proceso</th><th>Formulación</th></tr>
<tr><td>Transporte (Algoritmo discreto Lagrangiano)</td><td>

$$\vec{V}_t = \vec{V} + \vec{V}^{!}$$

Donde:

- $\vec{V_t}$: Velocidad de deriva.
- $\vec{V}$: Velocidad media superficial debido a la combinación de los efectos del viento y la corriente.
- $\vec{V}!$: Fluctuación turbulenta – Dispersión horizontal de derrame de crudo.

</td></tr>
<tr><td>Hidrodinámica de la corriente</td><td>

- Ecuación de Continuidad

$$\frac{\partial Q}{\partial x} + \frac{\partial A}{\partial t} = 0$$

- Ecuación de Momentum

$$\frac{\partial Q}{\partial t} + \frac{\partial \left(Q^2 / A \right)}{\partial x} + gA\left(\frac{\partial y}{\partial x} - S_0 \right) + gAS_f = 0$$

Donde:

- *x*: Distancia en metros del tramo a evaluar.
- *t:* Tiempo en segundos
- *A:* Área de la sección transversal
- *y:* Es la profundidad del agua.
- *So:* Pendiente del fondo del río.
- *Sf:* Pendiente de la gradiente de energía.
- *g:* Aceleración de la gravedad.

</td></tr>
<tr><td>Advección</td><td>

$$\vec{V} = \alpha_w \vec{V}_w + \alpha_c \vec{V}_c$$

Donde:

- $\vec{V}$: Velocidad de deriva por advección.
- $\vec{V_w}$: Velocidad del viento a diez metros de la superficie del agua.
- $\vec{V_c}$: Velocidad media de la corriente respecto a su profundidad.
- $\vec{\alpha_w}$: Factor de empuje del viento sobre la capa de crudo derramada.
- $\vec{\alpha_c}$: Factor de empuje que mueve la capa de crudo debido a la velocidad media de la corriente.

</td></tr>
<tr><td>Dispersión mecánica</td><td>

Tabla. Ley de dispersión para derrames de crudo (Fay, 1971, Hoult, 1979 y Walkman et al., 1973)

<table>
<tr><th>Fase de Dispersión</th><th>Ancho, Le</th><th>Radio, Re</th></tr>
<tr><td>Gravedad-Inercia</td><td>$1.39\left(\Delta g A t^2\right)^{1/3}$</td><td>$1.14\left(\Delta g V t^2\right)^{1/4}$</td></tr>
<tr><td>Gravedad-Viscosidad</td><td>$1.39\left(\Delta g A^2 t^{3/2} \nu^{-1/2}\right)^{1/4}$</td><td>$0.98\left(\Delta g V^2 t^{3/2} \nu^{-1/2}\right)^{1/6}$</td></tr>
<tr><td>Tensión Superficial-Viscosidad</td><td>$1.43\left(\sigma^2 t^3 \rho_w^{-2} \nu^{-1}\right)^{1/4}$</td><td>$1.60\left(\sigma^2 t^3 \rho_w^{-2} \nu^{-1}\right)^{1/4}$</td></tr>
</table>

Nota: A=0.5 volumen del crudo/longitud del derrame. V= Volumen del derrame de crudo. $\Delta = 1 - \left(\rho_o / \rho_w \right)$.

ν = viscosidad cinemática del agua.

</td></tr>
<tr><td>Adherencia a orillas</td><td>

$$\forall_2 = \forall_1 e^{-k(t_2 - t_1)}$$

</td></tr>
</table>

<table>
<tr><th>Proceso</th><th>Formulación</th></tr>
<tr><td></td><td>
Donde:
• $\forall_1$ y $\forall_2$: Volúmenes de crudo en la frontera en los tiempos t_1 y t_2.
• k: $\left[-\ln\left(\tfrac{1}{2}\right)\right]/\lambda$ y λ es la vida media determinada por la siguiente tabla:
Tabla. Descripción de la ribera de ríos y sus parámetros (Torgrimson, 1980)
<table>
<tr><th>DESCRIPCION DE LA RIBERA</th><th>VIDA MEDIA</th></tr>
<tr><td></td><td></td></tr>
<tr><td>Cabo o Promontorio</td><td>1 hr</td></tr>
<tr><td>Plataforma anti-oleaje</td><td>1 hr</td></tr>
<tr><td>POCKET BEACH</td><td>1 dia</td></tr>
<tr><td>Playa de Arena</td><td>1 dia</td></tr>
<tr><td>Playa de Arena y Grava</td><td>1 dia</td></tr>
<tr><td>Playa de Arena y Guijarro</td><td>1 año</td></tr>
<tr><td>EXPOSED TIDE FLATS</td><td>1 hr</td></tr>
<tr><td>Ribera de roca protegida</td><td>1 año</td></tr>
<tr><td>SHELTERED TIDE FALT</td><td>1 año</td></tr>
<tr><td>Humedal Protegido</td><td>1 año</td></tr>
<tr><td>Tierra</td><td>1 año</td></tr>
</table>
</td></tr>
<tr><td>Evaporación</td><td>
$$F = \left(\frac{1}{C}\right)\left[\ln P_0 + \ln\left(CK_E t + \frac{1}{P_0}\right)\right]$$
Donde:
• t: Es el tiempo transcurrido.
• C: Es una constante de evaporación definida por la siguiente ecuación:
$$C = 1158.9API^{-1.1435}$$
Donde: API es la gravedad específica del crudo bajo la siguiente relación:
$$API = \frac{141.5}{SG} - 131.5$$
Donde: SG es el valor del volumen molar típico para combustibles pétreos que es $200*10^{-6}$ m^3/mol.
• K_E: Es el exponente de la evaporación definido por: $K_E = \frac{0.0025U^{0.7}Av}{RTV_0}$
Donde:
• U: es la velocidad del viento medida a diez metros sobre el nivel del mar.
• A: Área del derrame.
• R: Constante de los gases ideales.
• T: Temperatura del crudo.
• v: Volumen molar.
Además:
• P_o: Presión de vapor inicial en atmósferas a TE, es decir, a temperatura de la superficie, que se define como:
$$\ln P_o = 10.6\left(1 - \frac{T_o}{T_E}\right)$$
Donde:
To: Es la temperatura de referencia o de ebullición del crudo definida por:
</td></tr>
</table>

<table>
<tr><th>Proceso</th><th>Formulación</th></tr>
<tr><td></td><td>

$$T_o = 542.6 - 30.275\,API + 1.565\,API^2 - 0.03439\,API^3 + 0.0002604\,API^4$$

</td></tr>
<tr><td>Emulsificación</td><td>

$$\frac{dF_{wc}}{dt} = K_{wc}(U_{10}+1)^2\left[\frac{1-F_{wc}}{OC_1}\right]$$

Donde:

- F_{wc}: Fracción de agua en aceite
- K_{wc}: Coeficiente de emulsificación de $2x10^{-6}$ para petróleo liviano o 4.5 x 10^{-6} para petróleos pesados.
- U^{10}: Velocidad del viento.
- OC^1: es 0.7 para petróleo crudo liviano y 1.15 para crudos pesados.

</td></tr>
<tr><td>Disolución y dispersión</td><td>

$$\frac{dm}{dt} = KA_sC_0e^{-\infty}$$

Donde:

- m: Masa de la mancha.
- K: Coeficiente de transferencia de la mancha.
- As: Área superficial de la mancha.
- α: Exponente de decaimiento.
- Co: Solubilidad del petróleo en agua fresca.

Los valores de K, Co y α fueron determinados por Yapa (1994) para crudos pesados y ligeros así:

Tabla. Constantes de disolución

Tipo de Crudo	C_o	*K*	KC_o	∞
Pesado	7.88	0.002335	0.0184	0.423
Ligero	21.3	0.041502	0.884	2.380

Fuente: Yapa P. y Shen H. 1994. "Modelling River Oil Spill: A Review, Journal of Hydraulic Reserch."

</td></tr>
<tr><td>Correcciones de la viscosidad</td><td>

Por contenido de agua:

$$\mu = \mu_o \exp\left[\frac{2.5F_{wc}}{1-0.65F_{wc}}\right]$$

Donde:

- μ_o: Es la viscosidad inicial del petróleo
- F_{wc}: Contenido de agua.

Por evaporación:

$$\mu = \mu_o \exp\left[C_2 * F_{evap}\right]$$

Donde:

- F_{evap}: Fracción evaporada de la mancha
- C2: Varía entre 1 y 10, siendo 1 para combustibles livianos.

Por temperatura:

$$\mu = \mu_o \exp\left[8770T_K^{-1} - 29.4\right]$$

Donde:

- Tk es la temperatura ambiental.

</td></tr>
<tr><td>Correcciones por la densidad</td><td>

$$\rho = F_W\rho + (1-F_{evap})\rho_o$$

Donde:

- F_{evap}: Fracción evaporada de la mancha

</td></tr>
</table>

Proceso	Formulación
	• F_w: Contenido de agua. • ρ_o: Densidad del petróleo. • ρ: Densidad del agua.

Tabla 2. *Formulaciones empleadas en el análisis del comportamiento de derrames de hidrocarburos en ríos*
Fuente: Zapata C. et al.,2002

De otro lado, en un esfuerzo por estimar el comportamiento de los incendios de hidrocarburo que ocurren durante un derrame, Imazu & Nishino (2018) proponen considerar dicho derrame como un conjunto de partículas circulares que flotan ardiendo, cuyas posiciones temporales dependen del campo de velocidades y trayectorias de la inundación y que su ignición inicial es impredecible. La zona de combustión y el flujo de calor radiante para cada una de las partículas incendiadas se estima de manera homogénea e independiente de la dirección del viento, desde el centro de la partícula de hidrocarburo y cesa cuando el espesor de cada partícula del derrame alcanza 1 mm.

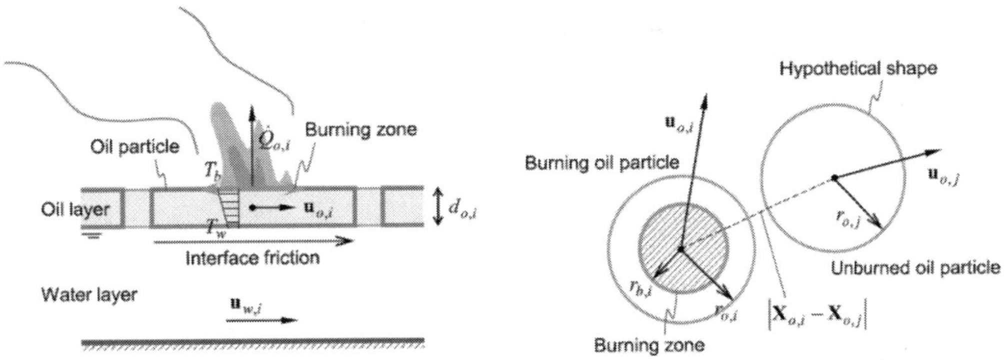

Imagen 14. *Esquema de un modelo de propagación de incendios por un derrame de hidrocarburo*
Fuente: T. Nishino, Y. Imazu, 2018.

2.1 Tratamiento del riesgo NaTech con enfoque de seguridad de procesos

Las capas de protección aplicables a los equipos y procesos expuestos a las fuerzas externas ejercidas por los fenómenos naturales, de acuerdo con el concepto propuesto por CCPS (2001), como barreras de seguridad proporcionan soluciones técnicas para evitar, prevenir o mitigar las pérdidas de contención dentro de un amplio espectro de alternativas que van desde las barreras pasivas hasta las activas.

Lecciones aprendidas de eventos NaTech ocurridos en distintas partes del mundo, recogidas por Misuri et. al (2019) y diversos autores, ponen de manifiesto fallas comunes en las barreras de seguridad asociadas a desabastecimiento de agua de los sistemas de protección contra incendios (Steinberg & Cruz, 2004), sistemas de supresión de vapores a base de agua y espuma no disponibles (Girgin, 2011), sistemas de contención de derrames colmatados (eMars), almacenamiento de químicos de refrigeración no disponibles (Additives for Polymers, 2017), válvulas de accionamiento manual inundadas e inaccesibles (NTSB, 1996) y generadores de energía de respaldo completamente inundados (Labib and Harris, 2015), entre otros.

De acuerdo con Necci A., Krausmann E. (2021), algunos ejemplos para tratar el riesgo asociado a las amenazas naturales pueden ir desde barreras físicas cómo: construcción de diques, construcción de canales artificiales, estabilización de riberas, compactación y estabilización de suelos, reorganización de instalaciones industriales y una correcta localización de la planta, mientras que en términos de medidas procedimentales, la definición de roles y funciones, así como la lista de acciones y tiempos de actuación, contribuirán a la reducción del riesgo.

Ahora bien, cuando sea posible anticiparse al impacto del evento natural en la instalación, a través de pronósticos y predicciones, la implementación de medidas como los apagados de emergencia, evacuaciones de personal no vital operacionalmente, aseguramiento de objetos flotantes y de equipos, instalación preventiva de diques y barreras de contención, además de la comunicación permanente con las autoridades locales y organismos de socorro, junto con el monitoreo de la evolución de la amenaza, serán determinantes en la actuación frente a este tipo de escenarios.

En un escenario NaTech relacionado con la afectación de tanques de almacenamiento de crudo debido a eventos hidrológicos extremos, es crucial centrarse en la seguridad de procesos e identificar los riesgos asociados al evento amenazante y su evolución. Aquí hay algunos aspectos específicos que podrían incluirse como medidas de tratamiento del riesgo:

Respecto al fenómeno amenazante:

- Instalación y monitoreo de sistemas de alerta temprana multicriterio, que permitan evaluar en tiempo real, los pulsos de lluvia detonantes y los niveles y la velocidad de flujo en la corriente amenazante.
- Analizar y caracterizar hidrológica e hidrodinámicamente el comportamiento de la corriente bajo escenarios climáticos extremos, de manera que pueda identificarse la mancha de inundación, las velocidades de flujo, las áreas impactadas por la creciente y los equipos potencialmente expuestos al fenómeno hidrológico.
- Monitorear periódicamente los cambios en la morfodinámica fluvial de la corriente, identificando procesos de erosión y/o sedimentación, que dieran paso cambios en la trayectoria de flujo y el posible incremento del nivel de riesgo.

Respecto al equipo y proceso expuesto:

- Realizar análisis de riesgos de proceso (ARP) involucrando análisis cuantitativos del riesgo (ACR), de forma que se pueda identificar los diferentes sucesos finales o consecuencias (efectos) asociados al fallo individual de la vasija y bajo un efecto dominó.
- Diseño inherentemente seguro del tanque, considerando la resistencia a las fuerzas externas ejercidas por el fenómeno natural amenazante.
- Identificación de equipos y/o unidades complementarias al recipiente que podrían ser afectadas por inundaciones.
- Instalación de mecanismos de anclaje en el tanque y unidades complementarias o acopladas a él.

- Dimensionamiento de los sistemas fijos de contención de derrames como diques, desnatadores y drenajes perimetrales.
- Con base en los resultados de la modelación hidrológica e hidrodinámica de la creciente, construir muros de contención para retener total o parcialmente la inundación.
- Diseño e implementación de mecanismos de apagado o de corte de procesos.
- Diseño e implementación de procedimientos y mecanismos de vaciado y/o trasiego de inventarios almacenados en los tanques expuestos.
- Diseño, construcción y mantenimiento de drenajes de agua lluvia de la instalación industrial.

Respecto al manejo de la emergencia:

- Actualización de planes de emergencia, considerando escenarios de riesgo NaTech.
- Con base en la modelación de consecuencias del evento, dimensionar la capacidad de respuesta (recursos) para atender escenarios de riesgo NaTech.
- Actualización o incorporación de procedimientos y estrategias de respuesta para el manejo de emergencias originadas por el evento NaTech.
- Diseño de estrategias de respuesta internas articulables con las estrategias de respuesta externa de entidades estatales o gubernamentales a nivel local, regional y nacional.
- Establecimiento de un procedimiento de escalamiento del manejo de la emergencia, desde lo operativo hasta instancias de crisis y continuidad de negocio.
- Diseño y desarrollo de programas de entrenamiento y capacitación a los equipos de respuesta interna y externa en el manejo del evento NaTech.
- Diseño y ejecución de un programa de simulacros y simulaciones con los equipos de respuesta internos y externos, en el manejo del evento NaTech.

2.2 Estudio de caso

Con el fin de evaluar los planteamientos metodológicos propuestos por los diversos autores relacionados en el numeral 2 del presente documento, se ha analizado el comportamiento de un hipotético evento NaTech en una instalación industrial que recibe, procesa, trata y despacha gas húmedo, ubicada en la planicie inundable de un río de gran caudal.

Como insumos para comprender el comportamiento de un fenómeno hidrológico extremo en el área de estudio, se empleó la recolección de información topográfica mediante tecnología LiDAR (*Light Detection and Ranging o Laser Imaging Detection and Ranging*[2]) capturada a través del uso de un vehículo aéreo no tripulado tipo drone, con tecnología RTK (imagen 15), que permitió la recolección de datos altitudinales y morfométricos de la planicie inundable, del cauce del cuerpo de agua (corregida batimétricamente) y de la superficie de la planta (modelo digital de superficie) (imagen 15 a, b y c), con rangos de precisión de ±0.1 m., tal como se relaciona en la tabla 3.

[2] ESRI. Disponible en: https://desktop.arcgis.com/es/arcmap/latest/manage-data/las-dataset/what-is-lidar-data-.htm

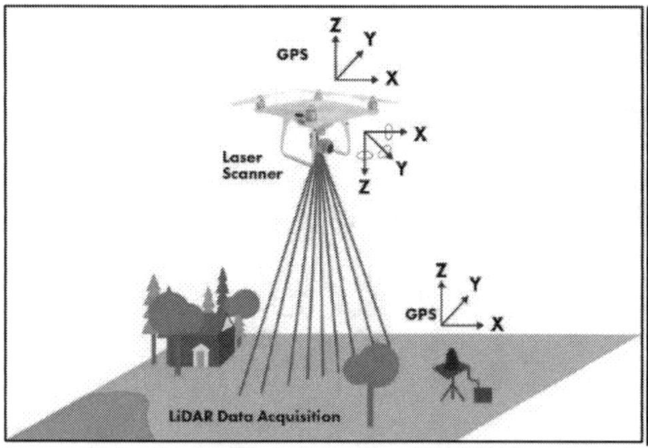

***Imagen 15**. Izq. Esquema ilustrativo de tecnología LiDAR con drone. **Fuente**: FlyGuys. Disponible en: https://flyguys.com/5-industry-use-cases-for-lidar/ Der. Levantamiento topográfico con tecnología LiDAR **Fuente**: SACS Consultores, 2022.*

Aeronave	
Fabricante	• DJI
Serie	• Phantom 4 RTK
Peso	• 1391 g.
Tamaño diagonal	• 350 mm.
Velocidad máxima de ascenso	• 6 m/s
Velocidad máxima de descenso	• 3 m/s
Velocidad máxima	• 58 km/h
Alcance máximo altitudinal	• 6000 m.
Límite de resistencia a la velocidad del viento	• 10 m/s.
Duración máxima de vuelo	• 30 minutos aproximadamente
Rango de temperatura ambiental de operación	• 0°C a 40°C
Rango de precisión	• RTK Conectado y funcionando apropiadamente: Vertical: ±0.1m – Horizontal: ±0.1m. • RTK desconectado: ±0.1m con posicionamiento visual - ±0.5m con posicionamiento GNSS. Horizontal: ±0.3m con posicionamiento visual - ±1.5m con posicionamiento GNSS.
GNSS	
Frecuencia única Alta sensibilidad GNSS	• GPS + BeiDou + Galileo (Asia); GPS +Glonass + Galileo (otras regiones)
Multifrecuencia – Multisistema de alta precisión RTK - GNSS	• GPS L1/L2; Glonass: L1/L2; BeiDou B1/B2; Galileo E1/E5 • Tiempo de ajuste inferior a 50 segundos.

Aeronave	
	• Precisión de posición: Vertical de 1.5 cm. + 1 ppm (RMS); Horizontal 1 cm + 1 ppm (RMS).
Cámara	
Sensor	• 1" CMOS; Píxeles efectivos: 20M
Lente	• FOV (Field of View) 84°, 8.8 mm (35 mm en formato equivalente: 24 mm.) f/2.8 – f/11, autoenfoque de 1 m - α
ISO Rango	• Video: 100 – 3200 (Auto), 100 – 6400 (Manual) • Foto: 100 – 3200 (Auto), 100 – 12800 (Manual)
Obturador mecánico	• 8 – 1/2000 s
Obturador electrónico	• 8 – 1/8000 s
Tamaño máximo de imagen	• 4864x3648(4:3); 5472x3648(3:2)

Tabla 3. *Características de la aeronave y de la información LiDAR capturada*

De otro lado, los datos correspondientes a la estimación hidrológica de la creciente en el cuerpo de agua, fueron obtenidos mediante técnicas de tratamiento estadístico y probabilístico de series de tiempo de caudal para obtener las magnitudes de dicho evento a un periodo de retorno (Tr) de 100 años, que, en términos de frecuencia, corresponde a un evento con una probabilidad de ocurrencia de 0.01.

Una vez estimada la magnitud del evento amenazante en función de su probabilidad de ocurrencia, se empleó un modelado computacional basado en el método de volúmenes finitos, para comprender hidráulicamente el comportamiento de la creciente en el cauce y en la planicie de inundación donde se localiza la planta de gas, a partir de la creación de un dominio de cómputo cuya geometría de modelado, permitió refinar la malla de cálculo hidráulico en una red de triángulos con precisiones de hasta 1 m, desarrollando el conjunto de ecuaciones de Navier – Stokes y asegurando el principio de la conservación de la masa a través de la malla de dicho dominio.

***Imagen 16.a**. Fotogrametría ortogonal de la planta de gas. **b**. Modelado digital de superficie + Fotogrametría de la planta de gas. **c**. Modelo digital de superficie de la planta de gas. Nota: En flecha roja, tanque de almacenamiento afectado. **Fuente**: SACS Consultores, 2022.*

Como resultado de la simulación, se obtuvo el campo de velocidades y profundidades de la creciente, que de acuerdo a lo hallado por V. Cozzani et al. (2010) en los registros de accidentes por inundaciones en tanques de almacenamiento y con base en lo planteado por Salzano et al., 2003 frente a las fuerzas que experimentan este tipo de equipos durante un evento hidrológico, se cumplen las condiciones de velocidades del agua superiores a 1 m/s (imagen 17, izq.) y profundidades de inundación mayores a 1 m. (imagen 17, der.), lo que supondría una afectación en los tanques de almacenamiento de hidrocarburo refinado existentes al interior de la instalación, objeto del presente análisis.

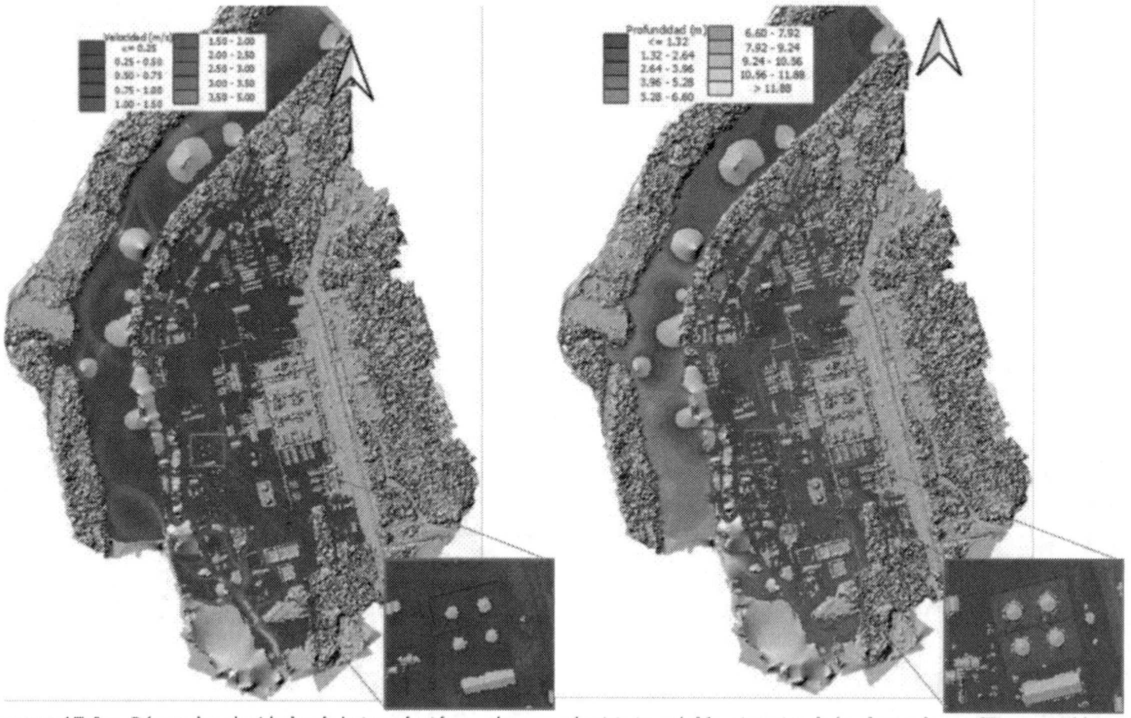

***Imagen 17**. **Izq**. Ráster de velocidades de la inundación en el cauce, planicie inundable e interior de la planta de gas (Zoom en clúster de tanques). Imagen 16. **Der**. Ráster de profundidades de la inundación en el cauce, planicie inundable e interior de la planta de gas (Zoom en clúster de tanques). **Fuente**: SACS Consultores, 2022.*

Los datos correspondientes a los tanques de almacenamiento de hidrocarburo (dimensiones y volumen), así como de las características del producto derramado y el balance de fuerzas resultantes entre la fuerza de empuje y la fuerza contrarrestante ejercida por el tanque y su contenido, se muestran en la Tabla 4.

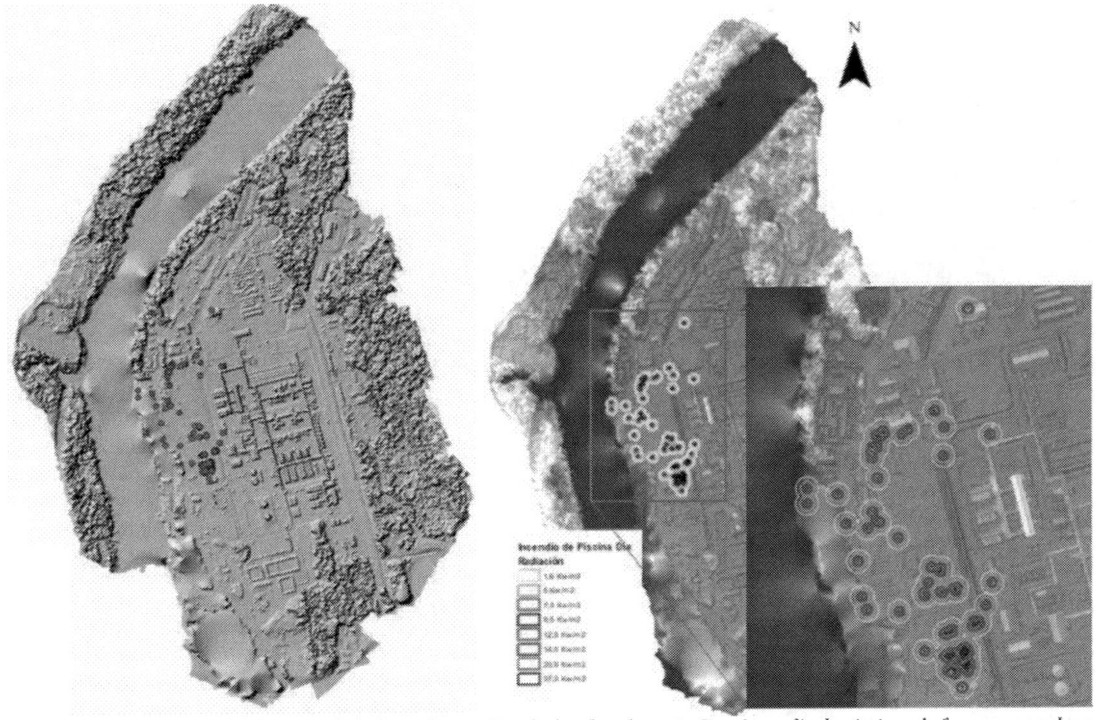

***Imagen 18. Izq**. Trayectoria del derrame de hidrocarburo – Partículas de color rojo. **Der**. Incendio de piscinas de fuego generadas por la ignición del derrame de hidrocarburo. **Fuente**: SACS Consultores, 2022.*

Dato	Valor
Tipo de tanque	Atmosférico - Vertical
Altura del tanque (m)	12.1
Diámetro del tanque (m)	10
Material del tanque	Acero
Módulo de elasticidad del acero (Pa)	2.1×10^{11}
Capacidad del tanque (bls.)	6000 aprox.
Producto	Diesel
Densidad del Diesel (kg/m^3)	850
Altura del dique (m)	0.8
Densidad del agua (kg/m^3)	$\simeq$1000
Velocidad del agua (m/s)	1.5
Profundidad del agua (m)	2
Coeficiente hidrodinámico de la inundación (adim.)	1.8
Frecuencia hidrológica del evento (Año^{-1})	1×10^{-2}
Presión hidrodinámica de la inundación (Pw_d) (Pas)*	19620
Presión hidrostática de la inundación (Pw_s) (Pas)*	3600
Presión total ejercida por la inundación ($Pw_d + Pw_s$) (Pas)*	23220

Dato	Valor
Presión de llenado del tanque (Contrarrestante) (Pf) (Pas)*	15134
Φ factor de nivel de llenado operativo del tanque*	0.15
Presión crítica del tanque (Pcr) (Pas)*	6760
Presión de resistencia del tanque (Pcr + Pf)*	21894
Tipo de falla del tanque	Pandeo
Volumen de derrame del hidrocarburo (bls.)	900

*Tabla 4. Resumen de datos de entrada aplicados en el presente estudio - *Obtenidos apliocando Landucci et al., 2012*

El resultado del modelamiento muestra que el derrame del hidrocarburo, al cabo de la primera hora de ocurrido el evento, queda contenido casi en su totalidad al interior de la planta de gas (imagen 18, izq.), mientras que una pequeña porción del derrame alcanza el cuerpo de agua aledaño. En relación con el modelo de piscina de fuego adoptado (aplicando Imazu & Nishino, 2018 & Det Norske Veritas (DNV), 2009), se observa que los charcos de fuego alcanzan varios equipos críticos o unidades de proceso de la instalación (imagen 18, der.), cuyas consecuencias potenciales, pueden llegar a ser la generación de un efecto dómino o escalamiento de la emergencia.

2.3 Conclusiones

La ocurrencia de fenómenos naturales, conforme avanzan los años, muestran una tendencia incremental no solo en su frecuencia, sino en su magnitud o severidad, abarcando zonas más extensas en la que, conforme avanza la industrialización, son ocupadas por instalaciones que reciben, procesan, almacenan y despachan sustancias peligrosas para la salud pública y el ambiente.

En ese orden de ideas, la exposición de equipos o unidades de proceso, son cada vez más frecuentes y los esfuerzos desde las autoridades regulatorias, hasta los propietarios del riesgo, parecieran no ser lo suficientemente robustos y rigurosos para establecer barreras y controles efectivos para hacerle frente a los eventos NaTech.

Aspectos como la variabilidad climática natural asociada al fenómeno ENSO (en algunas regiones) o la variabilidad inducida, como el efecto del cambio climático, han reflejado una estrecha correlación con la manifestación de fenómenos naturales más graves e intensos, lo que implica que los proyectos o activos industriales, necesariamente deberán incorporar este tipo de escenarios tendenciales, dentro del panorama de riesgos de la instalación.

Diversas investigaciones publicadas por distintos autores muestran aproximaciones para determinar el tipo de daño y la consecuencia potencial de un evento simplificado en un equipo expuesto, o el daño potencial desencadenado simultáneamente en un evento complejo tipo dominó, teniendo en cuenta el evento iniciador (inundación, tormenta eléctrica, sismos, etc.), la geometría y estructura del equipo expuesto y finalmente, el tipo de sustancia contenida.

En este sentido, herramientas computacionales para comprender la magnitud del impacto desencadenado en una pérdida de contención, juegan un papel crucial en el dimensionamiento tanto del escenario de riesgo, como del escenario de emergencia, lo que permitirá a las organizaciones, reducir la incertidumbre frente al comportamiento del evento y prepararse para actuar, mitigando el impacto de los sucesos NaTech en la instalación y sus alrededores.

Desde la gestión del riesgo con enfoque de gestión de seguridad de procesos, aspectos como la implementación de sistemas de alerta temprana (SAT) para detectar anticipadamente la ocurrencia del fenómeno natural que potencialmente puede desencadenar un accidente tecnológico, es vital para tratar y reducir el riesgo de la instalación, en tanto que medidas estructurales y organizacionales, como la conformación de diques y la adopción de un plan de emergencias, respectivamente, contribuirán significativamente a la mitigación de los impactos y afectaciones del evento NaTech.

3 Bibliografía y Referencias.

3.1 Bibliografía

Antonioni, G., Bonvicini, S., Spadoni G., Cozzani, V. (2009). *Development of a framework for the risk assessment of Na-Tech accidental events.* Journal Reliability Engineering and System Safety 94 (2009) 1442–1450.

Antonioni, G., Landucci, G., Necci, A., Gheorghiu, D., Cozzani, V. (2015). *Quantitative assessment of risk due to NaTech scenarios caused by floods.* Journal Reliability Engineering and System Safety 142 (2015) 334 - 345.

Cozzani, V., Campedel, M., Renni, E., Krausmann, E. (2010). *Industrial accidents triggered by flood events: Analysis of past accidents.* Journal of Hazardous Materials 175 (2010) 501–509.

Di Maio, F., Tonicello P., Zio E. (2022). *Modeling and Analysis Framework for Integrated Energy Systems Exposed to Climate Change-Induced NaTech Accidental Scenarios.* Sustainability 2022, 14, 786.

Jiang, P., Tong, S., Wang, Y., Xu, G. (2021). *Modelling the oil spill transport in inland waterways based on experimental study.* Environmental Pollution 284 (2021) 117473.

Jones, C., Sambanis, A., Kim, S., Cailas, M. (2020) *Chemical facility risks to natural flooding hazards in the United States.* Progress in Disaster Science 6 (2020) 100105

Khakzad, N., Van Gelder, P. (2017). *Vulnerability of industrial plants to flood natechs: A Bayesian network approach.* Reliability Engineering and System Safety (2017), doi: 10.1016/j.ress.2017.09.016

Landucci, G., Tugnoli, A., Antonioni, G., Cozzani, V. (2013). *Damage Models for Storage and Process Equipment Involved in Flooding Events.* Chemical Engineering Transactions, 31, 697-702 DOI: 10.3303/CET1331117.

Landucci, G., Antonioni, G., Tugnoli, A., Cozzani, V. (2012). *Release of hazardous substances in flood events: Damage model for Atmospheric storage tanks*. Journal Reliability Engineering and System Safety 106 (2012) 200 – 216.

Landucci, G., Necci, A., Tugnoli, A., Antonioni, G., Cozzani, V. (2014). *NaTech Scenarios Caused by Flooding: Evaluation of Accident Frequency by the Use of Fragility Models*. Chemical Engineering Transactions, 36, 427-432 DOI: 10.3303/CET1436072.

Landucci, G., Necci, A., Tugnoli, A., Antonioni, G., Cozzani, V. (2014). *Release of hazardous substances in flood events: Damage model for horizontal cylindrical vessels*. Journal Reliability Engineering and System Safety 132 (2014) 125 – 145.

Landucci, G., Antonioni, G., Necci, A., Cozzani, V. (2016). *Quantitative Risk Assessment of Cascading Events Triggered by Floods*. Chemical Engineering Transactions, 48, 901-906 DOI: 10.3303/CET1648151.

Misuri, A., Landucci G., Vivarelli, S., Bonvicini, S., Cozzani, V. (2019). Risk-based Vulnerability Analysis of Chemical Facilities Affected by Flooding. Chemical Engineering Transactions, 77. DOI: 10.3303/CET1977088.

Rojas, S. (2005). *Modelación de derrames de crudo en cauces – Aplicación al río Magdalena*. Tesis de grado. Facultad de ingeniería – Departamento de ingeniería civil y ambiental. Universidad de los Andes.

Salzano, E., Iervolino, I., Fabbrocino, G. (2003). *Seismic risk of atmospheric storage tanks in the framework of quantitative risk análisis*. Journal of Loss Prevention in the Process Industries 16 (2003) 403–409.

Yang Y., Chen G., Reniers G. (2019), *Vulnerability assessment of atmospheric storage tanks to floods based on logistic regression*. Journal Reliability Engineering and System Safety.

Yapa, P., Shen, H., Angammana, K. (1993). *Modeling oil spills in a river-lake system*. Journal of Marine Systems 4 (1994) 453-471.

Zapata, C., Calle, E., Parra, L. (2002). *Simulación del comportamiento fisicoquímico de derrames de hidrocarburos en ríos*. Escuela de geociencias y medio ambiente, Facultad de Minas. Universidad Nacional de Colombia Sede Medellín. *Avances en recursos hidráulicos 2002, Número 9* (pp. 7-18).

Zuluaga, S., Sánchez M., Ramírez O., Muñoz F. (2019). *Development of parametric fragility curves for storage tanks: A Natech approach*. Journal Reliability Engineering and System Safety 189 (2019) 1 – 10.

3.2 ***Referencias***

Cruz, W. (2020). Explotador RPAS/Profesional GIS. SACS Group. Adquisición de información topográfica mediante sobrevuelos LiDAR.

CENTRO DE MONITOREO PARA ANÁLSIS Y OPERACIÓN DE SEGURIDAD DE PROCESOS (MCPS) CON DATOS DEL DCS Y ACTIVIDADES PSM

Javier Acevedo Sandoval
HUNT LNG OPERATING COMPANY
Lima - Perú
jacevedo@huntloc.pe

Keywords: Gestión de Riesgos, Seguridad de Procesos, Sistema de Control Distribuido (DCS) y análisis de tendencias

Abstract

Las Plantas de Procesamiento de Gas y Refinerías implementan un Sistema de Control Distribuido (DCS) que sirve como interfaz para el personal operativo e ingeniería de procesos, permitiéndoles operar y analizar el rendimiento del proceso. El personal de Seguridad de Procesos también requiere una herramienta que les permita analizar las actividades diarias y datos del proceso desde una perspectiva de seguridad.

Con base en esta premisa, se introduce la metodología "CENTRO DE MONITOREO PARA ANÁLSIS Y OPERACIÓN DE SEGURIDAD DE PROCESOS CON DATOS DEL DCS Y ACTIVIDADES PSM - MCPS" como un enfoque dinámico y continuo para el análisis y control de la seguridad en los procesos.

El MCPS se diseña conforme a las actividades diarias establecidas por cada compañía en su Gestión de Seguridad de Procesos. Esto incluye datos del DCS que no necesariamente son monitoreados por los departamentos de Procesos, Control u otras áreas, pero que deben ser analizados por los responsables de Seguridad de Procesos.

El uso del MCPS facilita la identificación de desviaciones relacionadas con la seguridad de procesos, las cuales deben recibir una prioridad mayor para su abordaje en colaboración con los responsables de los elementos del PSM. Estas desviaciones también se presentan al comité PSM, donde se evalúan y proponen actividades complementarias, lo que resulta en un proceso dinámico y activo en términos de seguridad de procesos. La operación del centro de monitoreo de seguridad de procesos (MPCS) refleja la contribución de los diversos departamentos de una planta, promoviendo una cultura y un compromiso de ejecución en las actividades de PSM. Además, proporciona una visión del estado de las actividades diarias de seguridad de procesos.

1 INTRODUCCIÓN

El objetivo del presente artículo es detallar la metodología, del **"CENTRO DE MONITOREO PARA ANÁLSIS Y OPERACIÓN DE SEGURIDAD DE PROCESOS CON DATOS DEL DCS Y ACTIVIDADES PSM - MCPS"** adoptado en las operaciones, dando un alcance etapa por etapa de su elaboración, destacando sus aspectos más relevantes; así como, el desempeño que viene presentando desde su implementación y su importancia dentro del ámbito de seguridad de procesos aplicado en la Planta de Procesamiento de Gas de Peru LNG.

La metodología consta de siete etapas detalladas en la sección correspondiente. Su objetivo es desarrollar un Centro de Monitoreo de Seguridad de Procesos para ser usado por los responsables de la Seguridad de Procesos. Este instrumento facilita la detección temprana de cambios en las actividades de seguridad de procesos y respalda la coordinación efectiva de esfuerzos multidisciplinarios durante las operaciones, lo cual es fundamental para prevenir eventos de pérdida de control y tomar decisiones cotidianas con fundamentos sólidos.

Por tanto, el Centro de Monitoreo de Seguridad de Procesos - MCPS se viene implementando en la Planta de Procesamiento de Gas de Peru LNG desde 2023 y sigue en constante mejora gracias a la contribución de diversos especialistas de la planta, con el objetivo de mantener el enfoque en la seguridad de procesos.

2 METODOLOGÍA

2.1 ETAPAS PARA LA CREACIÓN DEL CENTRO DE MONITOREO DE SEGURIDAD DE PROCESOS - MCPS

Etapa 1: Seleccionar actividades de PSM que proporcionen Información de elementos de PSM a un primer nivel (macro)

Etapa 2: Seleccionar datos a monitorear del DCS de Planta. Información del DCS de planta que permita identificar desviaciones a un primer nivel (macro)

Etapa 3: Desarrollar Métrica PSM a través de Estandarización de KPI´s de Seguridad de Procesos a valores porcentuales para visión macro del sistema

Etapa 4: Identificar datos e integrarlos al centro de monitoreo

Etapa 5: Diseñar gráficas de primer nivel que faciliten la detección del inicio de desviaciones y la comprensión de sus relaciones mutuas.

Etapa 6: Analizar tendencias o patrones con gráficas de nivel macro y revisar gráficas de segundo y/o tercer nivel.

Etapa 7: Exponer desviaciones o patrones identificados en las tendencias y sus acciones correspondiente ante el comité PSM

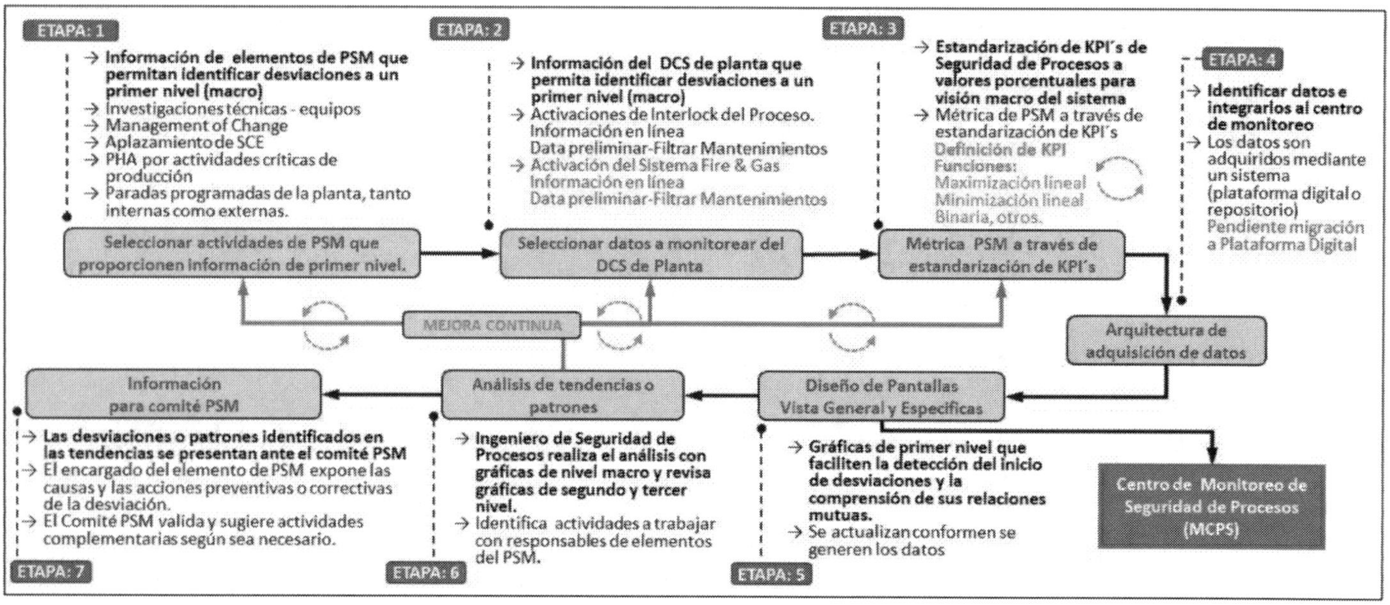

Figura 1. Diagrama de flujo de metodología para diseño de Centro de Monitoreo de Seguridad de Procesos - MCPS. Elaboración propia. Fecha de elaboración:04/03/2022

A continuación, se describe detalladamente cada una de las etapas.

Etapa 1: Seleccionar actividades de PSM que proporcionen Información de elementos de PSM a un primer nivel (macro).

En esta etapa, los encargados de la Seguridad de Procesos en cada organización deben identificar las tareas diarias que contribuyen a prevenir eventos de pérdida de control. Es esencial determinar qué datos a nivel macro pueden generar tendencias y permitir la detección de desviaciones que requieran atención. Se recomienda que los datos en esta sección sean genéricos para que puedan ser monitoreados en conjunto con otros. Por ejemplo, al representar gráficamente la cantidad diaria y acumulada de investigaciones técnicas, es posible detectar aumentos o disminuciones repentinas, lo cual sugiere problemas operativos. Asimismo, pueden identificarse patrones recurrentes que revelan el comportamiento de los equipos de proceso y la respuesta de los especialistas. Por consiguiente, la selección de los datos y el método de análisis variarán según las particularidades de cada empresa. Este enfoque permite adaptar las estrategias de seguridad de procesos a las necesidades específicas de cada organización, promoviendo así una gestión más efectiva de los riesgos y una mejora continua en el rendimiento operativo.

En esta sección se recomiendan las siguientes tendencias:

Investigaciones técnicas vs tiempo
Cambios en el Procesos vs tiempo
PHA por actividades críticas en el Proceso vs tiempo
Paradas programadas de la planta tanto internas como externas vs tiempo.

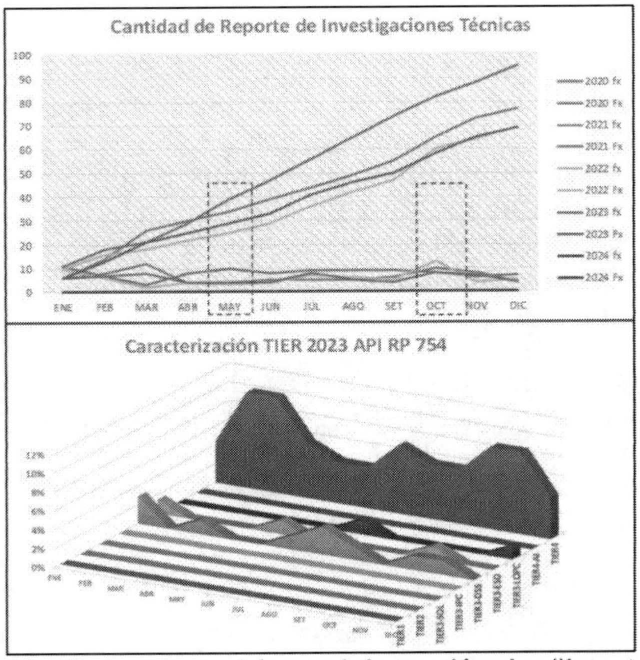

Figura 2. Aproximación de lo que podría ser información de diferentes operaciones y se muestran con el propósito de ilustrar lo que se podría obtener del análisis/procesamiento de los datos obtenidos. Elaboración propia 4/3/2024

Etapa 2: Seleccionar datos a monitorear del DCS de Planta. Información del DCS de planta que permita identificar desviaciones a un primer nivel (macro)

El Sistema de Control Distribuido (DCS) proporciona datos cruciales tanto para el personal operativo como para los equipos de ingeniería de procesos y control. Por lo tanto, es fundamental realizar un análisis exhaustivo en esta sección para determinar qué información debe ser monitoreada por los encargados de la seguridad de procesos. Es esencial identificar los parámetros y variables del DCS que son críticos para la seguridad de las operaciones y que pueden ofrecer indicadores tempranos de posibles riesgos o desviaciones en el proceso. Este análisis debe tener en cuenta la relevancia de la información, su precisión, su capacidad para prevenir incidentes y su utilidad para la toma de decisiones proactivas en materia de seguridad. La selección cuidadosa de los datos del DCS que se monitorearán permitirá una gestión eficaz de los riesgos y una respuesta rápida ante situaciones que puedan comprometer la seguridad del proceso.

El primer paso consiste en que cada empresa defina la accesibilidad de la información del Sistema de Control Distribuido (DCS) de la planta. Es importante reconocer que el DCS se instala en la Sala de Control Central, y configurar su disponibilidad a través de un software exclusivamente de visualización, sin acceso al control directo, facilita el análisis de la información por parte de diversos departamentos. En esta sección, se presupone que la información del DCS ya está accesible para su utilización. Este enfoque permite optimizar la rapidez y eficiencia en el análisis de datos, al mismo tiempo que garantiza la seguridad y la integridad del sistema de control centralizado. Es crucial establecer esta disponibilidad de información como base para las siguientes etapas del proceso de gestión de la seguridad de procesos.

Como ejemplo, es posible generar gráficos que representen los enclavamientos por unidad de proceso a lo largo del tiempo. Aunque esta información se monitorea típicamente por las áreas de control, procesos y operaciones, desde la perspectiva de la seguridad, permite detectar los enclavamientos realizados y determinar si fueron comunicados y revisados por los operadores de la sala de control. Llevar a cabo esta actividad es de gran relevancia, ya que alerta sobre las precauciones necesarias durante la operación de la planta. Estos gráficos proporcionan una visión clara de la gestión de enclavamientos y ayudan a garantizar que los procedimientos de seguridad sean seguidos de manera adecuada, minimizando así el riesgo de incidentes y promoviendo un entorno de trabajo seguro.

Etapa 3: Desarrollar Métrica PSM a través de Estandarización de KPI´s de Seguridad de Procesos a valores porcentuales para visión macro del sistema.

Para avanzar en esta etapa, es esencial que cada componente delineado en el Sistema de Gestión de la Seguridad de Procesos (PSM) de cada empresa cuente con sus indicadores clave de rendimiento (KPI) definidos por los responsables pertinentes, junto con sus rangos de aceptación correspondientes. Para facilitar la visualización rápida de toda esta información, se sugiere convertir los KPI en valores porcentuales, lo que permite obtener una medida de cada elemento del PSM y también una medida de referencia de todo el PSM.

La conversión de un KPI a valores porcentuales implica examinar el KPI y sus rangos de aceptación, y asignar la función matemática que mejor represente la situación, que puede ser de maximización lineal, minimización lineal, binaria, exponencial u otras. Esta práctica no solo proporciona una medición de rendimiento, sino que también ofrece una medida de tendencia para visualizar a nivel macro las actividades asociadas con cada elemento del PSM.

Este enfoque permite una evaluación más clara y efectiva del desempeño de cada componente del PSM, así como del PSM en su conjunto. Además, proporciona una base sólida para identificar áreas de mejora, tomar decisiones informadas y garantizar que se cumplan los estándares de seguridad y rendimiento establecidos.

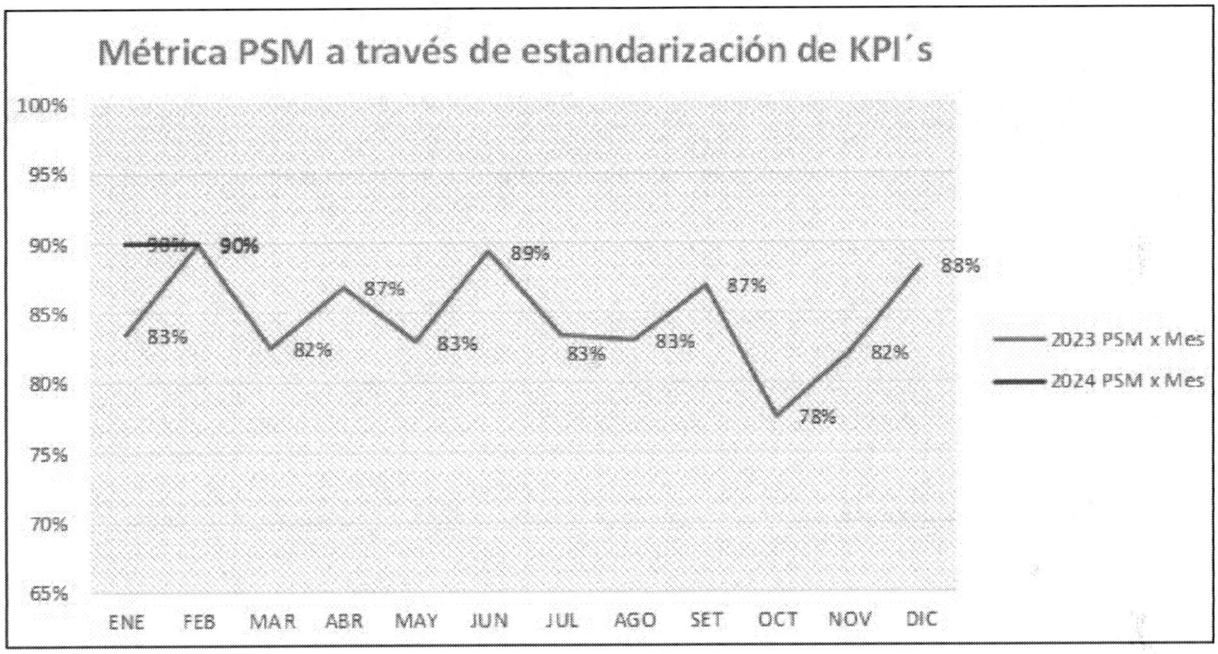

Figura 3. Aproximación de lo que podría ser información de diferentes operaciones y se muestran con el propósito de ilustrar lo que se podría obtener del análisis/procesamiento de los datos obtenidos. Elaboración propia 4/3/2024

Etapa 4: Identificar datos e integrarlos al centro de monitoreo.

Organizar y centralizar los ingresos de datos es crucial para facilitar su gestión y análisis posterior. La colaboración con el equipo de Tecnología de la Información es vital para implementar soluciones efectivas que se ajusten a las necesidades y capacidades de la empresa. Trabajar por fases permite dividir el proceso en tareas manejables y facilita el seguimiento y la evaluación del progreso. Al utilizar la metodología descrita en el documento como guía, se establece un marco estructurado que puede adaptarse y personalizarse según las circunstancias específicas de cada organización. Este enfoque sistemático contribuye a la creación de un sistema de gestión de datos sólido y eficiente, que sirve como base para la toma de decisiones informadas y la mejora continua del rendimiento operativo.

Etapa 5: Diseñar gráficas de primer nivel que faciliten la detección del inicio de desviaciones y la comprensión de sus relaciones mutuas.

En este momento del proceso, la contribución del equipo multidisciplinario es crucial para definir las gráficas de primer nivel y garantizar que el análisis inicial realizado por los responsables de la seguridad de procesos sea beneficioso para los distintos departamentos de la planta.

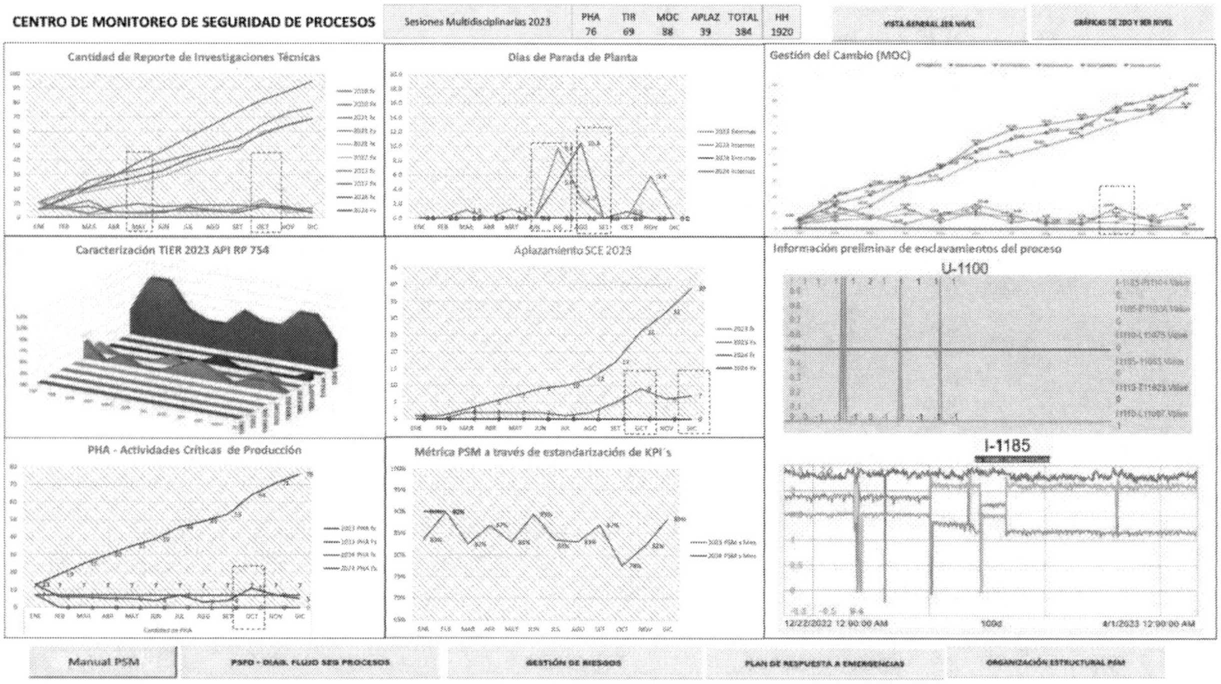

Figura 4. Aproximación de lo que podría ser información de diferentes operaciones y se muestran con el propósito de ilustrar lo que se podría obtener del análisis/procesamiento de los datos obtenidos. Elaboración propia 4/3/2024

Etapa 6: Analizar tendencias o patrones con gráficas de nivel macro y revisar gráficas de segundo y/o tercer nivel.

El análisis de tendencias se fundamenta en la detección de desviaciones en las gráficas de nivel inicial, además del pensamiento crítico del personal de seguridad de procesos para identificar cualquier otra desviación que necesite ser abordada. También implica la definición de qué gráficas o información de niveles secundario y terciario serán necesarias para comprender completamente la situación.

En esta fase, si es necesario, se inicia la colaboración de un equipo multidisciplinario para implementar acciones preventivas o correctivas que contribuyan al control de la desviación. Es fundamental que este equipo esté compuesto por profesionales con diversas habilidades y conocimientos, lo que permite abordar la situación desde diferentes perspectivas y garantizar la efectividad de las medidas adoptadas.

El análisis de tendencias y la colaboración multidisciplinaria son aspectos clave para mantener un enfoque proactivo en la gestión de la seguridad de procesos y garantizar la integridad y eficiencia de las operaciones en la planta.

Etapa 7: Exponer desviaciones o patrones identificados en las tendencias y sus acciones correspondiente ante el comité PSM

Las desviaciones o patrones observados en las tendencias son presentados ante el comité PSM, donde el responsable del elemento de PSM explica las causas subyacentes y propone las medidas preventivas o correctivas correspondientes. El comité PSM, por su parte, revisa y valida las propuestas, y sugiere actividades adicionales según sea necesario para abordar de manera efectiva las desviaciones identificadas.

Este proceso refleja la importancia de la comunicación transparente y la colaboración dentro del comité PSM. Al analizar las desviaciones y evaluar las acciones propuestas, el comité puede identificar áreas de mejora y tomar decisiones informadas para mitigar los riesgos y mejorar continuamente la seguridad y el rendimiento del proceso. Además, esta interacción permite aprovechar la experiencia y el conocimiento colectivo del equipo, lo que contribuye a un enfoque integral y proactivo en la gestión de la seguridad de procesos.

3 REVISIÓN Y MEJORA DEL CENTRO DE CONTROL DE SEGURIDAD DE PROCESOS

La evaluación del centro de control de seguridad de procesos es un proceso en constante evolución y adaptación. Se sugiere que se siga mejorando en base a las contribuciones proporcionadas tanto por el personal de la planta como por el comité de PSM.

Esta recomendación subraya la importancia de mantener un enfoque flexible y receptivo hacia los aportes y sugerencias de aquellos que están directamente involucrados en las operaciones diarias y en la gestión de la seguridad de procesos. Al estar abiertos a las opiniones y experiencias del personal de la planta y del comité de PSM, se pueden identificar oportunidades de mejora y optimización en el centro de control. Esto asegura que el centro de control esté alineado con las necesidades y desafíos específicos de la planta, así como con las mejores prácticas de seguridad de procesos. La revisión dinámica y continua del centro de control permite que siga siendo una herramienta eficaz para monitorear y mantener la seguridad en las operaciones de la planta.

4 CONCLUSIONES

- La metodología para diseñar un Centro de Monitoreo de Seguridad de Procesos (MCPS) motiva al personal de la planta a definir o redefinir sus actividades diarias en materia de seguridad de procesos.

- El sistema de control distribuido (DCS) de una planta proporciona datos para ser analizados desde una perspectiva de Seguridad de Procesos

- La estandarización de KPIs facilita su integración en un sistema de medición porcentual, lo que permite un análisis macro de su funcionalidad a través del centro de monitoreo de Seguridad de Procesos.

- La operación del centro de monitoreo de seguridad de procesos (MPCS) refleja la contribución de los diversos departamentos de una planta, fomentando una cultura y compromiso de ejecución en las actividades de PSM. Además, ofrece una visión del estado de las actividades diarias de seguridad de procesos.

- En este diseño se incluirá, información del sistema de detección de fuego y gas.

- El modelo sirve como base para migrar a una plataforma digital que integre toda la información.

5 REFERENCIAS

Online Reference:
OHSA Law & Regulations. (2011). http://www.osha.gov/law-regs.html . Accessed 30 January 2020.

Métricas Predictivas y Retrospectivas para Seguridad de ProcesosWWW.AICHE.ORG › SITES › DEFAULT › FILES › DOCS › PAGES
https://www.aiche.org/sites/default/files/docs/pages/Process%20Safety%20Leading%20and%20Lagging%20Metric%20CCPS%20espa%C3%B1ol.pdf

Cosechando la Potencia de la Inteligencia Artificial y Tecnologías Avanzadas para la Seguridad de Procesos.

César Augusto Puma Medina, Elec. Eng. MSc. Ph.D-DBA
Lima, Perú
cesar.puma.medina@gmail.com

Palabras clave: RBPS, IA, cybersecurity, PSM, Prompt, HAZOP-IA

Resumen

La seguridad de procesos en la industria es una prioridad crítica. La gestión efectiva de los riesgos es esencial para garantizar operaciones seguras y prevenir incidentes que puedan resultar en daños humanos, ambientales y económicos significativos. En este contexto, la integración de tecnologías avanzadas, particularmente la inteligencia artificial (IA) y otras herramientas tecnológicas, ha demostrado ser un componente clave para mejorar la seguridad de procesos.
La inteligencia artificial se ha convertido en una fuerza impulsora detrás de la revolución tecnológica en la seguridad de procesos. Al aprovechar la potencia de la IA, las organizaciones pueden anticipar y prevenir riesgos, optimizar sus operaciones y garantizar un entorno de trabajo más seguro. Este resumen se adentrará en el impacto de la IA y las tecnologías avanzadas en la seguridad de procesos, destacando sus beneficios, desafíos y aplicaciones.

La IA tiene un impacto significativo en la seguridad de procesos, en particular en los siguientes aspectos:

- La IA puede analizar grandes conjuntos de datos para identificar patrones y tendencias, lo que permite anticipar riesgos y prevenir incidentes antes de que ocurran. Los algoritmos de aprendizaje automático pueden detectar anomalías en los procesos y activar alarmas en tiempo real.
- La automatización y la IA pueden optimizar los procesos operativos, reduciendo el riesgo de errores humanos y mejorando la eficiencia. Esto se traduce en un ahorro de costos y una mayor productividad.
- La IA puede ayudar a evaluar y gestionar riesgos de manera más efectiva al analizar datos complejos y proporcionar información en tiempo real. Esto es crítico para las operaciones industriales, donde los riesgos pueden cambiar rápidamente.
- Los sistemas basados en IA pueden proporcionar análisis y recomendaciones basadas en datos, lo que permite a los tomadores de decisiones disponer de información más sólida para tomar medidas preventivas o correctivas.

La IA se puede aplicar en diversas áreas de la seguridad de procesos, incluyendo:

- La IA se utiliza para anticipar fallos en equipos y sistemas como los Gemelos digitales, lo que permite realizar mantenimiento predictivo en lugar de reactivo, reduciendo el tiempo de inactividad no planificado.
- La IA puede monitorear continuamente la integridad de los sistemas y detectar fugas o emisiones no deseadas, lo que contribuye a la gestión ambiental y la seguridad.
- Los sistemas de IA pueden evaluar riesgos de seguridad en tiempo real y tomar decisiones rápidas, como cerrar válvulas o desviar flujos en caso de situaciones de emergencia, principalmente a través de HAZOP activos.
- La IA puede detectar y prevenir ataques cibernéticos en sistemas de control industrial, protegiendo la integridad de los procesos y la seguridad de la información a través del Cyber-HAZOP.
- La IA se utiliza para ajustar los parámetros de operación en tiempo real para maximizar la eficiencia y minimizar los riesgos a través del monitoreo de SOL (safety operating limits).

A pesar de los beneficios, la implementación de la IA en la seguridad de procesos también presenta desafíos y consideraciones importantes. Algunos de estos incluyen:

- La recopilación masiva de datos en entornos industriales puede plantear problemas de privacidad y seguridad. La gestión de datos sensibles es esencial.
- La confiabilidad de los sistemas basados en IA es crítica. Los sistemas deben ser probados, monitoreados y mantenidos regularmente.
- Las organizaciones deben cumplir con regulaciones y normativas, lo que a menudo implica la implementación de medidas adicionales para garantizar la seguridad y la privacidad de los datos.
- La capacitación y competencia del personal en la operación de sistemas de IA es fundamental para garantizar una implementación efectiva y segura.
- La IA complementa a los trabajadores humanos, pero no los reemplaza. La colaboración efectiva entre humanos y sistemas de IA es esencial.

Para comprender mejor la aplicación de la IA en la seguridad de procesos hemos examinado estudios de casos en diversas industrias. Para implementar con éxito la IA en la seguridad de procesos, se deben seguir varios pasos clave, que incluyen:

- Identificar áreas críticas donde la IA puede tener un impacto significativo.
- Reunir datos relevantes y asegurar su calidad.
- Crear algoritmos de IA y modelos que sean apropiados para los datos y los procesos.
- Probar los sistemas en entornos controlados antes de la implementación en el mundo real.
- Capacitar a los empleados para trabajar junto a los sistemas de IA.
- Realizar un seguimiento constante de la eficiencia y la seguridad de los sistemas de IA.

La implementación exitosa de la inteligencia artificial y tecnologías avanzadas en la seguridad de procesos es un campo en constante evolución. Si bien existen desafíos significativos, los beneficios de la IA en la seguridad de procesos son innegables. A medida que continuamos avanzando hacia un mundo más tecnológico y sostenible, la IA seguirá desempeñando un papel fundamental en la protección de la vida, el medio ambiente y los recursos económicos. La cosecha de su potencia en la seguridad de procesos es un imperativo para el futuro.

1 Introducción

1.1.La inteligencia artificial y su relevancia en la seguridad de procesos

La inteligencia artificial (IA) ha emergido como una poderosa herramienta en la era moderna, transformando radicalmente diversas industrias a través de su capacidad para simular procesos cognitivos humanos y realizar tareas complejas de manera automatizada. En el ámbito industrial, la aplicación de la IA ha revolucionado la forma en que se abordan los desafíos de seguridad, especialmente en sectores donde la gestión de riesgos es crítica, como la industria química, petrolera y manufacturera.

1.2.Relevancia en la seguridad de procesos:

La seguridad de procesos es un aspecto vital en cualquier entorno de trabajo donde se manipulan materiales peligrosos, operan equipos críticos o se enfrentan a condiciones ambientales extremas. La IA ofrece un enfoque innovador y eficiente para abordar estos desafíos, al permitir la identificación temprana de riesgos, la predicción de fallas en equipos, la optimización de procesos y la toma de decisiones informadas en tiempo real.

Mediante el uso de algoritmos avanzados y análisis de datos, la IA puede analizar grandes cantidades de información proveniente de sensores, sistemas de monitoreo y bases de datos históricas para detectar patrones, tendencias y anomalías que podrían indicar potenciales peligros o situaciones de riesgo inminente. Esto permite a las empresas anticiparse a posibles accidentes, implementar medidas preventivas y minimizar los riesgos para el personal y el entorno.

En resumen, la inteligencia artificial está transformando la seguridad industrial al proporcionar herramientas y técnicas innovadoras para prevenir accidentes, mejorar la eficiencia operativa y proteger los activos de las empresas. Su capacidad para procesar y analizar grandes volúmenes de datos en tiempo real la convierte en un elemento indispensable en la gestión proactiva de riesgos en entornos industriales complejos.

2 Antecedentes y contexto

2.1 Desafíos comunes en la gestión de la seguridad de procesos

Los desafíos comunes en la gestión de la seguridad de procesos pueden variar dependiendo de la industria y el tipo de operaciones que se estén llevando a cabo. Sin embargo, algunos desafíos son ampliamente reconocidos en diversos entornos industriales. Aquí tienes algunos de los desafíos más comunes:

a) **Complejidad de los procesos:** Los procesos industriales modernos suelen ser altamente complejos, con múltiples variables interrelacionadas. Gestionar la seguridad en este contexto puede ser desafiante debido a la gran cantidad de factores a tener en cuenta.

b) **Identificación y evaluación de riesgos:** Determinar todos los posibles riesgos asociados con un proceso industrial y evaluar su probabilidad y gravedad puede ser complicado. A menudo, los riesgos pueden ser subestimados o pasados por alto si no se realiza una evaluación exhaustiva.

c) **Cambio en las condiciones operativas:** Los procesos industriales pueden experimentar cambios en las condiciones operativas debido a factores como fluctuaciones en la demanda, variaciones en la materia prima o cambios en el equipo. Estos cambios pueden introducir nuevos riesgos o modificar la evaluación de los riesgos existentes.

d) **Cumplimiento normativo y regulaciones:** Cumplir con las normativas y regulaciones locales, nacionales e internacionales relacionadas con la seguridad de procesos puede ser un desafío, especialmente si las regulaciones son complejas o cambian con frecuencia.

e) **Cultura de seguridad:** Establecer y mantener una cultura de seguridad sólida en toda la organización puede ser difícil. Es crucial que todos los empleados estén comprometidos con la seguridad y se sientan responsables de identificar y reportar posibles riesgos.

f) **Capacitación y competencia del personal:** Asegurar que el personal esté adecuadamente capacitado y tenga las habilidades necesarias para operar de manera segura los equipos y procesos es fundamental para prevenir accidentes.

g) **Gestión del cambio:** Implementar cambios en los procesos o en los equipos puede introducir nuevos riesgos si no se gestionan adecuadamente. Es importante realizar evaluaciones de riesgos y planificar cuidadosamente cualquier cambio para minimizar el impacto en la seguridad.

h) **Comunicación y coordinación:** Garantizar una comunicación efectiva y una coordinación adecuada entre todos los niveles de la organización es esencial para garantizar que la información relevante sobre seguridad se comparta y se actúe en consecuencia.

Abordar estos desafíos requiere un enfoque holístico y proactivo hacia la gestión de la seguridad de procesos, que incluya la implementación de sistemas de gestión de seguridad robustos, la capacitación continua del personal y una cultura organizacional que priorice la seguridad en todo momento.

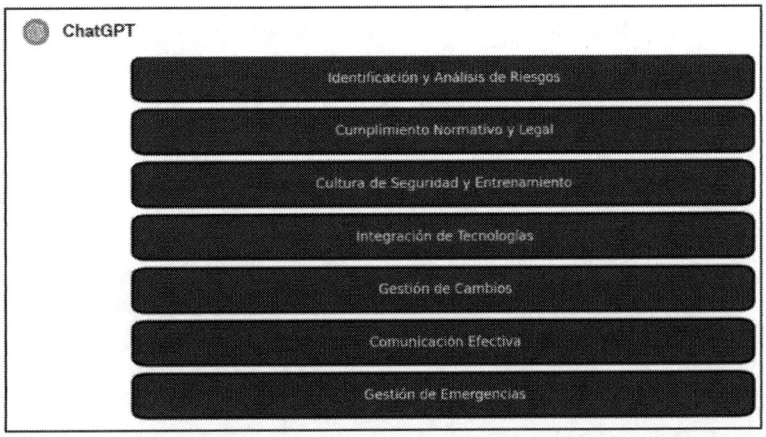

Fig. 01.- Desafíos comunes en la gestión de la seguridad de procesos (generado desde Prompt)

2.2 Métodos tradicionales y sus limitaciones

Los métodos tradicionales de Gestión de la Seguridad de Procesos (PSM) y Análisis de Peligros de Proceso (PHA) han servido como pilares fundamentales para evaluar y gestionar los riesgos en la industria de procesos. Estos métodos incluyen herramientas como HAZOP (Hazard and Operability Study), What-If Analysis, FMEA (Failure Mode and Effects Analysis) y LOPA (Layer of Protection Analysis), entre otros. Cada uno de estos enfoques tiene fortalezas específicas, como la capacidad de sistematizar el pensamiento crítico y la evaluación detallada de los peligros potenciales y sus controles asociados.

Sin embargo, estas metodologías tradicionales también presentan limitaciones:

- **Dependencia de la Experiencia Humana:** Los métodos tradicionales dependen en gran medida del conocimiento y la experiencia de los expertos que llevan a cabo el análisis. Esto puede introducir sesgos subjetivos y limitar la capacidad de identificar peligros no evidentes o poco frecuentes.

- **Manejo de Grandes Volúmenes de Datos:** Las técnicas convencionales pueden ser ineficientes para manejar y analizar la gran cantidad de datos generados en los procesos modernos, lo que puede llevar a una visión incompleta de los riesgos.

- **Actualización y Repetición:** Los análisis tradicionales son estáticos y no se actualizan automáticamente con cambios en las condiciones del proceso o con nueva información, requiriendo revisiones manuales periódicas que pueden ser laboriosas y lentas.

- **Integración de Datos Dispersos:** La integración de información de diversas fuentes para una visión holística del riesgo puede ser difícil de lograr con métodos tradicionales, que a menudo operan en silos.

- **Respuesta en Tiempo Real:** Los métodos tradicionales no están diseñados para proporcionar respuestas en tiempo real a condiciones cambiantes, lo que es esencial para la intervención oportuna y la mitigación de emergencias.

Además, el PHA tradicional puede no detectar interacciones complejas entre diferentes sistemas o el impacto acumulativo de pequeñas desviaciones, lo cual puede llevar a una subestimación de los riesgos emergentes. También, al depender de la experiencia humana, pueden existir sesgos cognitivos y limitaciones en la capacidad de procesar grandes cantidades de datos o identificar sutilezas dentro de esos datos.

Frente a estas limitaciones, la Inteligencia Artificial (IA) ofrece un enfoque avanzado que puede superar varios de estos obstáculos:

- **Análisis de Datos Mejorado:** La IA puede procesar y analizar grandes conjuntos de datos para identificar tendencias y patrones que podrían pasar desapercibidos para los humanos.
- **Modelos Predictivos:** Los sistemas de IA pueden desarrollar modelos predictivos que anticipan fallos o eventos peligrosos antes de que ocurran, permitiendo intervenciones proactivas.
- **Modelado Predictivo Avanzado:** Utilizando técnicas como los gemelos digitales, la IA puede simular escenarios de riesgo y predecir las consecuencias de posibles fallos, lo que permite una mejor planificación y mitigación de riesgos.
- **Aprendizaje Continuo:** La IA se beneficia de algoritmos de aprendizaje automático que se actualizan continuamente a medida que se recopilan nuevos datos, manteniendo el análisis de riesgos relevante y actualizado.
- **Automatización de Tareas Repetitivas:** Reducción de Carga de Trabajo a Al automatizar parte del análisis de riesgos, la IA puede reducir la carga de trabajo de los equipos de seguridad, permitiéndoles concentrarse en tareas de mayor valor.

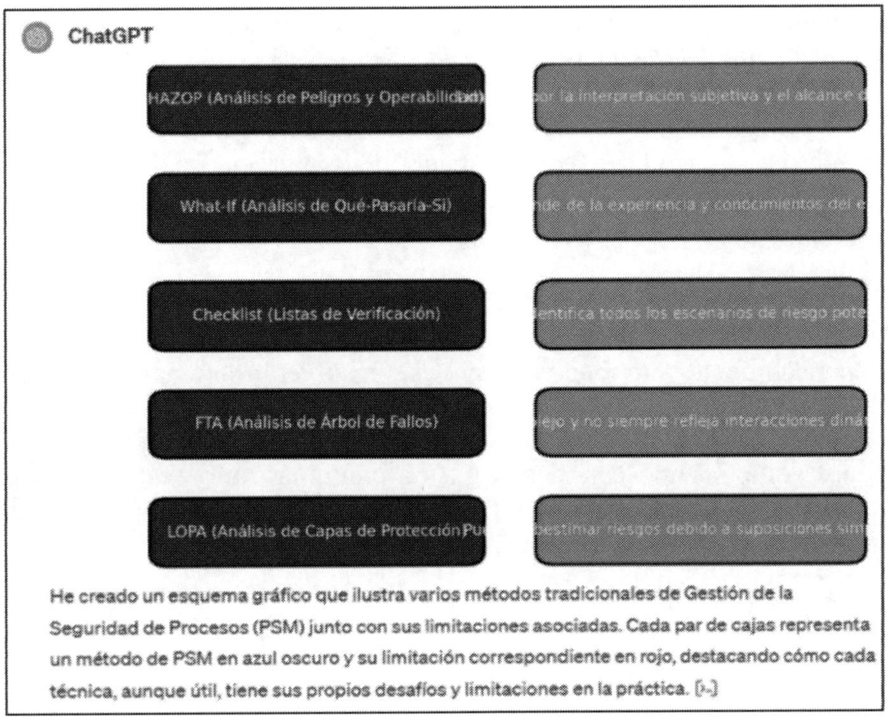

Fig. 02.- Limitaciones de las herramientas PHA (generado desde Prompt)

2.3. Conceptos básicos de inteligencia artificial

La inteligencia artificial (IA) es un campo de la informática que se enfoca en desarrollar sistemas y tecnologías que pueden realizar tareas que normalmente requieren de la inteligencia humana. Esto incluye actividades como el reconocimiento de patrones, la toma de decisiones, la resolución de problemas y el aprendizaje.

Las principales ramas de la inteligencia artificial incluyen:

a) **Aprendizaje automático (Machine Learning):** Es una subdisciplina de la inteligencia artificial que se centra en el desarrollo de algoritmos y modelos que permiten a las computadoras aprender y mejorar su rendimiento en una tarea específica a través de la experiencia. Los algoritmos de aprendizaje automático pueden entrenarse con datos para identificar patrones y tomar decisiones sin intervención humana explícita.

b) **Redes Neuronales Artificiales:** Inspiradas en el funcionamiento del cerebro humano, las redes neuronales artificiales son modelos matemáticos que pueden aprender de conjuntos de datos y realizar tareas complejas de reconocimiento de patrones. Consisten en una red de nodos interconectados que procesan y transmiten información de manera similar a las neuronas en el cerebro.

c) **Procesamiento del lenguaje natural** (Natural Language Processing, NLP): Se refiere al conjunto de técnicas utilizadas para que las computadoras comprendan, interpreten y generen lenguaje humano de manera natural. Esto incluye tareas como el análisis de sentimientos, la traducción automática, la generación de texto y la respuesta a preguntas.

d) **Visión por computadora (Computer Vision):** Es una rama de la inteligencia artificial que se enfoca en desarrollar sistemas capaces de interpretar y comprender la información visual del mundo real. Esto incluye la detección de objetos, el reconocimiento facial, la segmentación de imágenes y la clasificación de imágenes.

e) **Sistemas expertos:** Son programas informáticos diseñados para emular el comportamiento y el conocimiento de un experto humano en un campo específico. Utilizan reglas lógicas y bases de conocimiento para tomar decisiones y resolver problemas complejos.

Estas son solo algunas de las ramas principales de la inteligencia artificial, pero el campo es amplio y diverso, y sigue evolucionando con el tiempo. La combinación de estas técnicas y enfoques permite desarrollar sistemas cada vez más sofisticados y capaces de realizar una amplia gama de tareas inteligentes.

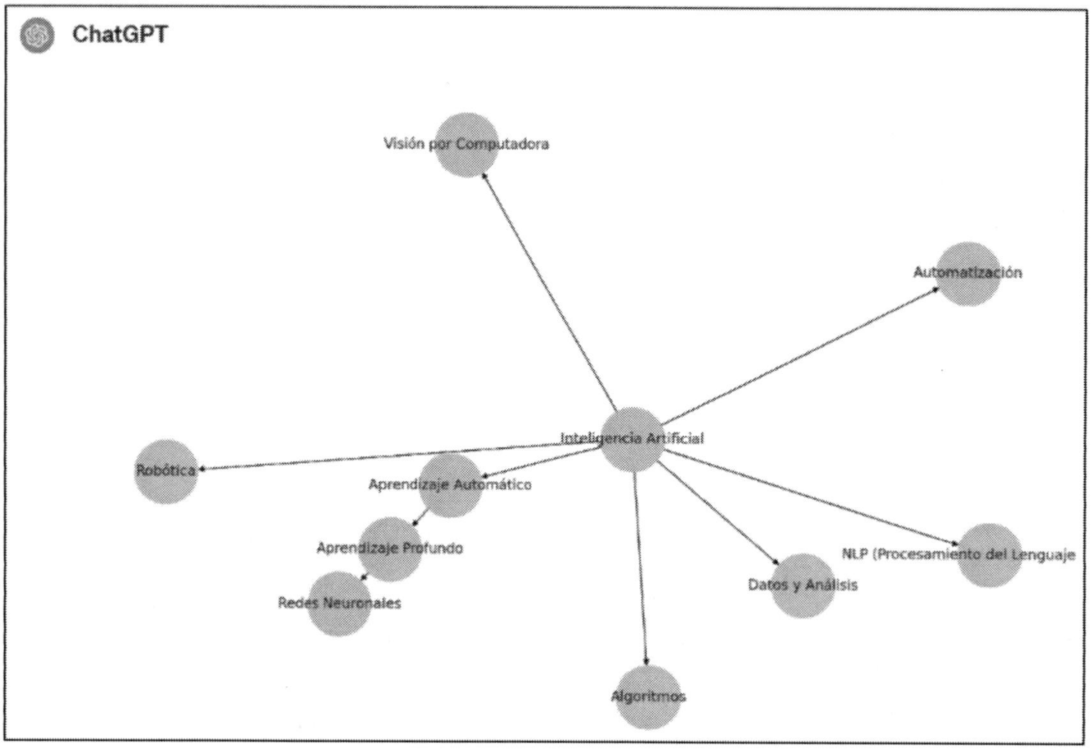

Fig. 03.- Gráfico que ilustra los conceptos básicos de la Inteligencia Artificial (IA)s (generado desde Prompt)

2.4. Ejemplos de aplicaciones en otras industrias

a) **Salud**:
   - Diagnóstico médico asistido por IA: Utilizando algoritmos de aprendizaje automático, se pueden desarrollar sistemas que analizan imágenes médicas (como radiografías, resonancias magnéticas o tomografías) para ayudar a los médicos a detectar enfermedades o anomalías.
   - Personalización del tratamiento: La IA puede analizar datos médicos de pacientes para predecir la eficacia de diferentes tratamientos y recomendar opciones personalizadas.

b) **Automotriz**:
   - Conducción autónoma: Las empresas automotrices utilizan sistemas de IA para desarrollar vehículos autónomos que pueden conducir de forma segura y autónoma en diferentes condiciones de tráfico y entornos.
   - Mantenimiento predictivo: Los fabricantes de automóviles pueden utilizar IA para predecir fallos en los vehículos mediante el análisis de datos de sensores y diagnósticos remotos, permitiendo la programación proactiva de mantenimiento.

c) **Finanzas**:
   - Análisis de riesgos y fraude: Los bancos y las instituciones financieras utilizan IA para analizar grandes volúmenes de datos y detectar patrones sospechosos que podrían indicar fraudes o riesgos financieros.
   - Gestión de inversiones: Los algoritmos de aprendizaje automático se utilizan para analizar tendencias del mercado, predecir cambios en los precios de las acciones y tomar decisiones de inversión.

d) **Retail**:
   - Recomendaciones de productos: Las plataformas de comercio electrónico utilizan sistemas de recomendación basados en IA para sugerir productos a los clientes en función de sus historiales de compra y comportamientos de navegación.
   - Gestión de inventario: Los algoritmos de IA pueden analizar datos de ventas históricos y pronósticos de demanda para optimizar los niveles de inventario y reducir los costos de almacenamiento.

e) **Educación**:
   - Tutoría personalizada: Los sistemas de tutoría virtual utilizan IA para adaptar el contenido educativo a las necesidades y habilidades individuales de los estudiantes, proporcionando una experiencia de aprendizaje más personalizada.
   - Evaluación automática: Los sistemas de evaluación automatizada pueden utilizar IA para analizar y calificar automáticamente trabajos escritos, pruebas y exámenes, ahorrando tiempo a los educadores y proporcionando retroalimentación inmediata a los estudiantes.

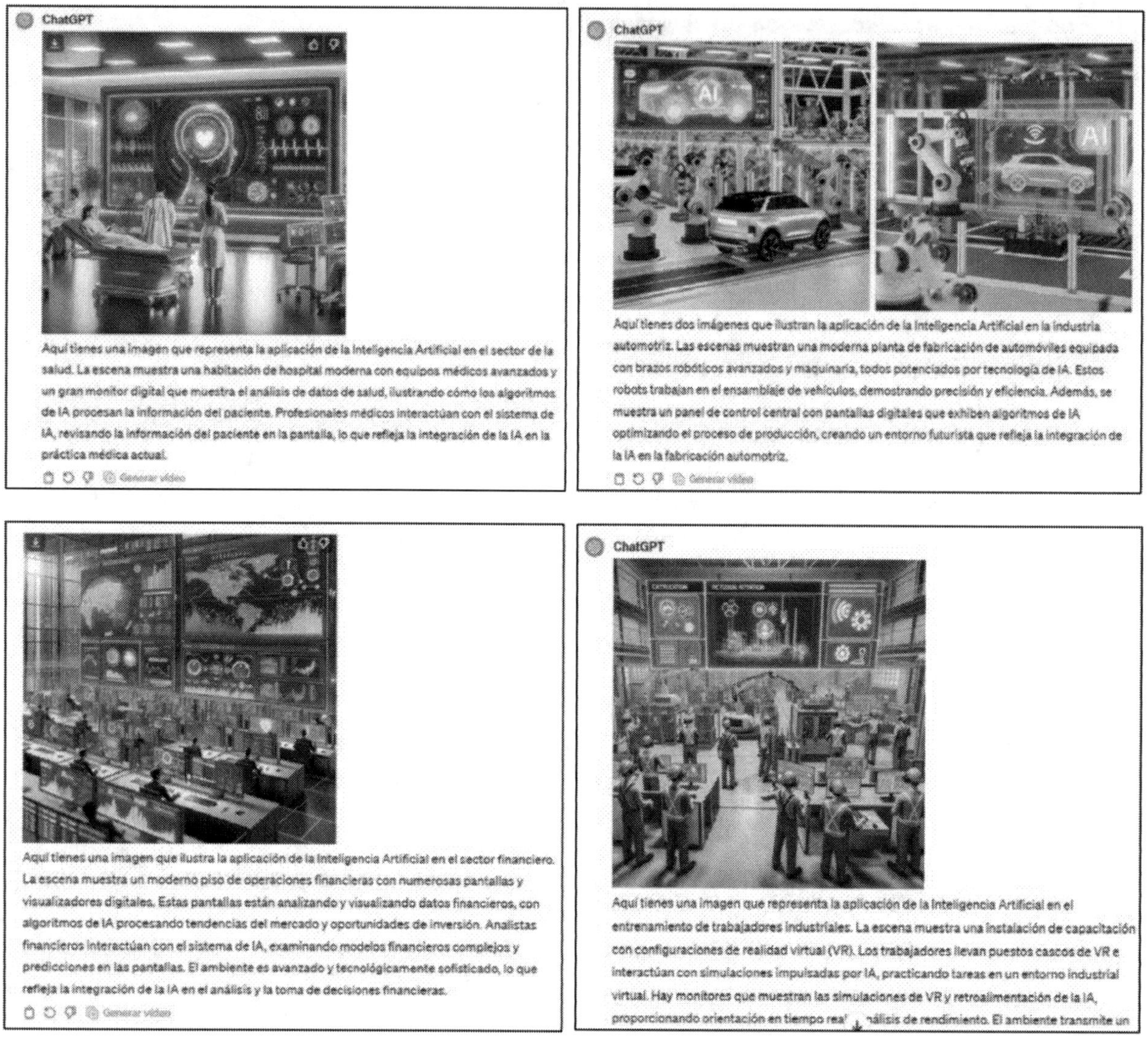

Fig. 04.- Ejemplos de aplicaciones IA en otras industrias (generado desde Prompt)

Estos son solo algunos ejemplos de cómo la inteligencia artificial está transformando diferentes industrias, pero su potencial es vasto y sigue expandiéndose a medida que se desarrollan nuevas aplicaciones y tecnologías.

3 Aplicaciones de IA en la seguridad de procesos

3.1 Análisis de riesgos: Métodos basados en IA para identificar y evaluar riesgos

El Análisis de Peligros y Operabilidad (HAZOP, por sus siglas en inglés: Hazard and Operability Analysis) es una técnica bien establecida en la gestión de riesgos de procesos industriales. Si bien tradicionalmente se ha realizado de manera manual, en los últimos años se ha explorado cómo la inteligencia artificial puede mejorar y agilizar este proceso. Aquí te explico cómo se pueden utilizar métodos basados en IA en el HAZOP:

a) **Automatización del proceso de análisis:** Los sistemas basados en IA pueden automatizar gran parte del proceso de análisis HAZOP. Esto implica la implementación de algoritmos de aprendizaje automático que son entrenados con datos históricos y normativas de seguridad para identificar posibles peligros y desviaciones en los procesos industriales.

b) **Identificación de desviaciones y escenarios de riesgo:** Utilizando técnicas de procesamiento de lenguaje natural (NLP) y análisis de datos, los sistemas de IA pueden analizar documentos y descripciones de procesos para identificar automáticamente desviaciones potenciales y escenarios de riesgo en los procesos industriales.

c) **Priorización de riesgos:** Los algoritmos de IA pueden ayudar a priorizar los riesgos identificados durante el análisis HAZOP, asignando niveles de riesgo basados en la probabilidad y la gravedad de los posibles incidentes. Esto permite a las organizaciones enfocar sus recursos en abordar los riesgos más críticos primero.

d) **Generación de recomendaciones y soluciones:** Los sistemas basados en IA pueden generar recomendaciones y soluciones para mitigar los riesgos identificados durante el análisis HAZOP. Esto puede incluir sugerencias para la implementación de controles de ingeniería, cambios en los procedimientos operativos o mejoras en el diseño del proceso.

e) **Actualización continua del análisis de riesgos:** Los sistemas de IA pueden monitorear de manera continua los procesos industriales y realizar análisis de riesgos en tiempo real. Esto permite una actualización constante del análisis HAZOP a medida que cambian las condiciones operativas o se introducen nuevos riesgos en el proceso.

En resumen, los métodos basados en IA pueden mejorar significativamente el proceso de análisis de riesgos HAZOP al automatizar tareas repetitivas, identificar desviaciones y escenarios de riesgo de manera más eficiente, priorizar los riesgos de manera adecuada y proporcionar recomendaciones para mitigar los riesgos identificados. Esto permite a las organizaciones mejorar la seguridad de sus operaciones industriales y reducir el riesgo de incidentes.

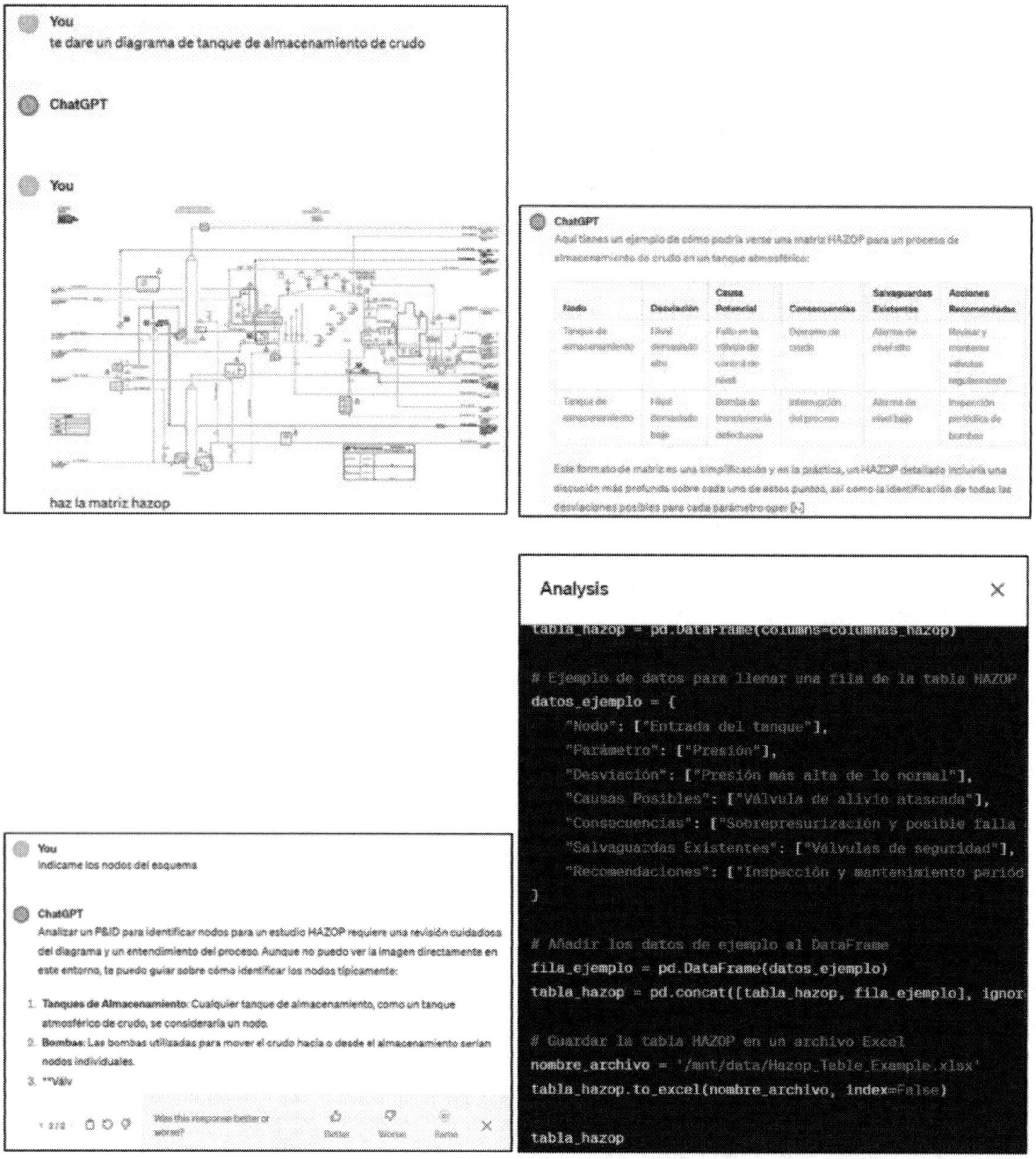

Fig. 05.- Ejemplos de aplicaciones IA en HAZOP (generado desde Prompt)

3.2 Predicción de fallos: Modelos predictivos para prevenir accidentes

La inteligencia artificial (IA) puede desempeñar un papel crucial en la predicción de fallos en equipos y sistemas industriales, lo que a su vez ayuda a prevenir accidentes. Aquí te explico cómo la IA puede contribuir en este sentido:

a) **Análisis de datos históricos:** Los sistemas de IA pueden analizar grandes volúmenes de datos históricos sobre el funcionamiento de equipos y sistemas industriales. Esto incluye datos de sensores, registros de mantenimiento, registros de operaciones, entre otros. Al procesar esta información, los algoritmos de IA pueden identificar patrones y tendencias que podrían indicar un deterioro o un posible fallo inminente en el equipo.

b) **Modelado predictivo:** Basándose en los datos históricos, la IA puede desarrollar modelos predictivos que estimen la probabilidad de que ocurra un fallo en el equipo en un momento dado. Estos modelos pueden tener en cuenta una variedad de factores, como la edad y el estado del equipo, las condiciones operativas, los patrones de mantenimiento y otros factores relevantes.

c) **Detección de anomalías:** Los algoritmos de IA pueden identificar anomalías o desviaciones inesperadas en el funcionamiento del equipo en tiempo real. Esto puede incluir cambios en los patrones de comportamiento de los sensores, variaciones en los niveles de rendimiento o cualquier otro signo de que el equipo podría estar experimentando problemas.

d) **Alertas tempranas y recomendaciones de acción:** Cuando se detecta una anomalía o se identifica un riesgo potencial de fallo, los sistemas de IA pueden generar alertas tempranas y recomendar acciones preventivas. Esto puede incluir la programación de mantenimiento preventivo, la inspección adicional del equipo, la implementación de controles de ingeniería adicionales o cualquier otra medida necesaria para mitigar el riesgo de un fallo.

e) **Aprendizaje continuo:** Los sistemas de IA pueden mejorar su capacidad predictiva con el tiempo a medida que se recopilan más datos y se obtiene retroalimentación sobre la precisión de las predicciones. Esto permite un aprendizaje continuo y la refinación de los modelos predictivos para hacerlos más precisos y efectivos en la prevención de fallos.

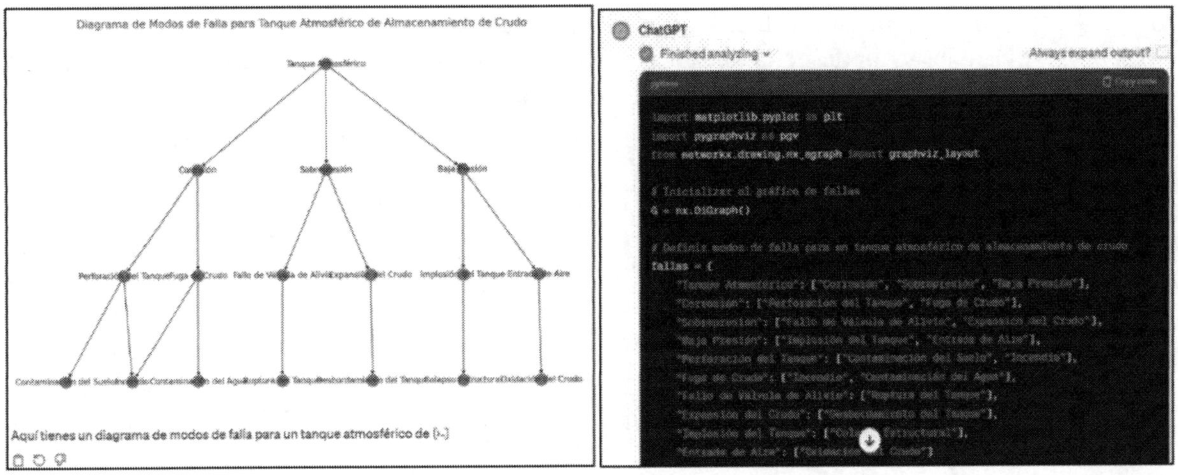

Fig. 06.- Ejemplos de Modelos predictivos para prevenir accidentes (generado desde Prompt)

En resumen, la IA puede desempeñar un papel crucial en la predicción de fallos en equipos y sistemas industriales al analizar datos históricos, desarrollar modelos predictivos, detectar anomalías en tiempo real, generar alertas tempranas y recomendar acciones preventivas. Esto ayuda a prevenir accidentes y a mejorar la seguridad en entornos industriales.

Por supuesto, aquí presento algunos modelos predictivos específicos basados en inteligencia artificial que se utilizan para prevenir accidentes en diversas industrias:

i. **Mantenimiento predictivo:**
- Utiliza algoritmos de aprendizaje automático para predecir cuándo es probable que ocurra un fallo en equipos industriales.
- Analiza datos de sensores, registros de mantenimiento y otros datos operativos para identificar patrones que indiquen un posible deterioro o fallo inminente.
- Permite programar intervenciones de mantenimiento preventivo de manera proactiva, reduciendo el riesgo de fallos no planificados y accidentes asociados.

ii. **Predicción de accidentes de tráfico:**
- Utiliza modelos de aprendizaje automático para prever la probabilidad de que ocurran accidentes de tráfico en diferentes áreas y condiciones.
- Analiza datos históricos de accidentes, condiciones del tráfico, clima y otros factores relevantes para identificar patrones y tendencias que puedan indicar áreas de riesgo.
- Permite a las autoridades y a los conductores tomar medidas preventivas, como la implementación de medidas de seguridad adicionales o la modificación de rutas de viaje, para reducir el riesgo de accidentes.

iii. **Modelos de pronóstico de riesgos naturales:**
- Utilizan datos históricos y en tiempo real, como imágenes satelitales, sensores meteorológicos y registros de eventos pasados, para predecir la probabilidad de desastres naturales, como tormentas, inundaciones o terremotos.
- Utilizan algoritmos de aprendizaje automático para identificar patrones y tendencias en los datos que puedan indicar la probabilidad de eventos extremos.
- Permiten a las autoridades y a las comunidades tomar medidas preventivas, como la evacuación anticipada o la construcción de infraestructuras de protección, para reducir el impacto de los desastres naturales.

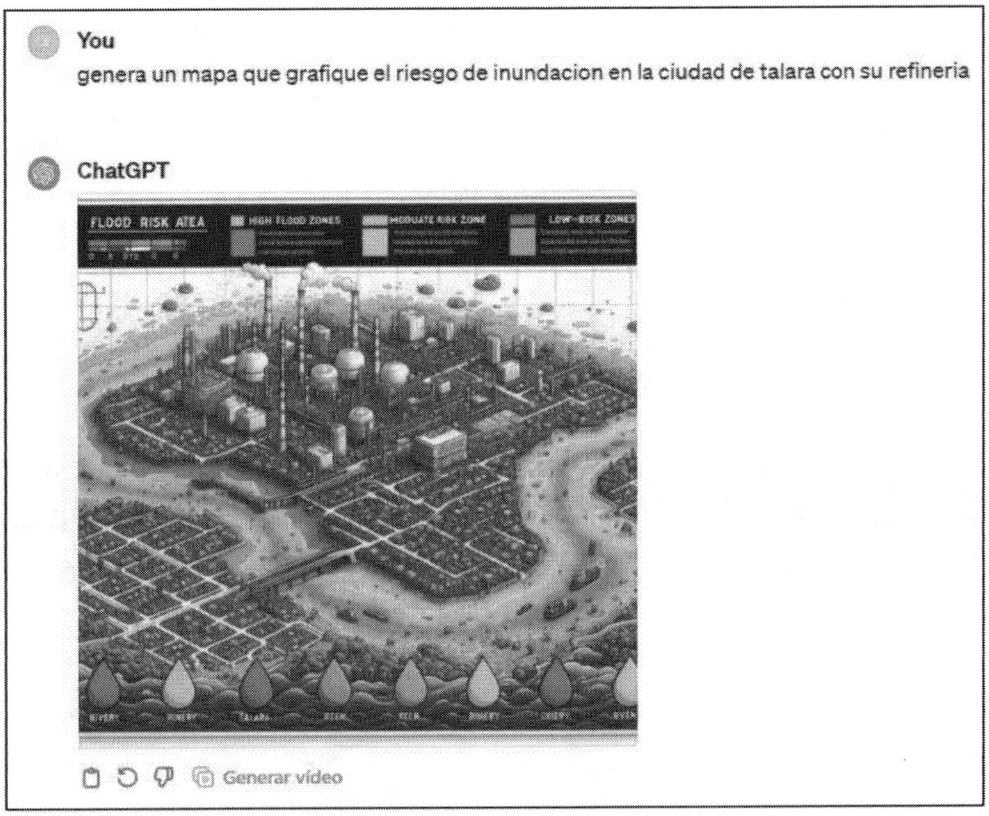

Fig. 07.- Modelo de pronóstico de riesgos naturales (generado desde Prompt)

iv. **Detección de anomalías en sistemas industriales:**
- Utiliza técnicas de análisis de datos y aprendizaje automático para detectar anomalías en el funcionamiento de equipos industriales, como variaciones inesperadas en los niveles de presión, temperatura o flujo.
- Permite identificar posibles problemas o fallos en los equipos antes de que ocurran, lo que permite tomar medidas preventivas para evitar accidentes y paradas no planificadas en la producción.

Estos son solo algunos ejemplos de modelos predictivos basados en inteligencia artificial que se utilizan para prevenir accidentes en diversas industrias. Cada uno de estos modelos puede adaptarse y personalizarse según las necesidades específicas de cada entorno industrial para mejorar la seguridad y proteger los activos y el personal.

3.3 Optimización de procesos: Mejora de la eficiencia y seguridad.

La Inteligencia Artificial (IA) juega un papel crucial en la optimización de procesos para mejorar la eficiencia y la seguridad, especialmente en entornos industriales y químicos. Su aplicación en estos campos es diversa y en constante evolución, ofreciendo nuevas oportunidades para el análisis de datos, la automatización y la toma de decisiones. A continuación, detallo cómo la IA contribuye a este objetivo:

3.3.1 Análisis Predictivo y Preventivo

- **Detección de Patrones y Anomalías:** La IA, especialmente el aprendizaje automático, puede analizar grandes cantidades de datos operativos para identificar patrones y anomalías que podrían indicar ineficiencias o riesgos potenciales.
- **Mantenimiento Predictivo:** Utilizando técnicas de machine learning, se pueden predecir fallos en el equipo antes de que ocurran, permitiendo un mantenimiento más efectivo y preventivo, lo que reduce las interrupciones no planificadas y mejora la seguridad.

Fig. 08.- Modelo de gemelo digital V.4.0 (generado para visor de realidad ampliada)

3.3.2 Optimización de Procesos en Tiempo Real

- **Control de Procesos:** Sistemas de IA pueden ajustar automáticamente las condiciones de operación en tiempo real para maximizar la eficiencia, como ajustar temperaturas, presiones o velocidades de flujo.
- **Respuesta Dinámica:** La IA permite una respuesta rápida a cambios en las condiciones de operación, ajustando los procesos para mantener la eficiencia y seguridad.

Fig. 09.- Modelo Agregado de gemelos digitales (DTA)

3.3.3 Automatización Mejorada

- **Robótica y Automatización Inteligente:** En la industria química, la IA facilita la automatización de tareas peligrosas o repetitivas, reduciendo el error humano y mejorando la seguridad laboral.
- **Sistemas Autónomos:** Vehículos y drones autónomos pueden ser utilizados para inspecciones y monitoreo de condiciones en áreas peligrosas o de difícil acceso.

3.3.4 Toma de Decisiones Basada en Datos

- **Optimización de Recursos:** La IA puede analizar patrones de consumo y producción para optimizar el uso de materias primas y recursos energéticos.
- **Decisiones Estratégicas:** Herramientas de IA pueden ayudar en la planificación a largo plazo, prediciendo tendencias del mercado y ayudando a tomar decisiones estratégicas más informadas.

3.3.5 Mejora de la Seguridad y Cumplimiento Normativo

- Simulaciones y Modelado de Riesgos: La IA permite simular diferentes escenarios de riesgo, ayudando a prevenir accidentes y a cumplir con regulaciones de seguridad y ambientales.

- Monitoreo Continuo: Sistemas de IA pueden monitorear continuamente las operaciones para garantizar que se mantengan dentro de parámetros seguros y reglamentarios.

3.3.6 Formación y Capacitación

- Realidad Virtual y Aumentada: La IA combinada con VR/AR puede crear simulaciones de entrenamiento realistas para preparar al personal para situaciones de emergencia, mejorando la preparación y la seguridad.

En resumen, la IA no solo mejora la eficiencia y productividad en la industria química y de procesos, sino que también juega un papel fundamental en la mejora de la seguridad, reduciendo riesgos y mejorando la protección tanto de los trabajadores como del medio ambiente. La clave está en la implementación cuidadosa y considerada de estas tecnologías, garantizando que complementen y mejoren los sistemas y procesos existentes.

3.4 Detección de anomalías: Identificación de patrones anómalos.

La identificación de patrones anómalos utilizando Inteligencia Artificial (IA) es una aplicación poderosa en diversos campos, desde la detección de fraudes financieros hasta el mantenimiento predictivo en la industria. A continuación, se presenta algunos ejemplos destacados de su aplicación, junto con fuentes y datos relevantes:

3.4.1 Mantenimiento Predictivo en la Industria

- **Aplicación**: Monitoreo del estado de maquinaria y equipos en la industria manufacturera o energética.
- **Datos**: Utiliza datos de sensores como vibración, temperatura y sonido.
- **Fuente**: Un estudio en "Journal of Manufacturing Systems" mostró cómo el aprendizaje automático puede predecir fallas en equipos industriales analizando patrones de vibración.
- **Resultado**: Mejora en la eficiencia operativa y reducción de tiempos de inactividad.

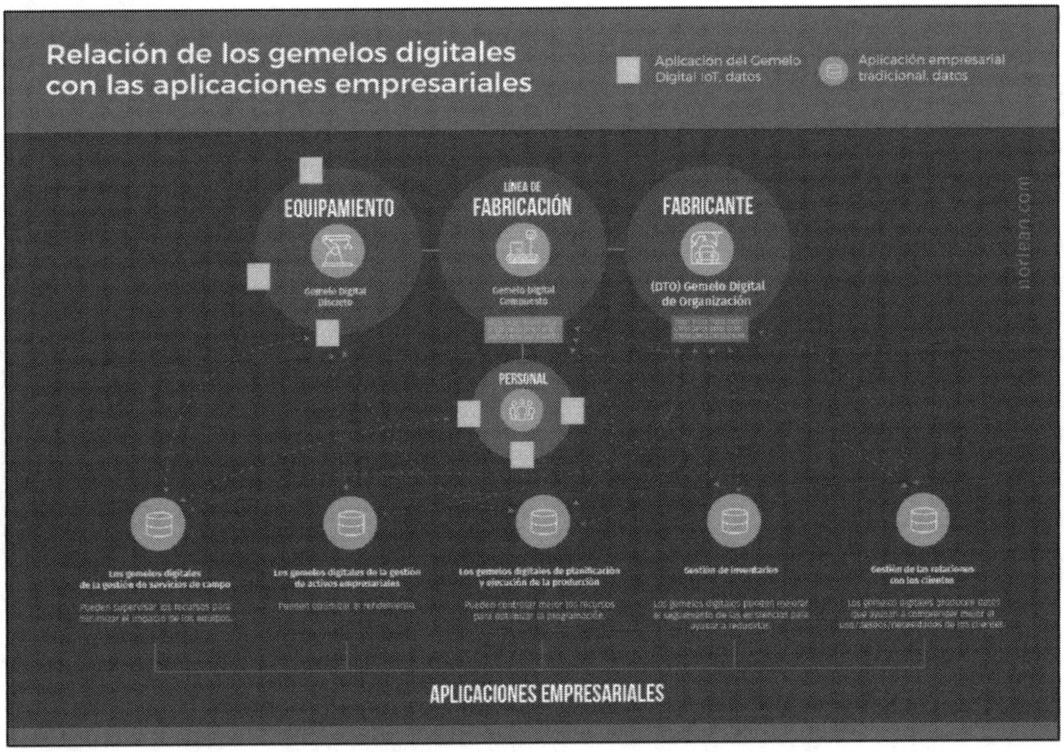

Fig. 10.- Relación de gemelos digitales

3.4.2 Detección de Fraude Financiero

- **Aplicación**: Identificación de actividades sospechosas en transacciones bancarias o tarjetas de crédito.
- **Datos**: Análisis de patrones de transacciones, incluyendo montos, frecuencia y ubicaciones.
- **Fuente**: Un informe de "IEEE Transactions on Neural Networks and Learning Systems" detalla el uso de redes neuronales para detectar patrones de fraude.
- **Resultado**: Reducción significativa en la incidencia de fraudes financieros.

3.4.3 Diagnóstico Médico

- **Aplicación**: Diagnóstico precoz de enfermedades como el cáncer a través de imágenes médicas.
- **Datos**: Imágenes de resonancias magnéticas (MRI), tomografías computarizadas (CT) o radiografías.
- **Fuente**: Investigaciones publicadas en "Nature Medicine" han demostrado la eficacia de la IA en la detección de patrones anómalos en imágenes médicas que indican la presencia de tumores.
- **Resultado**: Detección más rápida y precisa de enfermedades, mejorando las posibilidades de tratamiento exitoso.

3.4.4 Monitoreo Ambiental

- **Aplicación**: Detección de cambios anómalos en el medio ambiente, como la contaminación del aire o del agua.
- **Datos**: Datos de calidad del aire, niveles de contaminantes, imágenes satelitales.
- **Fuente**: Un estudio en "Environmental Science & Technology" utilizó IA para identificar patrones de contaminación en cuerpos de agua.
- **Resultado**: Mejora en la respuesta a problemas ambientales y en la gestión de recursos naturales.

3.4.5 Seguridad Cibernética

- **Aplicación**: Identificación de amenazas de seguridad, como ataques de malware o intrusiones en redes.
- **Datos**: Datos de tráfico de red, registros de acceso, patrones de uso.
- **Fuente**: Investigaciones en "IEEE Transactions on Dependable and Secure Computing" han explorado el uso de IA para detectar anomalías que indican ciberataques.
- **Resultado**: Mejora en la detección y respuesta a amenazas de seguridad cibernética.

Estos ejemplos ilustran cómo la IA se ha convertido en una herramienta esencial para la identificación de patrones anómalos en una amplia gama de aplicaciones, contribuyendo significativamente a la mejora de la eficiencia, seguridad y capacidad de respuesta en varios sectores.

4 Casos de estudio y ejemplos

La implementación de soluciones de Inteligencia Artificial (IA) en empresas ha llevado a numerosos casos de éxito en diversos sectores. A continuación, presento un resumen de algunos casos reales, destacando tanto los logros como las lecciones aprendidas:

4.1 General Electric (GE) - Mantenimiento Predictivo

- **Implementación**: GE utiliza IA para el mantenimiento predictivo de su maquinaria y equipos industriales.
- **Éxito**: La implementación de IA ha llevado a una reducción significativa en el tiempo de inactividad y un mantenimiento más eficiente.
- **Lecciones Aprendidas**: La importancia de la calidad de los datos y la necesidad de integrar soluciones de IA con sistemas operativos existentes para lograr resultados óptimos.

Fig. 11.- Ilustración de aplicación IA en mantenimiento (generado desde Prompt)

4.2 American Express - Detección de Fraude

- **Implementación**: AmEx emplea algoritmos de IA para analizar transacciones y detectar posibles fraudes.
- **Éxito**: La capacidad de identificar transacciones fraudulentas con mayor precisión y rapidez, protegiendo tanto a la empresa como a sus clientes.
- **Lecciones Aprendidas:** La IA necesita ser constantemente actualizada y entrenada con nuevos datos para adaptarse a las cambiantes tácticas de fraude.

Fig. 12.- Ilustración de aplicación IA en detección de fraude (generado desde Prompt)

4.3 Siemens - Optimización de Procesos

- **Implementación:** Siemens utiliza IA en sus fábricas para optimizar procesos y mejorar la eficiencia operativa.
- **Éxito:** Mejora en la producción, reducción de costos operativos y mayor flexibilidad en la manufactura.
- **Lecciones Aprendidas:** La integración de sistemas de IA en ambientes de producción requiere un enfoque colaborativo entre ingenieros, técnicos y científicos de datos.

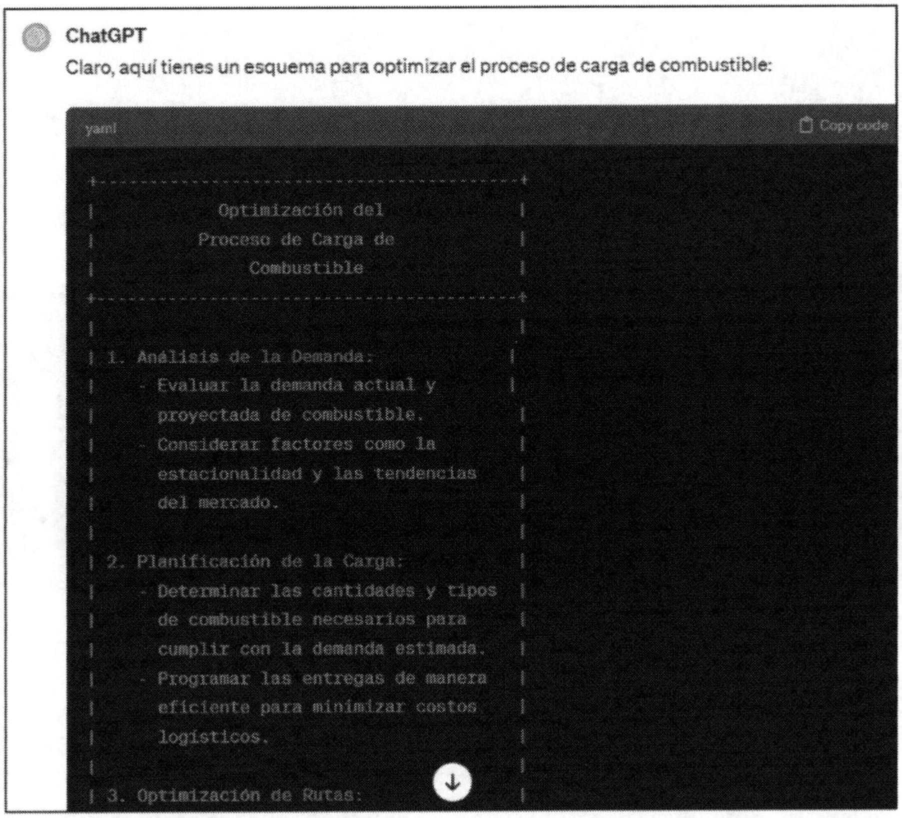

Fig. 13.- Ejemplo de mejora de proceso (generado desde Prompt)

4.4 IBM Watson Health - Atención Médica

- **Implementación**: IBM Watson se utiliza en el sector de la salud para análisis de datos médicos y apoyo en diagnósticos.
- **Éxito**: Ayuda a los médicos a tomar decisiones más informadas y proporciona diagnósticos más precisos.
- **Lecciones Aprendidas:** La IA en salud debe ser utilizada como una herramienta de apoyo, no como un sustituto del juicio clínico humano.

Fig. 14.- Imagen de atención de salud con análisis de datos por IA

4.5 Alibaba - Comercio Electrónico

- **Implementación:** Alibaba usa IA para personalizar la experiencia de compra, optimizar la logística y mejorar el servicio al cliente.
- **Éxito**: Incremento en las ventas, mejor satisfacción del cliente y eficiencia en la gestión de inventarios.
- **Lecciones Aprendidas:** La personalización basada en IA requiere un equilibrio cuidadoso entre relevancia y privacidad del usuario.

Fig. 15.- Modelo de negocios por IA

4.6 Tesla - Vehículos Autónomos

- **Implementación:** Tesla ha integrado IA en sus vehículos para facilitar características de conducción autónoma.
- **Éxito:** Avances significativos en la tecnología de vehículos autónomos, mejorando la seguridad y la experiencia de conducción.
- **Lecciones Aprendidas:** Los desafíos éticos y de seguridad son cruciales en la IA aplicada a vehículos autónomos, y es esencial una regulación y pruebas exhaustivas.

Fig. 16.- Interfaz de IA para vehículo Tesla

Estos casos ilustran cómo la IA puede transformar diferentes industrias, mejorando la eficiencia, la seguridad y la experiencia del cliente. Las lecciones aprendidas resaltan la importancia de la calidad de los datos, la necesidad de colaboración multidisciplinaria, la importancia de la ética y la privacidad, y la constante necesidad de adaptación y actualización de los sistemas de IA.

5 Beneficios, desafíos y el futuro

5.1 Beneficios potenciales: Reducción de accidentes, mejora de eficiencia,

La aplicación de la Inteligencia Artificial (IA) en la seguridad de procesos químicos conlleva beneficios potenciales significativos, particularmente en la reducción de accidentes y la mejora de la eficiencia operativa. Al integrar sistemas de IA, como el aprendizaje automático y el análisis predictivo, las instalaciones pueden anticipar y responder a condiciones anómalas antes de que se conviertan en incidentes. Esto se traduce en una disminución notable de eventos adversos, mejorando la seguridad general de los trabajadores y la protección del medio ambiente.

Desde una perspectiva de eficiencia, la IA optimiza las operaciones al monitorear y ajustar constantemente los parámetros del proceso para mantener la operatividad dentro de los márgenes óptimos. Esto no solo reduce el consumo de energía y materias primas, sino que también aumenta la calidad del producto y la consistencia del proceso. Además, al reducir el tiempo de inactividad a través del mantenimiento predictivo, la IA asegura una mayor productividad y menores costos operativos.

Por otro lado, la IA también permite una mejor gestión de datos, proporcionando insights valiosos a través del análisis de grandes volúmenes de información que de otro modo sería difícil de procesar manualmente. Esto conduce a una toma de decisiones más informada y basada en datos, lo que es esencial para una operación segura y eficiente en la industria química. Con la IA, las empresas pueden no solo responder a las condiciones actuales, sino también prever y planificar para el futuro, adaptándose a las tendencias emergentes y manteniendo la competitividad en un mercado en constante evolución.

Fig. 17.- Interfaz de modelamiento QRA (generado desde Prompt)

5.2 Desafíos y limitaciones: Implementación, costo, interpretación de resultados, etc.

La integración de la Inteligencia Artificial (IA) en la seguridad de procesos químicos no está exenta de desafíos y limitaciones, a pesar de sus numerosas ventajas. Estos desafíos pueden variar desde aspectos técnicos hasta preocupaciones operativas y éticas.

Uno de los principales desafíos técnicos es la calidad y cantidad de los datos necesarios para entrenar los algoritmos de IA de manera efectiva. Los sistemas de IA requieren grandes volúmenes de datos históricos de alta calidad para desarrollar modelos predictivos precisos. Sin embargo, en la industria química, a menudo hay una falta de datos históricos debido a la infrecuencia de ciertos eventos críticos, y la recopilación de datos puede ser dificultosa debido a la naturaleza compleja de los procesos químicos.

Además, la interpretación y el análisis de los datos presentan sus propias complejidades. Los procesos químicos son dinámicos y multifacéticos, y las variables que deben ser monitoreadas y analizadas son numerosas y a menudo interdependientes. Crear algoritmos que puedan entender y predecir el comportamiento de estos sistemas representa un desafío significativo de modelado.

Desde una perspectiva operativa, la integración de la IA en sistemas existentes puede ser complicada y costosa. La infraestructura de los sistemas de control industrial puede ser antigua y no siempre compatible con la tecnología más reciente. Esto puede requerir inversiones sustanciales en actualizaciones o reemplazos de equipos, así como en formación de personal para gestionar y mantener estas nuevas tecnologías.

La resistencia al cambio es otra barrera importante. La implementación de IA en la seguridad de procesos puede ser recibida con escepticismo por parte de los empleados que pueden temer por la seguridad de su empleo o que no confían en las decisiones tomadas por sistemas automatizados. Esto puede ser mitigado a través de una comunicación efectiva y programas de capacitación, pero sigue siendo una preocupación relevante.

- **Integración con Sistemas Existentes:** La implementación de soluciones de IA debe considerar la integración con sistemas y procesos existentes.
- **Fiabilidad y Transparencia:** Es crucial que los sistemas de IA sean fiables y sus decisiones y procesos sean transparentes y comprensibles para los operadores humanos.
- **Actualizaciones y Mantenimiento Continuo:** Los sistemas de IA requieren actualizaciones y mantenimiento regulares para asegurar su efectividad y precisión.

Finalmente, existen preocupaciones éticas y de privacidad relacionadas con la IA. En la industria química, donde se manejan sustancias potencialmente peligrosas, los errores en los sistemas de IA pueden tener consecuencias devastadoras. Por lo tanto, es crucial asegurar que los sistemas de IA sean transparentes, responsables y sujetos a una supervisión y regulación estrictas

5.3 Futuro de la IA en la seguridad de procesos

Como especialistas en Seguridad de Procesos, nuestra visión sobre el futuro de la Inteligencia Artificial (IA) en este campo debe ser eminentemente positiva y está marcada por un optimismo cauteloso. La IA tiene el potencial de revolucionar la forma en que abordamos la seguridad de procesos, al ofrecer mejoras significativas en la detección temprana de riesgos y en la toma de decisiones en tiempo real.

En los próximos años, anticipamos una adopción aún mayor de sistemas avanzados de IA que pueden analizar grandes conjuntos de datos operativos para identificar tendencias ocultas y predecir posibles incidentes antes de que ocurran. Esto llevará a una reducción de accidentes y a la minimización de tiempos de parada no planificados, lo cual es crítico en industrias donde los fallos pueden tener consecuencias desastrosas.

La IA también permitirá una gestión más eficiente de los recursos, optimizando el uso de energía y materiales y mejorando el rendimiento general de la producción. Los sistemas de control de procesos se beneficiarán de algoritmos autoajustables que mantendrán las operaciones dentro de los límites seguros, adaptándose a las fluctuaciones y minimizando la posibilidad de errores humanos.

Sin embargo, estos avances no están exentos de desafíos. La ciberseguridad se convierte en un tema crítico a medida que dependemos más de sistemas automatizados y conectados a la red. Además, la transparencia en la toma de decisiones de la IA y la necesidad de mantener la supervisión humana son esenciales para garantizar la confianza y la comprensión de los procesos automatizados.

En resumen, la IA es una poderosa aliada para el futuro de la seguridad de procesos químicos, pero su integración debe ser gestionada de manera estratégica, con una planificación cuidadosa y un enfoque en la fiabilidad, la seguridad y la sostenibilidad. La colaboración entre ingenieros de procesos, expertos en IA y operadores será clave para desbloquear el potencial completo de la IA en este campo.

Aquí damos una lista de algunas herramientas populares de Inteligencia Artificial (IA) que se utilizan en una variedad de aplicaciones:

1. **TensorFlow:** Una plataforma de código abierto desarrollada por Google para realizar tareas de aprendizaje automático y deep learning, como el reconocimiento de imágenes y el procesamiento del lenguaje natural.

2. **PyTorch:** Otro marco de aprendizaje profundo de código abierto que se utiliza ampliamente en la investigación y la producción de modelos de IA. Es conocido por su facilidad de uso y flexibilidad.

3. **Scikit-learn:** Una biblioteca de aprendizaje automático de código abierto para Python que proporciona herramientas simples y eficientes para la minería y análisis de datos, así como para la construcción de modelos de aprendizaje automático.

4. **Keras:** Una API de aprendizaje profundo de alto nivel que se ejecuta sobre TensorFlow, Theano o Microsoft Cognitive Toolkit. Es conocido por su facilidad de uso y su capacidad para crear modelos de IA rápidamente.

5. **OpenCV:** Una biblioteca de visión por computadora de código abierto que proporciona herramientas y algoritmos para procesar imágenes y videos, incluyendo reconocimiento facial, detección de objetos y seguimiento de movimiento.

6. **IBM Watson:** Una plataforma de servicios cognitivos en la nube que proporciona herramientas para crear aplicaciones de IA, incluyendo análisis de lenguaje natural, análisis de sentimientos y reconocimiento de voz.

7. **Microsoft Azure AI:** Una suite de herramientas y servicios de inteligencia artificial en la nube que incluye capacidades de aprendizaje automático, visión por computadora y procesamiento del lenguaje natural.

8. Amazon SageMaker: Un servicio de Amazon Web Services (AWS) que facilita la construcción, el entrenamiento y la implementación de modelos de aprendizaje automático a escala.

Estas son solo algunas de las herramientas más conocidas en el campo de la Inteligencia Artificial. Hay muchas otras disponibles, cada una con sus propias características y casos de uso específicos. Dependiendo de las necesidades de tu proyecto, es posible que encuentres útiles otras herramientas especializadas.

TEXTO	IMÁGENES	VIDEOS	AUDIO
ChatGPT	Midjourney	Runway	Murf
Claude	DALL-E 3	Pictory	Eleven Labs
Bing Chat	Leonardo	Veed	Otter
X Grok	Firefly	HeyGen	Speechify
CHATBOT	**MARKETING**	**PRESENTACIONES**	**DISEÑO**
Dante AI	AdCreative	Tome	Canva
Air	Claude	Simplified	Adobe Firefly
ChatSimple	Jasper	Beautiful.ai	Designs.ai
Zapier	SEMrush	Synthesia	Mokker
REDES SOCIALES	**INVESTIGACIÓN**	**REDACCIÓN PUBLICITARIA**	**PRODUCTIVIDAD**
Sprout Social	HARPA	Copy.ai	ChatGPT Plus
Emplifi	Perplexity	Grammarly	Taskade
Lately	Glasp	Wordtune	Audio Pen
Brandwatch	ChatGPT Plus	WriteSonic	Notion AI

Fig. 18.- Detalle de principales herramientas de IA según campo de aplicación

6 Conclusiones:

La incorporación de la Inteligencia Artificial (IA) y tecnologías avanzadas en la seguridad de procesos marca un hito transformador en la industria, ofreciendo beneficios significativos y abriendo nuevas fronteras en la gestión de riesgos. Estas tecnologías no solo realzan la capacidad de predecir y mitigar riesgos, sino que también catalizan un cambio fundamental en la eficiencia y la sostenibilidad operativa.

- En primer lugar, la IA, con su capacidad para analizar grandes volúmenes de datos y detectar patrones complejos, ha revolucionado la prevención de riesgos. Al anticipar incidentes antes de que ocurran, la IA mejora drásticamente la seguridad en los entornos de trabajo, lo que resulta en una reducción significativa de accidentes y daños ambientales. Esta previsión proactiva es un cambio radical respecto a los métodos reactivos tradicionales.
- Además, la automatización y la optimización de procesos impulsadas por la IA no solo incrementan la productividad y reducen costes, sino que también minimizan la posibilidad de errores humanos, tradicionalmente una de las mayores causas de accidentes en entornos industriales. La eficiencia mejorada en la operación y mantenimiento de equipos asegura un rendimiento óptimo y una mayor longevidad de los activos.
- Otro aspecto crucial es la toma de decisiones informadas. Los sistemas basados en IA proporcionan recomendaciones y análisis detallados, facilitando decisiones más rápidas y fundamentadas, esenciales en situaciones críticas. Esto representa un cambio significativo en la manera en que los operadores y gerentes interactúan con los sistemas de procesos, integrando conocimientos profundos y análisis predictivos en sus rutinas diarias.
- Sin embargo, estos avances no están exentos de desafíos. La ciberseguridad se convierte en una preocupación primordial a medida que los sistemas se vuelven más interconectados. Además, la necesidad de mantener una supervisión humana y garantizar la transparencia en las decisiones de la IA es crucial para mantener la confianza y comprensión en los procesos automatizados.
- En conclusión, la integración de la IA y las tecnologías avanzadas en la seguridad de procesos es un avance ineludible y vital. A medida que estas tecnologías continúen evolucionando, su aplicación inteligente y ética será clave para desbloquear todo su potencial, garantizando operaciones más seguras, eficientes y sostenibles en la industria del futuro.

7 Referencias (formato de ejemplo que se muestra a continuación).

Referencias en línea:

[1] https://www.unesco.org/es/artificial-intelligence/recommendation-ethics
[2] https://learn.microsoft.com/es-es/azure/cloud-adoption-framework/innovate/best-practices/trusted-ai
[3] https://www.mckinsey.com/capabilities/quantumblack/our-insights/leading-your-organization-to-responsible-ai/es-ES
[4] https://www.forbes.com/sites/bernardmarr/2022/06/20/the-best-examples-of-digital-twins-everyone-should-know-about/
[5] "Process Safety Progress Highlight: AI as an Ally in Hazard Analyses" de AIChE.

8 Bibliografía

1 Agrawal, A., Gans, J., & Goldfarb, A. (2018). Prediction Machines: The Simple Economics of Artificial Intelligence. Harvard Business Review Press.
2 Bostrom, N. (2014**).** ***Superintelligence: Paths, Dangers, Strategies***. Oxford University Press.
3 Bughin, J., Hazan, E., Ramaswamy, S., Chui, M., Allas, T., Dahlström, P., Henke, N., & Trench, M. (2018). ***Notes from the AI Frontier: Insights from Hundreds of Use Cases. McKinsey Global Institute***.
4 Crowl, D. A., & Louvar, J. F. (2019). ***Chemical Process Safety: Fundamentals with Applications***. (4th ed.). Prentice Hall.
5 Center for Chemical Process Safety (CCPS). (2017). ***Guidelines for Risk Based Process Safety***. Wiley-AIChE.
6 Davenport, T. H., & Ronanki, R. (2018). ***Artificial Intelligence for the Real World***. Harvard Business Review Press.
7 Domingos, P. (2015). ***The Master Algorithm: How the Quest for the Ultimate Learning Machine Will Remake Our World***. Basic Books.
8 Goodfellow, I., Bengio, Y., & Courville, A. (2016). ***Deep Learning***. MIT Press.
9 Kletz, T., & Amyotte, P. (2010). ***Process Plants: A Handbook for Inherently Safer Design***. (2nd ed.). CRC Press.
10 Lees, F. P. (2012). ***Lees' Loss Prevention in the Process Industries: Hazard Identification, Assessment and Control***. (4th ed.). Butterworth-Heinemann.
11 Lee, K.-F. (2018). ***AI Superpowers: China, Silicon Valley, and the New World Order***. Houghton Mifflin Harcourt.
12 Mannan, S. (Ed.). (2018). ***Guidelines for Process Safety Metrics.*** American Institute of Chemical Engineers
13 Mitchell, M. (2019). ***Artificial Intelligence: A Guide for Thinking Humans***. Pantheon.
14 Ng, A. (2018). ***Machine Learning Yearning***. DeepLearning.AI.
15 O'Neil, C. (2016). ***Weapons of Math Destruction: How Big Data Increases Inequality and Threatens Democracy.*** Crown.
16 Russell, S. J., & Norvig, P. (2016). ***Artificial Intelligence***: A Modern Approach (3rd ed.). Pearson.
17 Russell, S. J. (2019). ***Human Compatible: Artificial Intelligence and the Problem of Control***. Viking.
18 Schwab, K. (2017). ***The Fourth Industrial Revolution***. Crown Business.
19 Tegmark, M. (2017). ***Life 3.0: Being Human in the Age of Artificial Intelligence***. Knopf.

Cómo incorporar y analizar los modos no rutinarios en los PHAs de manera eficiente

(Arranque, parada y mantenimiento en línea)

Autor

Matías A. Massello – Process Safety Engineer
Process Improvement Institute, Inc. (PII)
e-mail: mmassello@piii.com

Co-autor

William G. Bridges – President
Process Improvement Institute, Inc. (PII)
e-mail: wbridges@piii.com

Abstract

Las Evaluaciones de peligros, también llamadas Análisis de Peligros del Proceso (PHA, por sus siglas en inglés), se han realizado con mejoras graduales durante más de cinco décadas. Durante este tiempo, se han desarrollado y perfeccionado métodos como HAZOP y What-If. Sin embargo, algunas debilidades identificadas hace 30 años aún existen en la mayoría de los PHAs realizados en todo el mundo. El punto más crítico es que la mayoría de los PHAs no analizan a fondo los errores que pueden ocurrir durante arranque, parada y otros modos de operación no rutinarios.

A pesar de que alrededor del 80% de los accidentes graves ocurren durante las operaciones no rutinarias, la mayoría de los PHAs se enfocan principalmente en la operación normal (por ejemplo, HAZOP de nodos de equipo) ignorando casi completamente los modos de operación más peligrosos. Lamentablemente el enfoque habitual utilizado para modo de operación continuo solo encuentra alrededor de 5 - 10% de los escenarios de accidentes que pueden ocurrir durante arranque, parada y mantenimiento en línea. Esto resulta en un gran porcentaje de potenciales accidentes sin identificar. Si el PHA no encuentra los escenarios que podrían ocurrir durante estas operaciones no rutinarias, la organización no sabrá qué salvaguardas son necesarias para controlar su riesgo.

Una de las razones por las que muchas empresas no realizan PHA de Procedimientos es porque creen que el tiempo requerido para dicho análisis será excesivo. El objetivo de este artículo es describir las mejores prácticas para: (1) la selección y clasificación de los procedimientos más críticos para el análisis, (2) decidir sobre el método para optimizar el tiempo invertido, y (3) agilizar la documentación de los resultados del PHA.

Introducción

La identificación de peligros y situaciones que podrían causar daño de algún tipo es un aspecto fundamental para tomar decisiones tanto en la vida cotidiana como en la industria de procesos. Para poder estar preparados para contrarrestar los efectos indeseados de las situaciones peligrosas, es necesario tener dichas situaciones identificadas: Si no esperamos (o ignoramos) la ocurrencia de una situación peligrosa en particular, que se produzcan (o no) sus efectos indeseados dependerá de condiciones y factores que no están bajo nuestro control (o lo que a veces denominamos como "suerte"). En el ámbito de la industria de procesos, la manera de identificar dichos peligros es mediante los Análisis de Riesgos/Peligros de los Procesos (PHA, Process Hazard Analysis).

Los Análisis de Riesgos/Peligros de los Procesos han estado presentes en la industria de procesos por más de medio siglo, y cobraron mayor importancia a partir de accidentes trágicos como los de Flixborough (1974), Seveso (1974) y Bhopal (1984). Estos accidentes tuvieron una relevancia tal que fueron los disparadores de la regulación en Europa y Estados Unidos sobre Seguridad de Procesos (*Directivas Seveso* y *US OSHA PSM Standard*). A su vez, la industria ha ido desarrollando las Mejores Prácticas para la identificación de peligros, las cuales fueron condensadas por el CCPS en el libro *Guidelines for Hazard Evaluation Procedures*, cuya primera edición fue en 1985, y posteriormente actualizado y reeditado en 1992 y 2010 [1].

A pesar de que los PHA no son un tema nuevo, siguen existiendo muchas deficiencias a la hora de realizarlos. El punto más crítico es que la mayoría de los PHAs no analizan a fondo los errores que pueden ocurrir durante arranque, parada y otros modos de operación no rutinarios.

La importancia de los PHA de modos no rutinarios

Sin duda alguna, la principal deficiencia en los PHAs es la falta de análisis de **todos** los modos de operación. Prácticamente la totalidad de los PHAs realizados se centran únicamente en el modo de operación continua (estado estacionario). Suelen incluirse algunos escenarios de otros modos de operación, pero depende de que algunos de los miembros del equipo los mencione ya que no se realiza una búsqueda sistemática para tratar de identificarlos.

Los modos de operación que actualmente **no son analizados** debidamente son:

- Arranque/Puesta en marcha
- Parada normal
- Parada de emergencia
- Mantenimiento en línea
- Otras tareas no estacionarias regidas por procedimientos

La pregunta que podría surgir es: *"Si estos modos de operación están presentes mucho menos tiempo que la operación normal, ¿qué tan grave puede ser no analizarlos en los PHA?"*. Para responder a esta pregunta y recalcar la importancia que tiene el análisis de todos los modos de operación, sólo basta con mirar las estadísticas de los accidentes ocurridos en la industria:

> *~80% de los accidentes investigados por el CSB fueron durante modos de operación no rutinarios o en procesos/tareas batch [2]*

Dicha estadística coincide con lo manifestado informalmente a Process Improvement Institute por miles de estudiantes (de diversas empresas) en los cursos de PSM, Investigación de Incidentes o PHA.

Es decir que, al no analizar adecuadamente dichos modos de operación durante los PHA, **no se identifican de manera apropiada los escenarios que causan ~80% de los accidentes!** Resulta evidente de por qué esta es la principal deficiencia actual.

Cómo realizar PHA de modos no rutinarios

Varios analistas han tratado de identificar los peligros asociados a los modos de operación no rutinarios mediante el agregado de desviaciones genéricas tales como *"Desviaciones durante el arranque"* o *"Desviaciones durante el mantenimiento"* en cada Nodo/Sección. Desafortunadamente, este método sólo encuentra una fracción de los accidentes que podrían ocurrir durante estos modos de operación debido a que el equipo está enfocado en el modo de operación continua y no en los diferentes estados transitorios que se presentan durante los modos no rutinarios. Esta metodología suele encontrar únicamente escenarios asociados a accidentes/cuasi-accidentes experimentados en la instalación.

La evaluación de peligros de los modos de operación no rutinarios implica la **revisión de procedimientos** utilizando las técnicas de HAZOP, HAZOP simplificado o What-if para descubrir posibles escenarios de accidentes. El objetivo de esta revisión no es el de asegurar que los procedimientos son correctos y aceptables, sino es evaluar los escenarios de accidentes posibles si los procedimientos no se siguen. Incluso el mejor procedimiento puede no ser seguido por diversas razones.

> *El objetivo de la evaluación de peligros de los modos de operación no rutinarios (gobernados por procedimientos escritos) es el de asegurar que la organización tiene suficientes salvaguardas para el momento inevitable en que un paso sea realizado incorrectamente o sea omitido (ya sea inadvertidamente, por tomar "atajos" o por otras razones).*

Previo a cualquier análisis, la pregunta que debe hacerse es: ¿Son los procedimientos aptos para ser analizados? Si los procedimientos no son actualizados, correctos, claros y utilizados por los operadores, analizarlos representaría una pérdida de tiempo y, peor aún, la posibilidad de asumir que hay protecciones suficientes cuando no las hay o no identificar algunos escenarios. Si los procedimientos no son al menos un 90% exactos, el primer paso a realizar es el de desarrollar los procedimientos siguiendo las Mejores Prácticas[1] tan pronto como sea posible y luego realizar el PHA a estos nuevos procedimientos.

Es importante notar que para este análisis el rol del personal de Operaciones toma mayor relevancia. **Es necesaria la presencia dentro del Equipo de PHA de un Operador Senior y un Operador de campo**. Muchas veces sólo está presente un Supervisor/Superintendente que fue Operador durante varios años pero que quizás no está al tanto de todas las actividades o "atajos" que toman los operadores en el campo actualmente.

---

[1] Las Mejores Prácticas sobre escritura y formato de procedimientos pueden encontrarse en *"Mejores Prácticas para escribir procedimientos operativos y guías de solución de problemas"* [3]

En líneas generales, los PHA de modos no rutinarios consta de dos etapas principales:

- Selección y categorización de los Procedimientos a analizar
- Selección y aplicación de la metodología a utilizar

En ambas etapas hay aspectos claves para tener en cuenta para ser eficientes con el uso de los recursos (i.e. Tiempo utilizado). Dichas etapas se describen en las secciones siguientes.

Selección y categorización de los Procedimientos a analizar

Dependiendo del tamaño de la planta, podría haber varias decenas o centenas de procedimientos a analizar. Analizar TODOS los Procedimientos podría llevar mucho tiempo sin necesariamente obtener grandes dividendos por dicho esfuerzo. Es por eso por lo que es necesario utilizar algún criterio para decidir a cuáles dedicarles mayor cantidad de recursos (tiempo). El criterio que utilizar, tal cual ocurre en otros ámbitos de la Seguridad de Procesos, es el riesgo.

La selección y categorización de los Procedimientos a analizar se realiza junto a un Operador ya sea antes de las reuniones o junto a todo el Equipo de PHA luego de finalizar el análisis de los Nodos/Secciones del modo de operación continuo. Para cada Procedimiento se decidirá cualitativamente, en función de su complejidad y los peligros asociados, cuál es el riesgo del Procedimiento (Alto, Medio o Bajo). La Tabla 1 muestra un ejemplo.

Tabla 1. Ejemplo de categorización de Procedimientos

Num. Doc.	Título	Rev.	Riesgo
SOP 023	DESCARGA DE FOSFATO	09-20	BAJO
SOP 024	ARRANQUE DE LA SECCIÓN DE REACCIÓN	09-20	ALTO
SOP 025	PARADA DE LA SECCIÓN DE REACCIÓN	09-20	MEDIO
SOP 026	ARRANQUE DEL HORNO DE CARGA	02-20	ALTO
SOP 027	MONITOREO DEL SISTEMA DE VENTEO	09-20	BAJO

Una vez que los Procedimientos están categorizados, el Equipo se organiza para analizar aquellos Procedimientos con riesgo Alto y luego, los de riesgo Medio. Los de riesgo Bajo, no son analizados ya que el Equipo considera que no tiene escenarios relevantes de Seguridad de Procesos asociados. En promedio, PII encuentra que se deben analizar alrededor del 20 - 30% del total de Procedimientos.

Selección y aplicación de la metodología a utilizar

Hay tres metodologías principales para analizar los modos de operación no rutinarios:

- HAZOP (7 Palabras Guía)
- HAZOP simplificado (2 Frases Guía)
- What-if

A continuación, se describe cada una de ellas y, posteriormente, el criterio para decidir qué técnica utilizar.

HAZOP

La variante del HAZOP orientada a procedimientos es la forma más antigua de HAZOP (originada en ICI por la década del 60). Los creadores optaron por subdividir los dos tipos de errores humanos (omisión y comisión) generando las Palabras Guía que se describen en la Tabla 2.

Tabla 2. Palabras Guía de HAZOP para procedimientos

Tipo error	Palabra guía	Significado
Omisión	No	La intención específica de la instrucción no se realiza
	Parte de	Una parte de la intención de la instrucción no se realiza. Usualmente sólo aplica a tareas que implican 2 o más tareas prácticamente simultaneas (*"Abrir las válvulas A, B y C"*).
Comisión	Más	Incremento cualitativo. Se realiza demasiado de la intención de la instrucción (no aplica a tareas de Prender/Apagar o Abrir/Cerrar) o se hace demasiado rápido.
	Menos	Disminución cualitativa. Se realiza demasiado poco de la intención de la instrucción o se hace demasiado lento.
	Antes/Después	La instrucción se realiza fuera de la secuencia correcta
	Además	Se realiza otra acción además de la especificada en la instrucción (podría ser un atajo)
	Otro	Se selecciona, usa, lee, etc. el dispositivo incorrecto o se utiliza de otra manera. O se selecciona/agrega otro material.
-	Faltante	Un paso, instrucción o precaución está faltante en el procedimiento previo a este paso (Similar a *"Fuera de secuencia"*, pero el paso no está escrito).

Cada una de las Palabras Guía (Tabla 2) es aplicada a cada instrucción del procedimiento que implique una acción que modifique el estado del sistema. Es decir, todas las instrucciones del tipo Verificar/Asegurar/Chequear no son analizadas y son consideradas como "protecciones" para los otros pasos. En el caso de que el procedimiento no sigua las Mejores Prácticas [3] y cada instrucción no implique una única acción, el primer paso será subdividir las instrucciones en acciones individuales.

El desarrollo de la reunión es prácticamente idéntico a la de la revisión del modo continuo. Para cada desviación de la intención de la Instrucción/paso el equipo debe ir más allá de la causa obvia de *"Error operativo"* para identificar las Causas raíz asociadas a dicho error; como, por ejemplo: *"No hay suficiente énfasis en este paso durante el entrenamiento"*, *"El Operador es responsable de realizar dos tareas simultáneamente"*, *"El etiquetado de las válvulas en campo es confuso"*, o *"La lectura del indicador es difícil de leer porque está borrosa"*. La Palabra Guía *"Faltante"* permite identificar causas como *"No hay un paso escrito o entrenamiento formal para solicitar el permiso de trabajo en caliente previo a este paso"* o *"No hay un paso escrito o entrenamiento formal para abrir la válvula de descarga antes de poner en marcha la bomba"*.

HAZOP Simplificado

Un enfoque simplificado del HAZOP (descripto en los párrafos anteriores) consta en utilizar dos Palabras (Frases) Guía en vez de las siete Palabras Guía de la Tabla 2. Las dos Frases Guía que se utilizan se presentan en la tabla a continuación (Tabla 3).

Tabla 3. Frases Guía de HAZOP Simplificado para procedimientos

Frase guía	Significado
El paso/Instrucción no se realiza	El paso (o parte de él) no se realiza. Algunas posibles razones incluyen que el operador se olvidó de realizar el paso, no comprendió su importancia, o el procedimiento no incluía este paso.
El paso/instrucción se realiza de manera incorrecta	El Operador tiene la intención de realizar el paso (no omitirlo) sin embargo no lo realiza como es pretendido. Algunas posibles razones incluye

La primer Frase Guía está relacionado a las Palabras Guía de *"No"*, *"Parte de"* y *"Faltante"* de la Tabla 2, mientras que la segunda frase está relacionada a los errores de comisión abarcados en las Palabras Guía *"Más"*, *"Menos"*, *"Antes/Después"*, *"Además"* y *"Otro"* de la Tabla 2.

La aplicación y la dinámica de las reuniones de este método son equivalente a la variante de HAZOP con 7 Palabras Guía. Este enfoque ha demostrado ser útil en los casos que:

- El Procedimiento analizado no posea grandes peligros y/o no sea excesivamente complejo.
- El Líder del PHA tenga mucha experiencia con el enfoque previo y pueda compensar las deficiencias de este enfoque debidas a la simplificación.

What-if

El método de What-if para analizar modos de operación basados en procedimientos es un *brainstorming* libre sin la ayuda (o restricción) de las palabras guía. El equipo de PHA lee todo el procedimiento y luego responde la pregunta: *"¿Qué errores podrían llevar a una consecuencia de interés?"*. El equipo lista estos errores y luego analiza las consecuencias, causas, protecciones existentes y realiza recomendaciones de ser necesario. Este método no se aplica paso por paso, sino que se analiza la tarea completa (procedimiento) en conjunto.

Este método se describe en detalle en el libro del CCPS *Guidelines for Hazard Evaluation procedures* [4].

Selección de la metodología a utilizar

Las tres metodologías anteriores fueron descriptas en orden de rigurosidad; y a mayor rigurosidad, mayor cantidad de tiempo requerido. De manera análoga a lo que ocurre con la selección de la técnica para el modo de operación continuo el criterio básico de selección es simple: Cuanto mayor sean los peligros asociados al procedimiento y cuanto mayor sea su complejidad; se requerirá mayor rigurosidad en la técnica de análisis.

A modo general, la regla para decidir la metodología a utilizar es:

- **HAZOP (7 PG):** Para procedimientos con el potencial de consecuencias catastróficas y de muy alta complejidad. Típicamente no se encuentran este tipo de procedimientos en plantas químicas, petroquímicas, refinerías.
- **HAZOP Simplificado (2 FG):** Para procedimientos complejos en el cual los errores podrían tener consecuencias severas o muy costosas.
- **What-if:** Para procedimientos poco complejos.

La experiencia tanto del Líder como del personal de Operaciones juega un papel fundamental en los PHA de modos de operación no rutinarios tanto en la selección de qué procedimientos a analizar como en la elección de la metodología de análisis para cada uno de ellos.

La Figura 1 muestra la distribución típica del uso de cada una de las técnicas en un PHA de una planta química o de una refinería. La gran mayoría de los procedimientos son lo suficientemente simples o no tienen grandes peligros asociados como para justificar el uso de las técnicas menos rigurosas: What-if y HAZOP simplificado. El método HAZOP (7-8 Palabras Guía) prácticamente no se utiliza ya que la mayoría de las tareas no requiere tanto nivel de escrutinio para identificar los posibles escenarios.

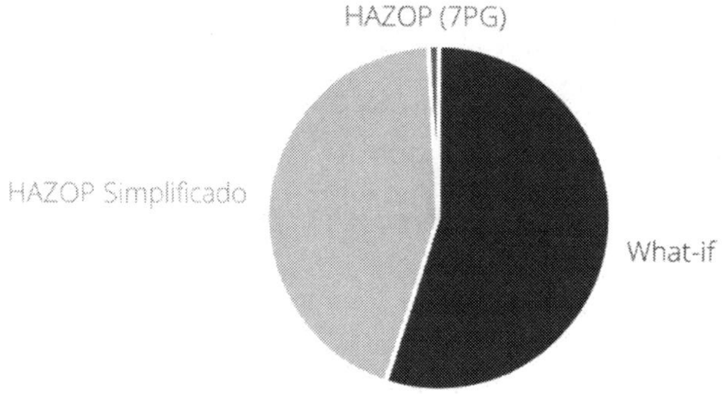

Figura 1. Distribución típica del uso de las técnicas de análisis de modos de operación no rutinarios

Resumen

La Figura 2 muestra un diagrama de flujo de la metodología para realizar PHA de los procedimientos de arranque, parada y mantenimiento en línea de manera eficiente.

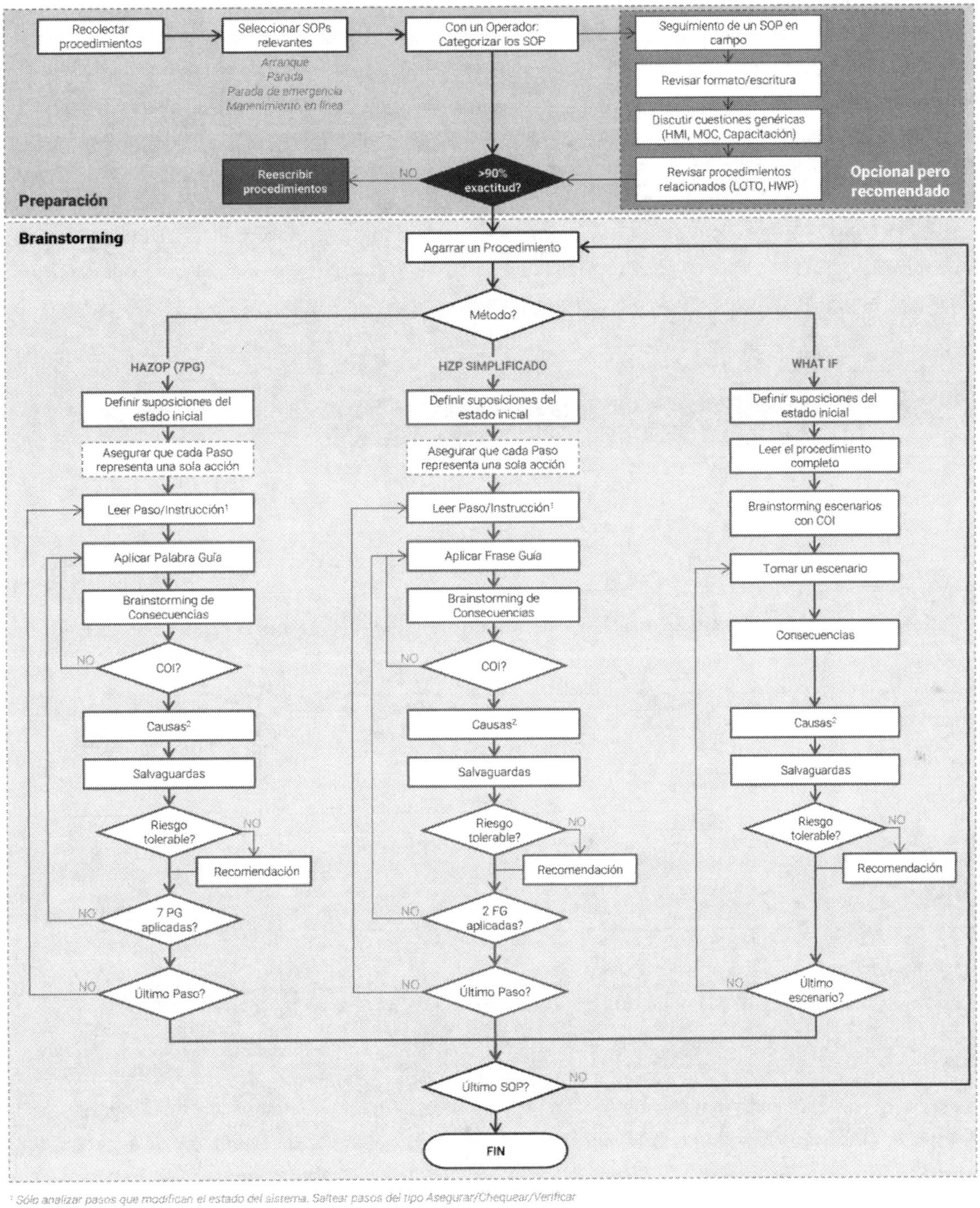

[1] Sólo analizar pasos que modifican el estado del sistema. Saltear pasos del tipo Asegurar/Chequear/Verificar

[2] Indagar más allá de la causa "Error humano" e identificar las cuestiones relacionadas a los Factores Humanos

COI: Consecuencia de Interés

Figura 2. Diagrama de flujo de un PHA de Procedimientos

Tiempo requerido

La mayoría de las empresas no realizan un análisis de los riesgos de los modos de operación de arranque, parada y mantenimiento en línea. La razón brindada normalmente es que el análisis de dichos modos de operación "lleva demasiado tiempo". La evaluación del modo normal de operación consume demasiado tiempo y la organización cree que ya no queda tiempo disponible para incluir el análisis de los modos de operación no rutinarios. Sin embargo, si se optimiza la forma en la que se realiza un análisis del modo continuo (i.e. utilizando las Mejores Prácticas [5]), la organización tendría tiempo suficiente para realizar el análisis de los modos no rutinarios. La Figura 3 muestra la distribución típica del tiempo de reunión promedio dedicado al análisis del modo continuo de operación y a los modos no rutinarios (arranque, parada, mantenimiento en línea).

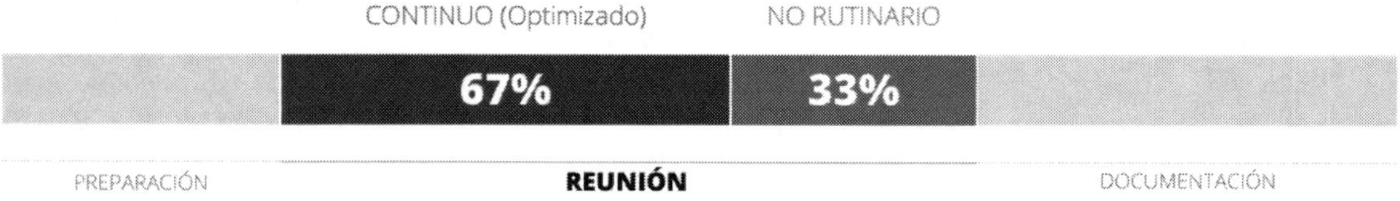

Figura 3. Distribución típica del tiempo requerido para Modo continuo y Modos no rutinarios

Ejemplos

La Tabla 4 muestra algunos ejemplos reales de los tiempos de reunión de varios PHAs.

Tabla 4. Ejemplos de distribución de tiempo de reunión

Planta	País	SOPs		Modo Continuo		Modos No rutinarios	
		Total	% Alto/Medio	Días	%	Días	%
2EH	KSA	87	23 %	14	70 %	6	30 %
Urea	KSA	141	33 %	9.5	72 %	3.5	28 %
PDH	KSA	84	25 %	10	71 %	4	29 %
Coker	USA	33	30 %	10	67 %	5	33 %
PP	KSA	174	27 %	10	62 %	6	38 %

Conclusión

La identificación de peligros juega un rol fundamental en la Gestión de Seguridad de Procesos:

Es imposible gestionar una situación de la que no se tiene conocimiento de su posible ocurrencia.

Teniendo en cuenta que la mayoría de los accidentes se producen en estados no estacionarios, es imperativo optimizar el análisis del modo continuo y dedicar tiempo al análisis de los modos de arranque, parada y mantenimiento en línea. El análisis cualitativo de los procedimientos operativos no rutinarios es una herramienta poderosa para descubrir las fallas que podrían llevar a un humano a cometer un error y escenarios de potenciales accidentes durante todos los modos de operación. Independientemente de la técnica utilizada, es importante que los Equipos de PHA pregunten "¿Por qué un humano podría cometer un error?" cuando se identifique un "Error humano" como posible causa de un accidente. A su vez, seguir los lineamientos descriptos en este artículo permitirá un uso eficiente de los recursos (tiempo) para el análisis de los modos no rutinarios.

Acrónimos

CCPS: *Center for Chemical Process Safety*

CSB: *US Chemical Safety Board*

COI: Consecuencias de interés (*Consequence Of Interest*)

HAZOP: Estudio de peligros y operatividad (*Hazard and Operability study*)

OSHA: *US Occupational Safety and Health Administration*

PHA: Análisis de Peligros de Procesos (*Process Hazard Analysis*)

PSM: Gestión de Seguridad de Procesos (*Process Safety Management*)

RBPS: Seguridad de Procesos basada en riesgo (*Risk Based Process Safety*)

SOP: Procedimiento operativo estándar (*Standard Operating Procedure*)

Referencias

[1] CCPS/AIChE, Guidelines for hazard evaluation procedures, 2nd ed., Wiley, 1992.

[2] CSB, «Completed Investigations,» [En línea]. Available: www.csb.gov/investigations/completed-investigations.

[3] W. G. Bridges y R. Tew, *Best practices for writing operating procedures and trouble-shooting guides,* San Antonio, TX: 13th Global Congress on Process Safety, 2017.

[4] CCPS/AIChE, Guidelines for hazard evaluation procedures, 3rd ed., Wiley, 2008.

[5] W. G. Bridges y R. Tew, *Optimizing qualitative Hazard Evaluations for maximized brainstorming,* Tampa, Florida: 5th Global congress on process safety, 2009.

Las Capas de Protección Independientes tipo SCAI, aquí y más allá del LOPA. ¿Cómo gestionar su integridad?

Romel Rodríguez
CSF Consultoría en Seguridad Funcional
Barcelona, Edo Anzoátegui - Venezuela
rodriguezrx@grupocsf.com

Andrei A Núñez Z
CSF Consultoría en Seguridad Funcional
Barcelona, Edo Anzoátegui - Venezuela
nuneza@grupocsf.com

Carlos A Guilarte Rodriguez
CSF Consultoría en Seguridad Funcional
Barcelona, Edo Anzoátegui - Venezuela
guilartec@grupocsf.com

Keywords: Ciclo de Vida de Seguridad, Función de Seguridad, Función Instrumentadas de Seguridad (SIF), IPL, LOPA, SCAI, Seguridad Funcional, SIL, SIS.

Abstract

Los Análisis de Capas de Protección (LOPA: Layer Of Protection Analysis) se han convertido en una de las herramientas más utilizadas en la industria de procesos para identificar, analizar y evaluar las funciones de seguridad necesarias para alcanzar los criterios de riesgo de una instalación.

Un LOPA nos permite (a) evaluar las funciones de seguridad requeridas y (b) verificar si existen suficientes capas de protección acreditables para evitar o mitigar un posible evento peligroso y (c) determinar si una de ellas es una Función Instrumentada de Seguridad (SIF) y su SIL asociado. Así, al realizar un LOPA, debemos verificar que capas de protección cumplen con los requisitos para ser consideradas una capa de protección independiente (IPL: Independent Protection Layer) y acreditables (en términos de independencia, efectividad, especificidad y auditabilidad entre otros aspectos)

Identificar las capas de protección independiente (IPL: Independent Protection Layer) durante un LOPA es de suma importancia, pero si las IPL nos son validadas y mantenidas apropiadamente, el riesgo estimado puede incrementarse y los accidentes pudieran presentarse de forma más frecuente de lo que fue previsto durante el análisis.

Trabajar bajo el marco de la normativa IEC 61511 e ISA 84.91.01 le permite a una organización enfocarse en mantener la confiabilidad de los equipos e instrumentos considerados críticos a la hora de prevenir o mitigar accidentes. Este mismo tipo de vigilancia activa debe ser implementada para los otros medios reducción de riesgo, no solo las SIF. Todas IPL debe ser diseñadas, instaladas, validadas, mantenidas y operadas de forma adecuada. Debemos estar conscientes que alcanzar la Seguridad Funcional depende de ello.

Este trabajo aborda particularmente los tópicos a tomar en cuenta cuando se debe gestionar la integridad de IPL tipo SCAI (Safety Controls, Alarms, and Interlocks) que son susceptibles de sufrir una degradación de su desempeño en forma muy rápida.

1 Introducción

El riesgo se ha definido como una medida de daño a las personas, ambiente o pérdidas económicas en términos tanto de la frecuencia de ocurrencia como de la severidad de un evento no deseado. Por lo tanto, cualquier esfuerzo para reducir el riesgo que surge del diseño y de la operación de una instalación de procesos puede enfocarse en reducir la probabilidad de materialización del evento no deseado (frecuencia de incidentes) como en minimizar la magnitud de sus efectos (en caso de que se ocurra), o alguna combinación de ambos. En general, las estrategias para reducir el riesgo se pueden clasificar en una de las siguientes categorías [1]:

- Diseño inherente seguro; que busca eliminar el peligro mediante el uso de materiales y condiciones de proceso que no sean peligrosos (por ejemplo, sustituir un solvente inflamable por agua).
- Control Administrativo: utilizar procedimientos operativos, procedimientos de respuesta a emergencias y otros enfoques de gestión para prevenir incidentes o minimizar los efectos de un evento no deseado.
- Control Pasivo: eliminar o minimizar el peligro mediante características de diseño de procesos y equipos que no eliminan el peligro, pero sí reducen la frecuencia o las consecuencias de la materialización del evento no deseado sin la necesidad de que ningún dispositivo funcione activamente (por ejemplo, el uso de equipamiento para excederlas condiciones de presión de diseño del proceso).
- Control Activo: utilizar controles, interlocks de seguridad y sistemas de parada de emergencia para detectar desviaciones de procesos potencialmente peligrosas y tomar medidas correctivas.

La filosofía de diseño seguro basado en capas o también conocida como la defensa en profundidad [2], es una filosofía de seguridad que implica el uso de sucesivas medidas compensatorias (a menudo llamadas salvaguardas, barreras o capas de protección) para prevenir o minimizar los efectos de un evento no deseado si ocurre un mal funcionamiento en una instalación que maneja sustancias peligrosas. Las capas deben, en la medida de lo posible, ser independientes, lo que significa que la falla de una de ellas no afecta la eficacia de otras barreras para realizar su función. La filosofía garantiza que la seguridad no dependa totalmente de ningún elemento individual del diseño, construcción, mantenimiento u operación de la instalación.

La filosofía de diseño seguro basado en capas es fácilmente representada a través del modelo del queso suizo [2], en este se puede observar que el riesgo de la materialización de un accidente es dinámico, es decir, dependerá de la integridad de las capas en el tiempo (crezcan los agujeros como manifestación de degradación), o que la independencia entre las capas no se pueda asegurar (falla más de una capa a la vez) y que estos procesos lleven a que las capas permita la progresión del peligro (se alineen los agujeros). Ver figura 1.

Figura 1. Modelo del queso suizo y filosofía de diseño seguro basado en capas Basado en IOGP [4]

2 Proceso de Asignación de las Funciones de Seguridad a las Capas de protección

Dentro del Ciclo de Vida descrito en el estándar IEC/ISA 61511 [5] [6] se da cuenta de cómo se debe en primera instancia realizar un Análisis de Peligros y Riesgos que da somo resultado un listado de Funciones de Seguridad, entendiendo estas como: *"función a ser implementada por una o más capas de protección, la cual está destinada a lograr o mantener un estado seguro para el proceso, con respecto a un evento peligroso específico"*. Luego, se debe hacer una revisión sistemática del diseño del proceso para verificar, mediante análisis, que existen suficientes capas de protección para mantener a raya esos peligros específicos. Este proceso se denomina asignación de las funciones de seguridad a las capas de protección, que común y erróneamente se suela llamar asignación del SIL. La idea central es poder verificar si existen suficientes capas y luego establecer si una de esas capas es un Función Instrumentada de Seguridad (SIF) y proceder a determinar su desempeño en términos del nivel de integridad requerido (SIL: Safety Integrity Level). Por tanto, alcanzar la seguridad no solo dependerá de la SIF, sino que muy probablemente tenga una mayor dependencia de otras múltiples capas de protección.

Ahora, esas capas de protección deben contar con unas ciertas características para que puedan ser consideradas adecuada para poder reducir el riesgo.

3 Análisis de Capas de protección (LOPA: Layer Of Protection Analysis)

LOPA es una técnica simplificada (semicuantitativo) de valoración de riesgo en el que un equipo multidisciplinario confirma si existen suficientes capas de protección. Es una de las herramientas más utilizadas para identificar, analizar y evaluar las funciones de seguridad de una instalación.

Durante el LOPA se realiza una evaluación de riesgo, que generalmente usa órdenes de magnitud para indicar la frecuencia del evento iniciador asociada a un peligro en específico, la severidad de las consecuencias y la probabilidad de falla de las capas de protección independientes (IPL) para establecer el riesgo de un escenario. Por lo general no es necesario ser tan preciso (por lo de los órdenes de magnitud) pero no existe impedimento matemático para refinar el cálculo.

Analizar el efecto de las capas de protección en la reducción de riesgo de un escenario dependerá de la efectividad de estas, en términos de reducción de riesgo, y determinar si éstas son suficientes para prevenir o mitigar un posible evento peligroso asociado a un peligro en específico. En este proceso se requiere de un examen riguroso que permita emitir un juicio sobre la calidad de la capa de protección y poder acreditarle la condición de Capa de Protección Independiente (IPL).

Podemos definir una IPL como un *"dispositivo, sistema o acción capaz de reducir el riesgo mediante control, prevención o mitigación, independientemente del evento iniciador o de otras capas de protección"*

Según cómo generan la reducción del riesgo, pueden ser clasificadas en:

- **Preventivas:** diseñadas para prevenir o evitar la propagación de un escenario peligroso. Por ejemplo: Un lazo del BPCS, una Alarma con intervención del Operador, etc.
- **Mitigación:** diseñadas para minimizar la consecuencia de un escenario peligroso. Por ejemplo: Diques de contención, SF&G, etc.

Para poder acreditar una IPL la misma debe cumplir cabalmente con las siguientes características [7]:

- **Específica:** debe ser diseñada específicamente para prevenir o mitigar las consecuencias del escenario peligroso evaluado. Por ejemplo: ejemplo: Una PSV para sobre presión, un dique para sobre llenado.
- **Independiente:** debe ser independiente del evento iniciador y de otras capas de protección asociadas al escenario peligroso evaluado y su evolución. La independencia se analiza de forma específica para cada escenario, es decir dos capas pueden ser independientes en un escenario, pero no serlos para otro. Por ejemplo, una alarma y una SIF pueden ser independientes para un escenario de aumento de presión, pero no serlo si el escenario de análisis incluye corte de energía eléctrica y ambas capas comparten suministro de energía. Por otra parte, una PSV es diversa e independiente de los mecanismos de falla de un lazo de control.
- **Efectiva:** debe ser efectiva para la prevención o mitigación del riesgo en análisis, es decir relación directa con el evento iniciador y su evolución. Por ejemplo: será distinta la efectividad de una PSV diseñada para caso fuego, que, para alivio térmico o descarga bloqueada, su capacidad de despejar el problema y cambiar la dinámica del proceso en un tiempo adecuado será muy diferente.
- **Confiable:** debe asegurarse que operará según fue diseñada para un periodo de tiempo específico. Que puede asignársele una probabilidad de falla (o de éxito) tal que pueda reducir el riesgo, es decir que puede calcularse o estimarse. Por ejemplo: Cada vez que se prueba una PSV se hace parte de una estadística de falla que permite calcular la probabilidad de falla a partir de las tasas de falla que puedan documentarse de esas pruebas.

- **Auditable:** debe existir evidencia documentada que confirme la efectividad de la capa de protección para prevenir o mitigar el escenario peligroso evaluado y adicionalmente debe confirmarse el diseño, instalación, mantenimiento y pruebas para demostrar que la misma cumple con la reducción de riesgo establecida. Por ejemplo: Los resultados de las pruebas Pre Pop de las PSV según el programa de manteniendo mayor de la instalación.
- **Control de Acceso**: Adicionalmente deben utilizarse controles administrativos y físicos que permitan reducir cambios no intencionales o no autorizados. Por ejemplo: Control de acceso físico a los gabinetes de sala de control, o protección por password para la calibración de un instrumento o para modificar un setpoint.
- Control de Cambios: además se debe revisar, documentar y aprobar modificaciones que no sean reemplazo de tipo (uno a uno) antes de su implementación. Si las capas de protección no son validadas, probadas y mantenidas apropiadamente, el riesgo puede incrementarse y los accidentes pudieran presentarse de forma más frecuente de lo que fue previsto. Por ejemplo: La realización de un proceso de análisis de impacto que derive en la necesidad del cambio del setpoint en un disparo (PSV) o en una alarma que este controlado administrativamente bajo el procedimiento de Manejo del Cambio de la organización.

En este sentido, se debe garantizar que los escenarios asociados a una IPL estén bien definidos y que la incorporación de cada IPL sea coherente con la desviación del proceso asociada, todo esto con el fin de que operen según es requerido, cuando es requerido, para prevenir el desarrollo o mitigar las consecuencias de uno o varios escenarios peligrosos determinados.

Aunque el Ciclo de Vida de Seguridad de los SIS no da requisitos específicos sobre el diseño, tratamiento de las IPL, si indica que todo lo que es considerado como otro medio de reducción de riesgo debe seguir las pautas de las mejores prácticas de diseño y operación según la tecnología de cada IPL. Sería el equivalente a una declaración de un elemento crítico para la seguridad (SCE: Safety Critical Element) [8].

4 Mejores Prácticas y Requisitos de los Estándares

Aunque el Ciclo de Vida de Seguridad de los SIS definido en IEC/ISA 61511 [5] [6] no da requisitos específicos sobre el diseño, si indica que todo lo que es considerado como otro medio de reducción de riesgo debe seguir las pautas de las mejores prácticas de diseño y operación según la tecnología de cada IPL y que a cada una de ellas debe hacérsele el seguimiento correspondiente en la fase de operación y mantenimiento para garantizar que la reducción de riesgo estimada inicialmente se mantiene en el tiempo. Sería el equivalente a una declaración de un elemento crítico para la seguridad (SCE: Safety Critical Element) [8].

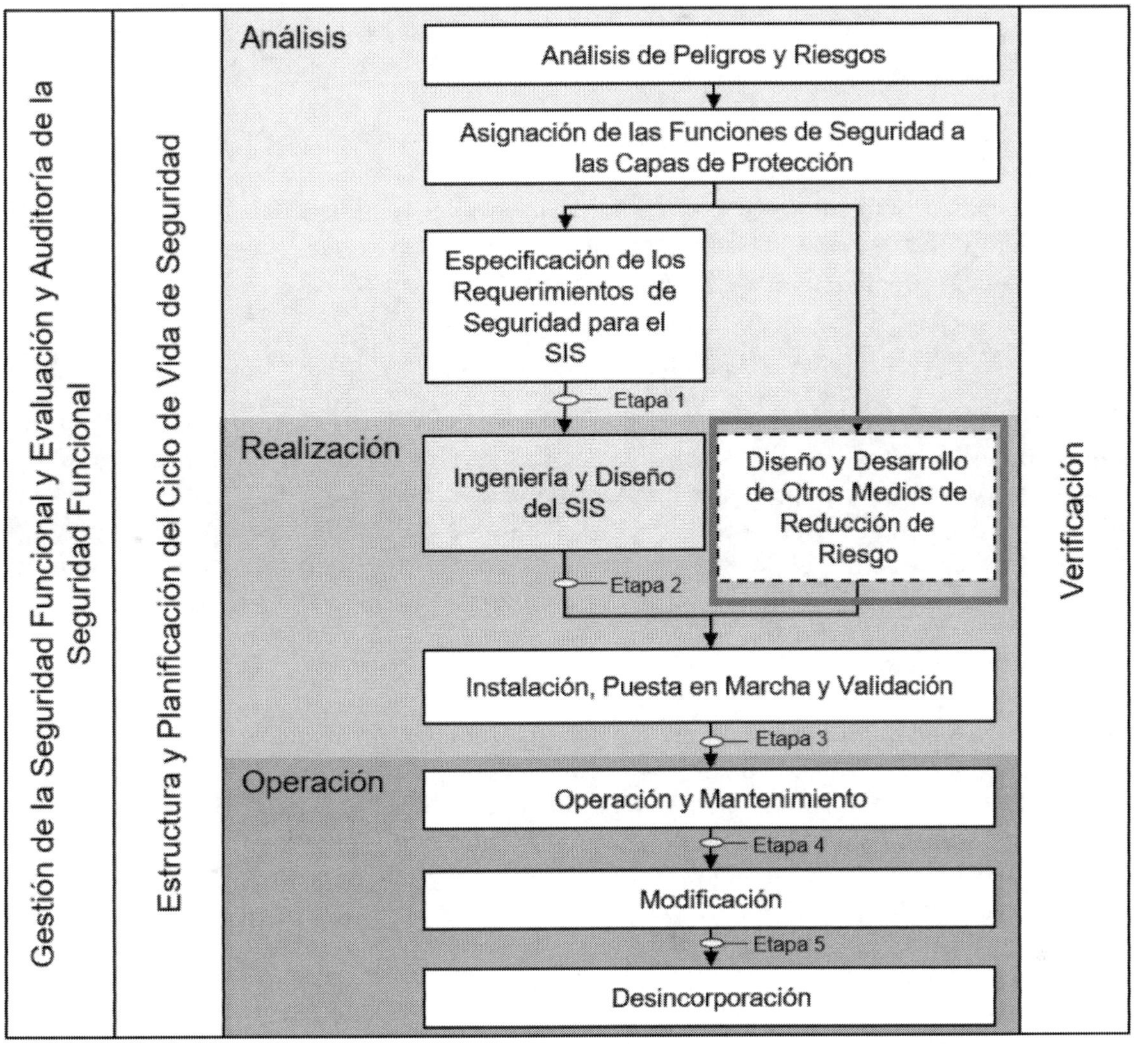

Figura 2. Ciclo de Vida de Seguridad de los Sistemas Instrumentados de Seguridad

En línea con lo anterior, regulaciones como OSHA [9], establecen en forma general que "*el Empleador deberá determinar y documentar que los equipos han sido diseñados, mantenidos, inspeccionados, probados y están operando en una manera segura*", entre ellos cualquier dispositivo o sistema que se considere tienen un papel importante en la seguridad y en el proceso de reducción de riesgo.

En forma completaría el estándar ANSI/ISA 84.91.01 [10], que inicialmente fue necesaria para aclarar el alcance de OSHA [9], sobre los sistemas instrumentados relacionados con la seguridad en la industria de procesos, fue actualizada para dar respuestas a los aspectos de Integridad Mecánica (IM) (inspección, mantenimiento, prueba y documentación) del programa de Gestión de Seguridad de Procesos relativo a estos sistemas.

En este estándar estos dispositivos o sistemas automatizados / instrumentados relacionados con la seguridad fueron denominados Sistemas de Control de Seguridad, Alarmas e Interlocks o SCAI por sus siglas en inglés (Safety Control, Alarms and Interlocks) y establece que los mismos, deben ser:

- Identificados de manera que se distingan de otros sistemas instrumentados.
- Instalados según las especificaciones de diseño y del fabricante.
- Inspeccionados, mantenidos y probados en intervalos regulares según el plan de IM y siguiendo las buenas prácticas de ingeniería.
- Adecuadamente documentados (diseño, mantenimiento y operación).

5 IPL tipo SCAI

SCAI o controles de seguridad, alarmas e interlocks son salvaguardas de seguridad de procesos implementadas con instrumentación y controles, utilizadas para lograr o mantener un estado seguro para un proceso y requeridas para proporcionar reducción de riesgos con respecto a un evento peligroso específico (ANSI/ISA 84.91.01 [10], CCPS [11] [12]). Normalmente cuando cumplen con las características de IPL se les puede acreditar un Factor de Reducción de Riesgo no mayor a10 (FRR= 10) es decir Probabilidad de Falla en Demanda de 1/11 (PFD = 0,1).

Controles de seguridad

Normalmente operan para respaldar el control del proceso (o control regulatorio) y evitar desviaciones del proceso tomando medidas para mantener el proceso en el rango operativo normal. Estos controles normalmente están relacionados con lazos de control del BPCS que cumplen tareas control y de seguridad. Están no son diseñados ni gestionados de acuerdo con IEC/ISA 61511.

Alarmas de seguridad

Estos inician una alarma quc requiere que el operador tome medidas de acuerdo con un procedimiento de respuesta para detener la propagación del evento peligroso. Están son diseñados y gestionados de acuerdo con ANSI/ISA 18 [13].

Interlocks de seguridad

Estos toman medidas automáticas para lograr o mantener un estado seguro del proceso cuando una variable del proceso alcanza un límite definido fuera del entorno operativo normal. Están no son diseñados ni gestionados de acuerdo con IEC/ISA 61511. Estos interlocks suelen ser implementados en el BPCS y también pueden estar definidos como Permisivos de seguridad que son una condición dentro de una secuencia lógica que debe cumplirse antes de que se permita que la secuencia pase a la siguiente fase.

Sistema Instrumentado de Detección de Gas y Fuego (F&G)

Sistema compuesto por sensores, controlador de fuego y gas y elementos alarmas que ejecutan funciones de seguridad de detección, supervisión, control y alarma de fuego y gas, integra las señales de entrada y salida de campo y permite, de acuerdo con lo especificado en cada instalación, el monitoreo, control y alerta de eventos y siniestros en las instalaciones de procesos. Normalmente no incluye los sistemas de extinción/mitigación de incendio. Aunque por su carácter de mitigación suele dejarse de lado la acreditación como IPL.

Burner Management System (BMS)

Conjunto de dispositivos de campo, sistema lógico y elementos finales, dedicados a la seguridad del encendido, supervisión de llama y asistencia del operador en el arranque y la parada de equipos de fuego directo (hornos, calderas, etc.), para evitar su mal funcionamiento y posible el daño. Al ser un sistema normalmente dedicado, puede cumplir y gestionarse con IEC/ISA 61511.

Sistema Instrumentado de Seguridad (SIS)

El SIS, por lo general, está compuesto por varias Funciones Instrumentadas de Seguridad (SIF) y todos los servicios comunes y necesarios para su funcionamiento. Detectan una condición fuera de control y automáticamente llevan o mantienen el proceso a un estado seguro, responden a condiciones en las cuales la instalación puede ser peligrosa por sí misma, o en las que, si no se toma ninguna acción, podría eventualmente alcanzar una condición peligrosa. Se diseña y gestiona usando IEC/ISA 61511.

La siguiente figura muestra la clasificación general de los SCAI.

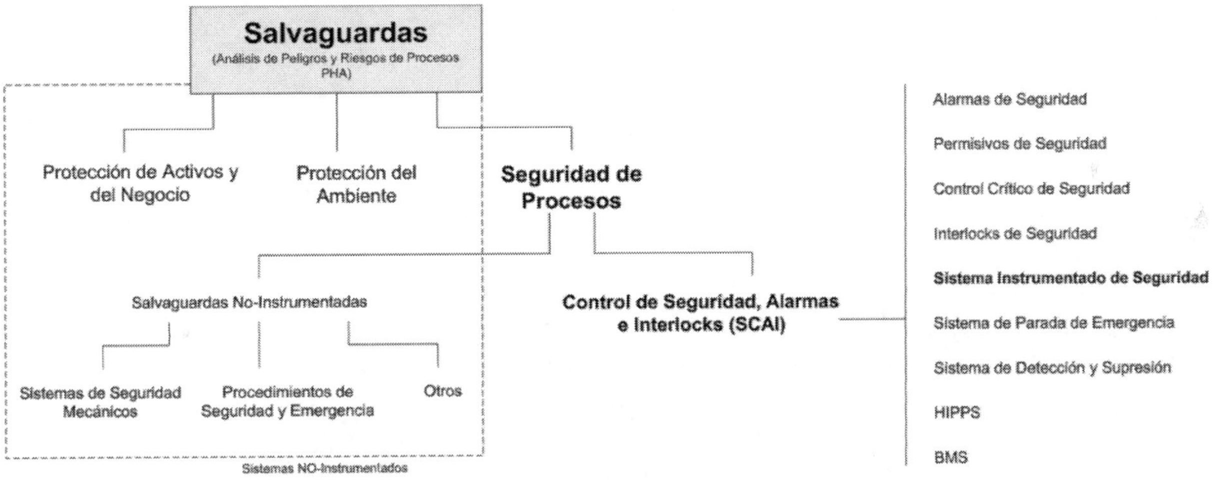

Figura 3. Clasificación SCAI según ANSI/ISA 84.91.01 [10].

Al igual que cualquier salvaguarda, barrera o capa de protección, un SCAI debe cumplir con todas las características de un IPL para que pueda ser calificada como tal.

6 El Problema

Al trabajar en el marco de la norma IEC/ISA 61511 podemos identificar los requisitos funcionales y de integridad para que las SIF logren la reducción de riesgo requerida, en conjunto las capas de protección acreditables como IPL.

El desempeño o SIL (Safety Integrity Level) que debe alcanzar una SIF, está directamente relacionado al buen funcionamiento de todas las IPL consideradas en cada escenario durante el LOPA. Pero, si las IPL nos son validadas, probadas y mantenidas apropiadamente, el riesgo estimado puede incrementarse y los accidentes pudieran presentarse de forma más frecuente de lo que fue previsto durante el LOPA

Situaciones comunes como cambios en el diseño, modificación del punto de ajuste de una alarma, eliminación de algún dispositivo, o la falta de mantenimiento pueden afectar la integridad de una IPL, creando una situación potencial de riesgo no cubierto (ni por el SIS ni por la capa que ha sido eventualmente degradada), como podemos apreciar en las siguientes figuras.

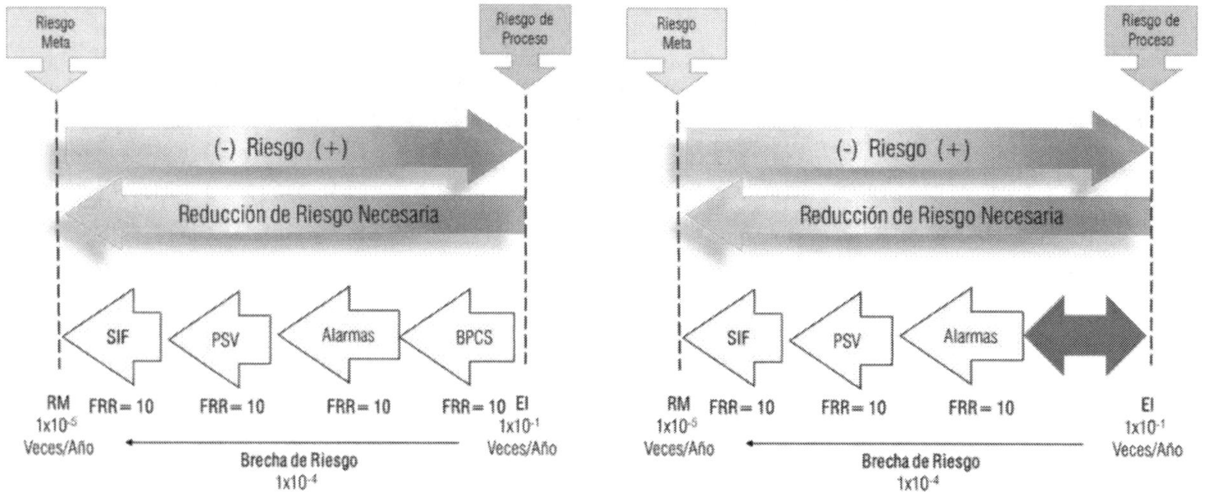

Figura 4. Distribución del riesgo entre IPL

Identificar una IPL durante las sesiones de trabajo del LOPA es de suma importancia, el grupo de análisis utiliza la información relacionada al proceso para identificar las capas de protección disponibles (P&ID, Filosofías de Control, Manuales, etc.), pero si no tiene el tiempo o los recursos para asegurar que cada IPL considerada cumplirá con los requisitos que la acreditan como tal fácilmente se puede sobre estimar las capacidades de una IPL o simplemente no documentarla adecuadamente para que se pueda hacer un seguimiento de su desempeño durante la vida útil de la instalación.

Vale la pena resaltar que es posible que el desempeño de IPL tipo SCAI tengan una importancia, en términos de reducción de riesgo, similar a otro tipo de IPL como una SIF o una PSV, por lo que no debe menospreciarse el impacto que tienen en el mantenimiento de la seguridad de la instalación.

A pesar de que el equipo que participa en un LOPA puede identificar las IPL que apliquen a cada escenario, dentro de sus responsabilidades no está el velar por su correcta implementación. Si no existen medios que permitan validar su funcionamiento es muy fácil exponer al personal y la instalación a un riesgo no considerado.

Como ya mencionamos, el estándar IEC/ISA 61511 no regula el diseño, mantenimiento y operación de las IPL (no SIF), pero en ella se establecen evaluaciones (o assessment) que deben ser realizadas durante el diseño, antes de la puesta en marcha y durante la operación en las que debemos considerarlas. El buen funcionamiento de las IPL depende de la forma en que cada organización maneje su sistema de integridad mecánica (por ejemplo, como parte la Gestión de Seguridad de Procesos (PSM)).

Como vemos, las IPL (en conjunto con las SIF) nos ayudan a mantener los niveles de riesgo dentro de valores tolerables al evitar (o mitigar) la propagación de un evento peligroso. Pero, debemos recordar que la correcta implementación de una IPL inicia desde el momento que documentamos las sesiones de un LOPA, y solo se alcanza cuando todo el personal encargado de su funcionamiento o involucrado en su mantenimiento se hace consciente que, además del proceso de selección, es importante gestionarlas, validarlas, probarlas, auditarlas y documentarlas apropiadamente a lo largo de su vida útil.

7 Integridad de IPL tipo SCAI

Una IPL tipo SCAI es un sistema instrumentado/ automatizado que debe responder a unas ciertas condiciones del proceso y en consecuencia debe generar acciones o alertas que lleven a que el proceso se mantenga o sea llevado a una condición segura, como vemos estamos hablando de un sistema que tiene una connotación relacionada con la seguridad funcional. Por tanto, es necesario crear los controles necesarios para que esa funcionalidad tenga ciertas garantías a lo largo de la vida útil de la instalación. Cualquier degradación en el desempeño de este tipo de IPL tiene una relación directa con el riesgo de la instalación y como son sistemas relativamente complejos (si lo comparamos con otras IPL como una PSV o un dique) requieren de nuestra atención en un marco gestionado.

7.1 Identificación de las IPL tipo SCAI

Como resultado del LOPA se debe obtener un listado de las SIF y de todas las IPL acreditadas incluyendo las IPL tipo SCAI. Se debe mantener una lista de equipos que identifique las IPL tipo SCAI mediante una designación única, como un número de etiqueta o identificación del equipo que sea rastreable hasta los requisitos de integridad mecánica necesarios para mantener la integridad y confiabilidad requeridas durante toda su vida útil y que se pueda ver en forma trazable a que escenario de riesgo o que peligro está ligada.

7.2 Especificación / Estándar de desempeño

Acto seguido a la acreditación de la IPL tipo SCAI en las sesiones del LOPA se debe desarrollar una especificación o estándar de desempeño (CCPS [12], Energy Institute [8]) que permita en todo momento verificar la idoneidad de la IPL para que cumplir satisfactoriamente con su propósito en cualquier instancia de su vida útil, esta debe incluir:

- Requerimientos de funcionalidad, integridad (disponibilidad y confiabilidad), supervivencia e interacciones y dependencia con otras IPL.
- Condiciones de funcionamiento satisfactorio para alcanzar o mantener el estado seguro del proceso.

- Requisitos de configuración, instalación y mantenimiento para lograr y mantener el rendimiento que se le ha acreditado.
- Modos de falla de la IPL tipo SCAI, medios utilizados para detectar estos modos de falla y respuesta esperada del sistema y del operador ante una falla detectada.
- Medidas de compensación necesarias para continuar las operaciones seguras en caso de ausencia de la IPL tipo SCAI en caso de falla.
- Condiciones requeridas para colocar en bypass el SCAI y cualquier medida de compensación que se implementará durante el bypass.
- Condiciones requeridas para el reinicio seguro del SCAI.

7.3 Consideraciones generales de diseño

Cada tipo de IPL tipo SCAI debe responder tanto a las especificaciones o estándar de desempeño como a su normativa particular para asegurar que son lo suficientemente robustas para soportar las tensiones reales del entorno operativo y proporcionar la integridad y confiabilidad requeridas. En la siguiente tabla se indican algunas de los estándares de diseño más relevantes relacionados con algunas de ILP tipo SCAI.

Tabla 1. Estándares de Diseño para IPL tipo SCAI

SCAI	Estándar de Diseño
Alarmas de seguridad	ANSI/ISA 18.2 2016 : Management of Alarms Systems for the Process Industries IEC 62682 : Management of Alarms Systems for the Process Industries EEMUA 191 2007 : Alarm Systems - A Guide to Design, Management and Procurement API RP 554 : Process Control Systems API RP 584 : Integrity Operating Windows ASM Guidelines : Abnormal Situation Management Consortium
Burner Management System (BMS)	NFPA 85: Boiler and Combustion Systems Hazards Code. 2019 NFPA 87 : Standard for Fluid Heaters. 2021 NFPA 86 : Standard for Ovens and Furnaces. 2019 ISA TR84.00.05 Guidance on the Identification of Safety Instrumented Functions (SIF) in Burner Management System (BMS). 2009 ISA S77.41 Fossil Fuel Power Plant Boiler Combustion Controls. 1992 API Recommended Practice 538: Industrial Fired Boilers for General Refinery and Petrochemical Service. 2015 API Recommended Practice 556: Instrumentation, Control and Protective Systems for Gas Fired Heaters. 2011 API Recommended Practice 573: Inspection of Fired Boilers and Heaters. 2010
Sistema Instrumentado de Detección de Gas y Fuego (F&G)	NFPA 72: National Fire Alarm and Signalling Code. 2007 NFPA 550: 2012 - Guide to the Fire Safety Concepts Tree NFPA 551: 2007 - Guide to the Evaluation of Fire Risk Assessments ISA TR84.00.07: Guidance on the Evaluation of Fire, Combustible Gas and Toxic Gas System Effectiveness 2018 IEC 60079-29-3: Guidance on functional safety of fixed gas detection systems BS 60080:2020: Explosive and toxic atmospheres — Hazard detection mapping — Guidance on the placement of permanently installed flame and gas detection devices using software tools and other techniques. 2020 SFPE Handbook of Fire Protection Engineering: 2002 3ra Edición

	SFPE Engineering Guide - Fire Risk Assessment: 2006
Sistema de Protección contra Sobrepresión (HIPPS: High Integrity Pressure Protection System)	API STD 521 Pressure-relieving and Depressuring Systems. 2014 RP 14C. Analysis, Design, Installation & Testing of Basic Safety Systems for Offshore Production Platforms. (2018). API RP 17O Subsea High Integrity Pressure Protection System (HIPPS). 2008 ASME Section VIII, Code Case 2211-1, 1995/1996 HSE Safety Instrumented Systems for the Overpressure Protection of Pipeline Risers. 2008
Sistema Instrumentado de Seguridad (SIS)	IEC 61511 (2016) - Functional Safety: Safety Instrumented Systems for the Process Sector. Edition 2.0. ANSI/ISA 61511-1 (2018) - Functional Safety –Safety Instrumented Systems for the Process Industry Sector.

7.4 Gestión de la Integridad Mecánica (IM) de IPL tipo SCAI

Mantener el desempeño de una IPL tipo SCAI requiere que la organización se prepare para gestionar la integridad de esta. La gestión de la integridad de los activos relacionados con las IPL tipo SCAI incluye una variedad de actividades, como inspección, calibración, mantenimiento preventivo, reparación/reemplazo y pruebas de prueba, gestión del cambio, de las competencias y de la documentación.

En el marco de la Gestión de la Seguridad de los Procesos, bien sea usando PSM de OSHA [9] o RBPS de CCPS [14], la Integridad Mecánica de IPL tipo SCAI es tan importante como la de cualquier otro sistema de la instalación. Entre las razones principales por las cuales se debe operar y mantener estos sistemas en forma adecuada podemos nombrar:

- Garantizar su operatividad para mantener controlado el nivel de exposición al riesgo en la instalación ante eventos no deseados.
- Cumplir con las regulaciones (OSHA / nacionales).
- Cumplir con los requerimientos de las normativas de la organización relacionadas a los SCAI (RAGAGEP declaradas en el Sistema de Gestión de la Seguridad de los Procesos (SGSP).
- Cumplir con los requerimientos de los fabricantes (garantías).
- Demostrar a través de testigos documentados que las IPL tipo SCAI son probadas y mantenidas en buenas condiciones.
- Detectar, corregir y evitar introducir fallas.
- Verificar que el programa de Integridad Mecánica (IM) y los Intervalos de Pruebas (TI) son suficientes para asegurar los requerimientos funcionales y de seguridad para toda la vida útil del equipamiento.
- Monitorear los mecanismos de degradación que puedan comprometer el desempeño las IPL tipo SCAI.
- Verificar que las premisas y requisitos de las especificaciones o estándares de desempeño son válidas en las condiciones operacionales y ambientales de la instalación.
- Recolectar datos del desempeño del equipamiento que forma parte de las IPL tipo SCAI para verificar que se cumple con los requerimientos de disponibilidad y confiabilidad.

- Proveer información para evaluar si el programa de Integridad Mecánica (IM) de sistemas automatizados o instrumentados relacionados con la seguridad y la reducción de riesgo es exitoso y soporta la mejora continua.
- Corregir las desviaciones del desempeño esperado de las IPL tipo SCAI.

Mantener la Integridad de los sistemas SCAI, es una tarea del día a día que se deben realizar en un marco armónicamente gestionado. Si bien no existen para todos los SCAI una normativa tan rigurosa como la que existe para los SIS la gestión de los SIS es un buen punto de inicio para la gestión de los SCAI. Entonces se deben definir para cada SCAI las actividades que permitirán operar, mantener, probar, inspeccionar, monitorear y controlar sus cambios además de mantener las competencias del personal a cargo. A continuación, se describen brevemente estas actividades:

Operación de los SCAI:

- Operación: incluye la interacción normal con los SCAI durante la operación, arranque, parada, inspecciones, documentación de fallas encontradas, inicio de ordenes de mantenimiento, reset, bypass y medidas de compensación en caso de degradación y la recolección de los datos asociados a esas actividades.
- Mantenimiento: incluye las inspecciones, mantenimiento preventivos y correctivos, pruebas funcionales y la recolección de los datos asociados a esas actividades.

Gestión de las actividades relacionas a los SCAI:

- Manejo del Cambio: incluye el manejo de las desviaciones y las solicitudes de cambios sobre los SCAI, análisis de impacto, procedimiento del cambio, aprobación del cambio y prueba de estos.
- Monitoreo: incluye el establecimiento de los indicadores de desempeño, las metas de los indicadores de desempeño, el análisis de los datos recolectados para verificar si el desempeño requerido es alcanzado.
- Gestión de Competencias: Mantener identificadas, asignadas, mantenidas y actualizadas las competencias de los responsables de los SCAI.
- Auditoría / Assessment: Realizar las revisiones periódicas que las actividades se estén realizando en un marco gestionado que esté vigente y en uso.

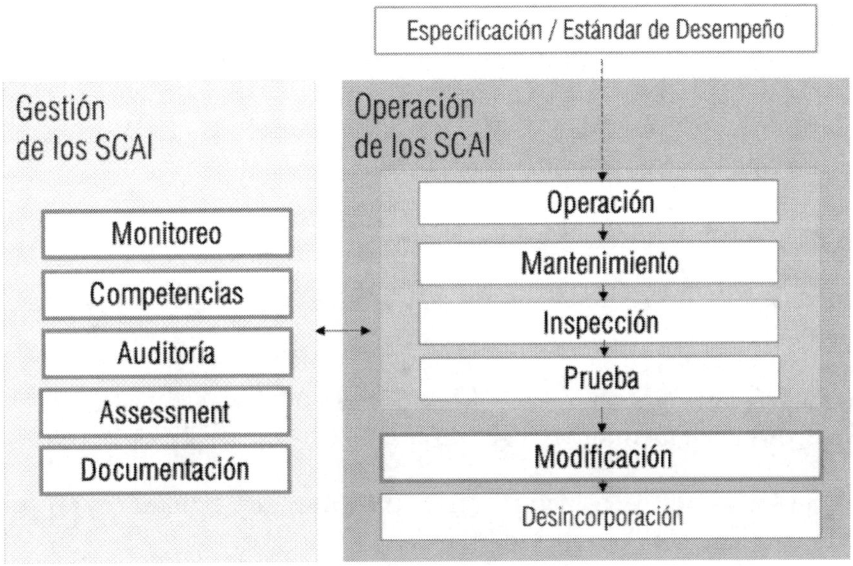

Figura 5. Actividades Principales del Ciclo de Vida de Seguridad del SIS en la Fase de Operación y Mantenimiento.

8 Conclusion

El marco de la Gestión de Seguridad de Proceso Basada en Riesgo implica la gestión de los activos que genera una reducción de riesgo y pueden evitar o mitigar un accidente mayor, por lo tanto, aquellas capas de protección que han sido identificadas y acreditadas como IPL tipo SCAI en un LOPA deben formar parte de este esfuerzo sin dejar de lado ninguna de ellas, ben sea por descuido, omisión o desconocimiento.

Garantizar la reducción de riesgo de la instalación nos obliga a definir un marco gestionado para cada IPL tipo SCAI, que parta de la base de una especificación o estándar de desempeño que pueda soportar las actividades de operación, mantenimiento, inspección, prueba, manejo del cambio, documentación, competencias.

9 Referencias

[1] CCPS. (2012). Guidelines for Engineering Design for Process Safety, 2nd Edition.

[2] IAEA. (1990). Application of the single failure criterion. IAEA safety series n°50-P-1. Vienna. IAEA, ISBN: 92-0-123790-1.

[3] Reason, James (1990). The Contribution of Latent Human Failures to the Breakdown of Complex Systems. Philosophical Transactions of the Royal Society of London. Series B, Biological Sciences. 327 (1241): 475–84.

[4] IOGP Report 556 (2023). Process safety - recommended practice on key performance indicators.

[5] International Electrotechnical Commission (IEC). IEC 61511 (2016) - Functional Safety: Safety Instrumented Systems for the Process Sector. Edition 2.0.

[6] International Society of Automation (ISA). ANSI/ISA 61511-1 (2018) - Functional Safety – Safety Instrumented Systems for the Process Industry Sector.

[7] CCPS. (2001). Layer of protection analysis: Simplified process risk assessment.

[8] Energy Institute. (2020). Guidelines for management of safety critical elements (SCEs). Third edition.

[9] Seguridad Ocupacional y Administración de Salud [OSHA]. (1992). CFR 29-part 1910.119 Process Safety Management of Highly Hazardous Chemicals. USA.

[10] ANSI/ISA S84.91.01 (2012). Identification and Mechanical Integrity of Safety Controls, Alarms, and Interlocks in the Process Industry.

[11] CCPS. (2015). Guidelines for Initiating Events and Independent Protection Layers in Layer of Protection Analysis.

[12] CCPS. (2007). Guidelines for Safe and Reliable Instrumented Protective Systems.

[13] ANSI/ISA 18.2. (2009), Management of Alarm Systems for the Process Industries.

[14] CCPS. (2010). Guidelines for Risk Based Process Safety.

Dispositivos de alivio en sistemas de hidrógenos. ¿Un peligro escondido?

Alejandro Jiménez
Engineering Systems Inc. (ESi)
6735 NW 36th Street, Suite 325 Miami, FL 33166
aljimenez@engsys.com
Center for Risk and Reliability, University of Maryland
0151 Glenn L. Martin Hall, Building 088
University of Maryland
College Park, MD 20742, USA

Keywords: Hidrogeno, Dispositivos de Alivio, Integridad Mecánica, Confiabilidad

Abstract

La industria de energía se encuentra en los últimos años en la búsqueda de una alternativa sostenible que permita reducir el impacto de las operaciones en el planeta y los seres humanos mientras mantenemos el consumo energético actual. Específicamente, se ha buscado mejorar la sostenibilidad de los procesos a través de mejoras tecnológicas buscando alternativas que reduzcan el impacto de carbono en la atmosfera.

Una de las alternativas que conjuntamente la industria ha estado apuntando es el uso de hidrogeno como medio de producción y almacenamiento de energía. Sin embargo, el hidrogeno posee peligros adicionales como los son una mayor facilidad para su ignición y la fragilización por hidrogeno (hydrogen embrittlement), los cuales tienen el potencial de causar una mayor cantidad de incendios y explosiones al visto hoy en día.

Debido a estos nuevos peligros, existe una incertidumbre en el riesgo presente cuando se aplican dispositivos de alivios tradicionales en sistemas de hidrogeno. Adicionalmente, a pesar de que estos se encuentran con el fin de mitigar una ruptura, se han observado incidentes donde dicho dispositivo ha sido el evento iniciador del incidente, generando la interrogante sobre donde nos encontramos en la balanza de riesgo.

Debido a esta incertidumbre, esta presentación mostrará un estudio documental realizado en conjunto con la Universidad de Maryland, donde se buscaron diversos incidentes relacionados con dispositivos de alivio en sistemas de hidrogeno, mostrando la distribución estadística más importantes de estos incidentes.

Durante la introducción de esta problemática, esta presentación realizará un resumen del incidente de Dimond Bar, el cual fuel el desencadenante de este estudio. Luego mostrará el estudio relazado, culminando con una comparación de los resultados con información publicada sobre la confiabilidad de los dispositivos de alivio.

La presentación culminará con una conclusión sobre actividades a realizar a miras hacia el futuro para reducir la incertidumbre del riesgo presente en la aplicación de dispositivos de alivio en sistemas de hidrogeno, buscando así la mejora continua, evitando un desbalance del riesgo, y buscando como fin último una transición a energías limpias y un mundo sostenible de la forma más segura posible.

Acrónimos

AIChE	American Institute of Chemical Engineers
CCPS	Center for Chemical Process Safety
H2Tool	Hydrogen Tools
HIAD	Hydrogen Incident and Accidental Database
KHK	High Pressure Gas Safety Institute of Japan
NTSB	National Transport Safety Administration
PHMSA	Pipeline and Hazardous Materials Safety Administration
PSM	Process Safety Management

1 Introducción

En los últimos, años la industria de energía ha estado en la búsqueda de una fuente de energía alternativa sostenible que reduzca el impacto ambiental de las operaciones mientras mantenga el consumo energético actual. Específicamente, se ha buscado mejorar la sostenibilidad de los procesos a través de mejoras tecnológicas buscando alternativas que reduzcan el impacto de carbono en la atmosfera. En este sentido, debido a su potencial producción limpia y uso flexible, el hidrogeno ha sido apuntado como una fuente alternativa de energía renovable que permitiría alcanzar dicha meta.

Sin embargo, esta transición trae la necesidad de establecer una infraestructura para el almacenaje, transporte y distribución de hidrogeno. Adicionalmente, el incremento de la demanda puede incrementar la frecuencia de eventos peligrosos, potencialmente impactando la seguridad de las comunidades y la percepción de la exposición al riesgo impactando significativamente el desarrollo, implementación, y la aceptación publica de tecnologías a base de hidrogeno.

Esta infraestructura de almacenaje, transporte y distribución de hidrogeno se encuentran equipado con dispositivos de alivios de presión para la protección en contra del aumento incontrolado de presión, el cual tiene la capacidad de generar una ruptura catastrófica provocando fugas de hidrógenos, incendios y/o explosiones. Adicionalmente, el hidrogeno posee peligros adicionales como los son una mayor facilidad para su ignición y la fragilización por hidrogeno (hydrogen embrittlement), los cuales tienen el potencial de causar una mayor cantidad de incendios y explosiones al visto hoy en día.

Debido a estos nuevos peligros, existe una incertidumbre en el riesgo presente cuando se aplican dispositivos de alivios tradicionales en sistemas de hidrogeno. A pesar de que estos se encuentran con el fin de mitigar una ruptura, se han observado incidentes donde dicho dispositivo ha sido el evento iniciador del incidente, generando la interrogante sobre donde nos encontramos en la balanza de riesgo.

Debido a esta incertidumbre y que uno de los elementos más importante de seguridad de procesos es el entendimiento de los peligros y riesgos[1] en conjunto con la ocurrencia de incidentes relacionados, como el Dimond Bar en California, se decidió realizar un estudio documental donde se buscaron diversos incidentes relacionados con dispositivos de alivio en sistemas de hidrogeno, buscando contribuciones estadísticas más importantes de forma que alimente el entendimiento de riesgo y peligro cuando se utilizan este tipo de dispositivos.

2 Metodología

Este trabajo se enfocó en la identificación de contribuciones y brechas en riesgo y confiabilidad de tecnologías de alivio en sistemas de hidrogeno. Con dicho objetivo en mente, se realizaron tres actividades principales: 1) la búsqueda de incidentes, 2) el análisis de data, y 3) el desarrollo de un modelo genérico de bowtie que permita alimentar estudios posteriores. La Figura 1 a continuación muestra la metodología general, mostrando los objetivos, métodos y herramientas usadas en cada etapa.

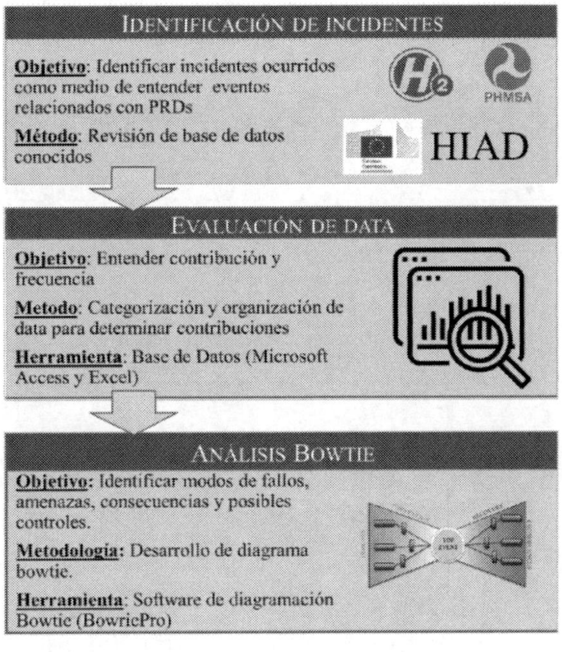

Figura 1. Metodologia

3 Incidentes

Para poder realizar un análisis de contribución adecuado se necesita la definición de una taxonomía adecuada. La Tabla 1muestra la taxonomía definida en este estudio.

Tabla 1. Taxonomía usada

Características de Incidentes	**Descripción**
Eventos	Describe la consecuencia final del incidente, p.ej., dispersión de gas, incendio, explosión, BLEVE/Burst
Tipo de sistema	Clasificación del sistema involucrado en el incidente, p.ej., Gas Comprimido, Liquido criogénico.
Evento iniciador PRD	Desglose de la data en eventos iniciados por fallos por Sistema de Alivio y eventos con fallos fuera del Sistema de Alivio pero que generaron incremento de presión, demandando activación del sistema de alivio.
Tipo de instalación	Indica tipo de instalación donde ocurrió el evento p.ej., Industrial, Comercial, terminal de llenado.
Ubicación o equipo involucrado	Describe el equipo especifica involucrado en el evento. P.ej., cilindros, tanque de almacenamiento, compresor.
Operación realizada durante incidente	Describe la operación que se encontraban realizando al momento del incidente, p.ej., loading/offloading, test, mantenimiento, transporte.
Tipo del PRD	Indica el tipo de dispositivo de alivio involucrado en el evento, p.ej., spring-operated pressure relief valve, burst disk
Modo de fallo	Describe la manera o forma el Sistema falla, p.ej., falla de operar, fuga externa, operación esporádica.
Mecanismo de fallo	Describe el proceso físico y/o químico que conduce al modo de fallo, p.ej., corrosión, fragilización, desgaste
Factor casual	Situación precursora del modo de falla que conduce al incidente, p.ej., mantenimiento inadecuado, instalación inadecuada, aplicación indebida de materiales.

La etapa de identificación de incidentes condujo a la identificación de 50 incidentes únicos, identificados en diversas base de datos globales y nacionales como H2Tools[2], HIAD[3], PHMSA[4], KHK[5], noticias y casos legales.

Estos 50 incidentes fueron categorizados y organizados con respecto a taxonomía mostrada en la Tabla 1. Las Figura 2, Figura 4, Figura 5, y Figura 6 muestran contribuciones de tipo de sistema, operaciones, modos de fallos, y evento iniciador respectivamente. A pesar de que la taxonomía definida muestra otros desgloses, estos presentan un resultado importante. En primer lugar, un resultado indirecto importante encontrado es que la mayoría de los incidentes no están siendo investigada a detalle. La descripción de la mayoría de los eventos se enfoca en causas inmediatas y no permite entender en detalle factores causales y mecanismo de falla; de igual forma no hay eventos casi accidentes (near missess) los cuales podrían enfatizar frecuencias de ocurrencia de eventos iniciadores y brechas.

En relación con la distribución encontrada, se muestra que la mayoría de los eventos (60%) ocurren en sistemas de gas comprimido. Esta contribución no era esperada ya que los sistemas de líquidos criogénicos tienen el potencial de generar mucha mayor presión al evaporar el hidrogeno. Sin embargo, esta observación puede ser el resultado de la confiabilidad del sistema de protección térmica y una mayor conciencia respecto a los peligros de líquidos criogénicos por parte de los operadores. Sin embargo, en este momento no existen estudios de cultura de seguridad en instalaciones de ambos sistemas. Por otro lado, vemos que la gran mayoría de los eventos resultaron en consecuencias relacionados a ignición (fuego y explosiones, 52%), menos de la mitad (42%) resultaron en dispersión de gases sin ignición. Sin embargo, esto puede ser resultado de no reportar dichos eventos por considerarlos no significantes o sin consecuencias. El hecho de la falta de datos que no permita validar dicha contribución demuestra la importancia de reportar todos los incidentes, e investigarlos adecuadamente como fuente de información fidedigna que alimente análisis y estudios de confiabilidad y riesgo.

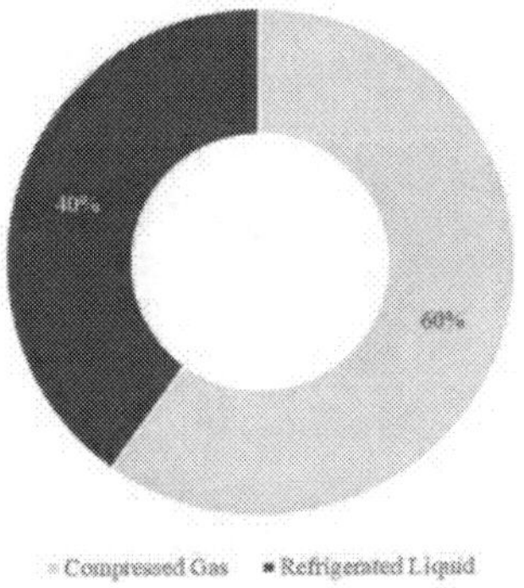

Figura 2. Tipo de Sistema

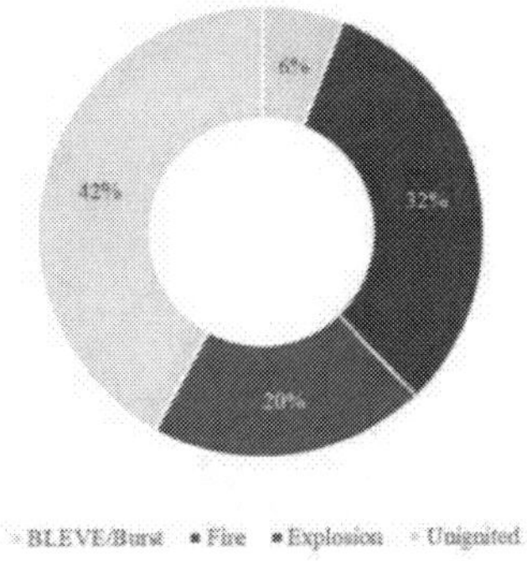

Figura 3. Eventos

Respecto a las operaciones realizadas durante el incidente, la data muestra una mayor contribución (36%) de operaciones de carga, descarga y llenado. Este resultado es esperado, sin embargo, la contribución encontrada en operaciones normales, transporte y equipos no en uso es mayor que la esperada. Esta contribución no esperada puede ser resultado de los nuevos mecanismos de degradación, pero la falta de detalle en la investigación no permite el desglose detallado con respecto mecanismo de fallo y factores causales.

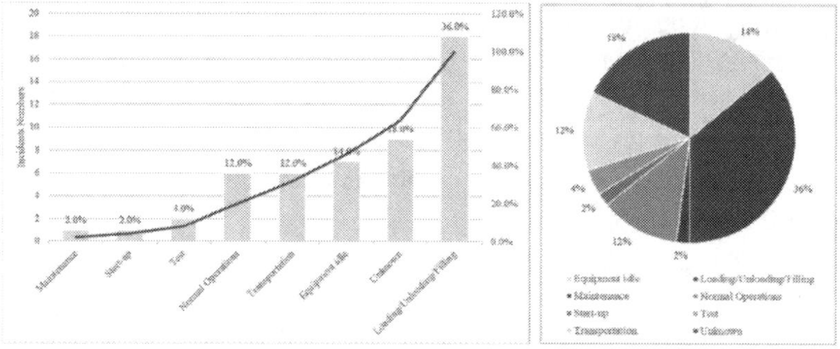

Figura 4. Operaciones

La distribución de modos de fallo indica que operaciones esporádicas son el modo de fallo más común (34%), esto puede alimentar la hipótesis anterior de una contribución por mecanismo de degradación nuevos, aunque actualmente no hay data detallada que pueda probar dicha hipótesis. Los otros modos de fallos importantes son desviaciones operacionales (24%) y fugas externas (20%). Sin embargo, cuando se reorganiza esta data para identificar los incidentes por eventos iniciadores, se muestra que la gran mayoría de eventos iniciadores son relacionados con fallos de sistemas de alivio (56%). Estos eventos corresponden en su mayoría, a la contribución cumulativa de operaciones esporádicas y fugas externas.

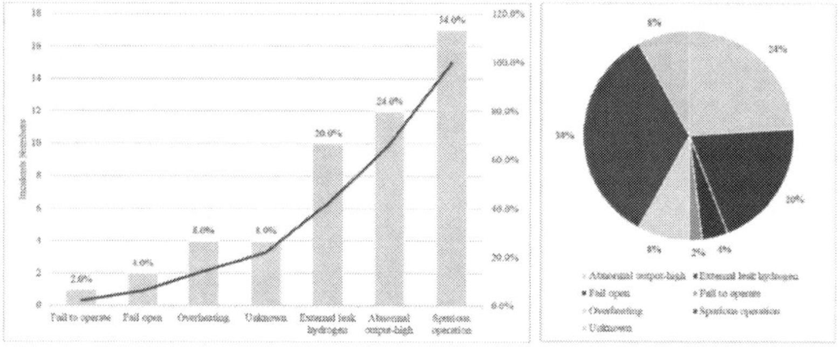

Figura 5. Modo de Falla

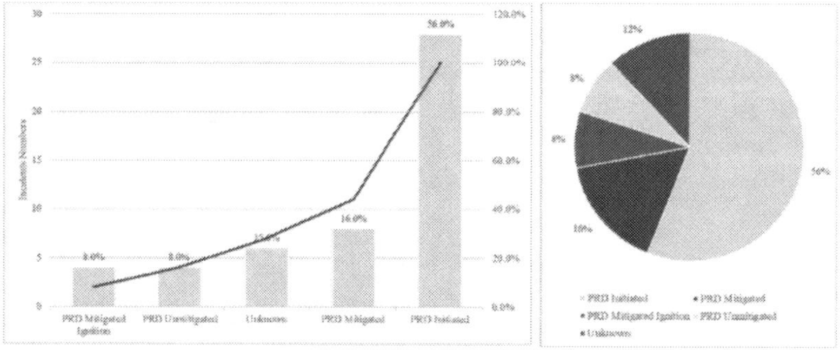

Figura 6. Eventos Iniciadores PRD

Adicionalmente, debido a que la fuente de información no es una base de datos de ingeniería de confiabilidad, y presenta los eventos de forma anecdótica, no todos los eventos presentan información sobre los factores causales. Sin embargo, utilizando los eventos que permitan inferir dichos factores fueron utilizados para indicar sobre la posibilidad de esto factores. Debido a la falta información, no se puede realizar contribución de los mismos, pero si se puede listar los mismos. Estos factores causales fueron definidos desacuerdo con elementos de gestión definidos por la guía de gestión de seguridad basada en riesgo de la CCPS[1] y presentados en la Tabla 2

Tabla 2. Factores Causales Identificados

No.	Factor Causal	Descripción
1	Control de Procura Inadecuada	Especificación de un componente adquirido es incompatible con el requerimiento técnico
2	Control de Calidad Inadecuado durante la fabricación	Fallo de control de Calidad durante el proceso de fabricación de componentes y sistemas generando un sistema o componente inadecuado para su uso previsto.
3	Falta o implementación inadecuada de programa de integridad de las instalaciones.	El elemento de integridad realiza actividades que permita que lo componentes y equipos se encuentren en condiciones adecuados para la aplicación prevista[1]. Este factor causal incluye falta e implementación inadecuada de actividades de inspección, pruebas, y/o mantenimiento.
4	Falta o inadecuado control de Calidad durante diseño e instalación.	Parte de la gestión de Confiabilidad es asegurar que los equipos y componentes sean adecuados para su uso previsto, esto incluye asegurar que dichos equipos y componentes estén adecuadamente diseñados e instalados para las condiciones de operación. En tal sentido, este factor causal incluye defectos de diseño e instalación que produzca los eventos observados.
5	Falta o implementación inadecuada de programa de control del cambio.	El control de cambio de realizar una revisión adecuada antes de la implementación de cualquier cambio, ya sea temporal o permanente, asegurando que dicho cambio no introduzca peligros adicionales o incremente el riesgo inadvertidamente[1]. Este factor causal indica que el proceso de gestión del cambio no fue realizado o fue realizado minuciosamente dejando brechas que generaron el incidente.
6	Falta o implementación inadecuada de preparación operacional (operational readiness)	Este elemento realiza una verificación de la operaciones asegurando que sus condiciones son adecuadas para el arranque[1]. Este factor causal incluye situaciones cuando este proceso de

No.	Factor Causal	Descripción
		verificación no se ha realizado, o se ha realizado ineficientemente dejando componentes en condiciones inadecuada para la operación.
7	Falta o inadecuados procedimientos operativos.	Los procedimientos operativos son instrucciones escritas que indican las actividades a realizar asegurando una operación segura[1]. Este factor causal incluye eventos donde no había procedimientos o estos no comunicaron adecuadamente las instrucciones de cómo realizar las operaciones.
8	Entendimiento de inadecuado de peligro y riesgo	El peligro y/o riesgo asociado no fue propiamente entendido, por lo tanto, los controles necesarios no fueron implementados. Esta falta de entendimiento fue alimentada por la no realización de estudios de identificación y análisis de riesgo o por una metodología inadecuada.
9	Gestión de emergencias inadecuado o ineficiente.	La gestión de emergencias es el elemento encargado de la planificación de emergencias potenciales, proveyendo recursos para su ejecución, y asegurando que el personal encargado se encuentre entrenado[1]. Este factor incluye la inexistencia o ineficiencia de dicho plan, la falta de recursos y/o la falta de entrenamiento de los respondedores.

4 Bowtie Genérico

Luego de entender la contribución de las diversas características de los incidentes, se desarrolló un modelo de bowtie utilizando dicha información. Este diagrama siguió las reglas del libro de bowtie del CCPS y el instituto de energía británico[6]. La idea con este modelo es el de presentar de forma gráfica los diversos aspectos generales que intervienen en incidentes que involucran dispositivos de alivio instalados en sistemas de hidrogeno. El propósito no es el de generar un modelo especifico, sino tener un modelo genérico que pueda ser tomado como base inicial en estudios más detallados, p.ej., arboles de fallos. En este sentido, solo se incluyeron eventos iniciadores relacionados con dichos dispositivos, p.ej., activación esporádicos, fuga externa. La Figura 7 muestra los elementos principales del diagrama, y la Figura 8, Figura 9, y Figura 10 muestra los detalles de cada barrera con sus factores de degradación y controles asociados. Finalmente, la Figura 11 muestra el diagrama completo.

Por otro lado, los modos de fallos relacionados con eventos iniciadores ajenos al sistema de alivio requieren un estudio más complejo que considere la interconexión entre dichos eventos y el dispositivo de alivio, p.ej., fallo de operación, fallo de cerrar, operación exitosa, por lo que estos no pueden ser efectivamente evaluado con un diagrama de bowtie genérico.

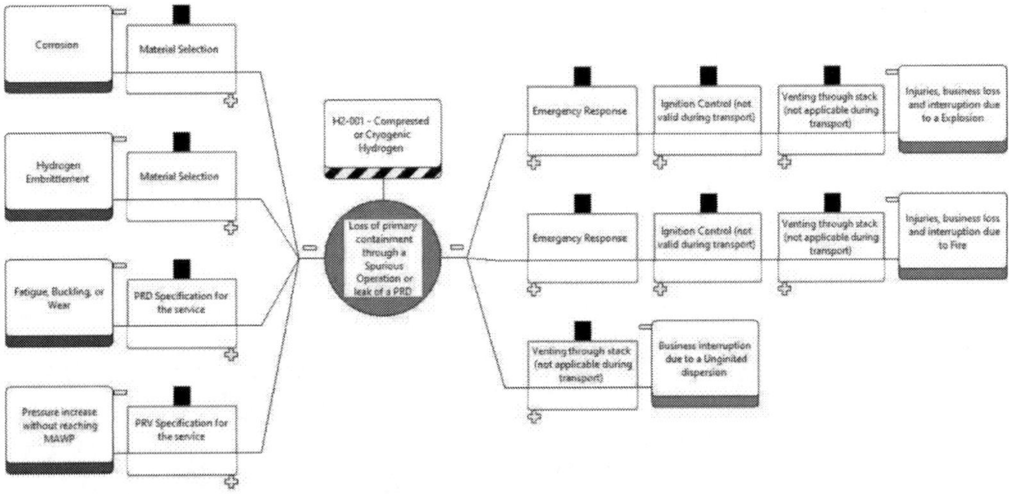

Figura 7. Bowtie genérica

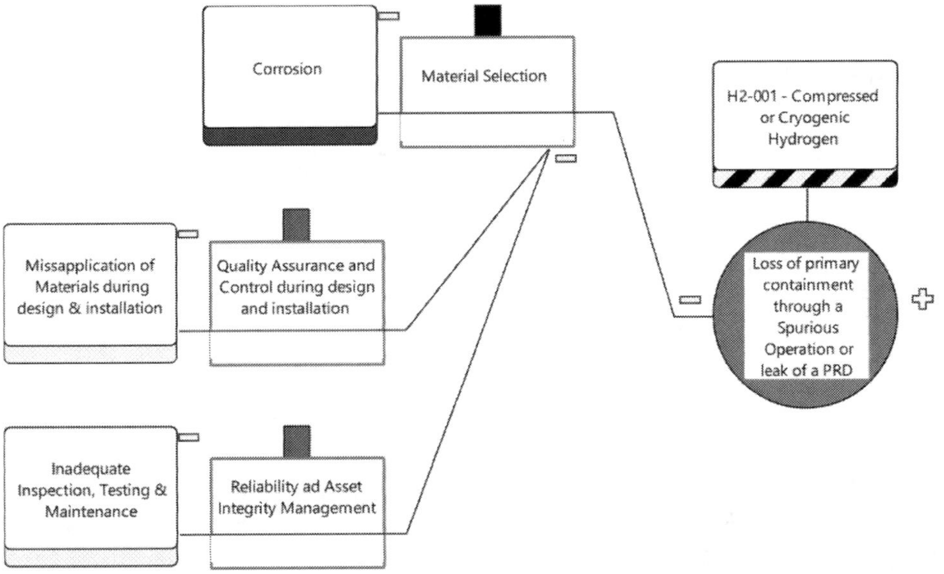

Figura 8. Barrera selección de material

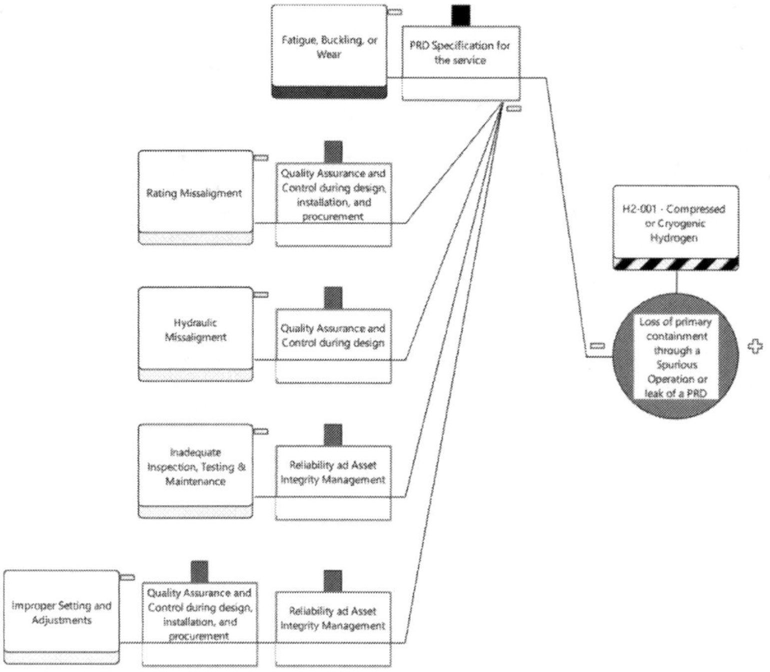

Figura 9. Barrera especificación de dispositivo

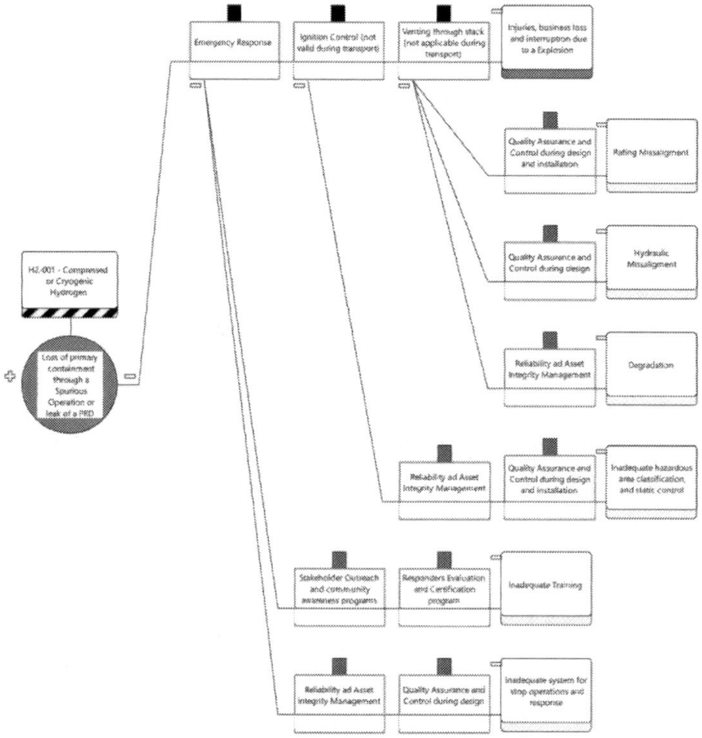

Figura 10. Barreras en consecuencias

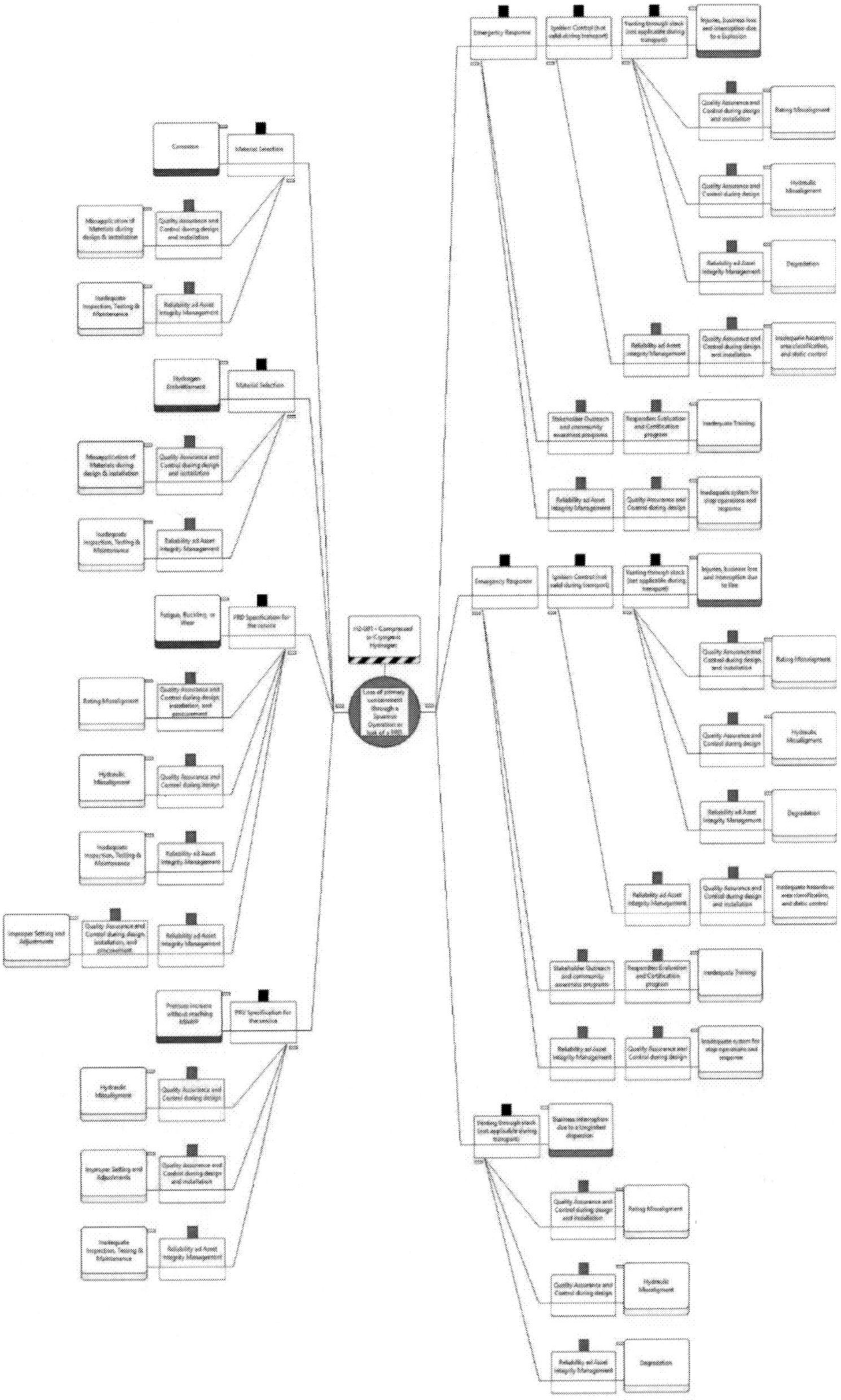

Figura 11. Diagrama completo

Este diagrama de bowtie muestra el sistema de control de calidad, de gestión de la integridad, y la preparación de operación antes del arranque en repetidas ocasiones. Esto demuestra su importancia, sin embargo, también muestra como el factor humano puede fácilmente afectar el riesgo de los mismos. Estas tres barreras (y elementos PSM) dependen directamente de actividades humanas que pueden introducir y propagar el error humano. En tal sentido, para asegurar que los dispositivos de alivio se encuentren en condiciones adecuados, se destaca la importancia del uso de los estándares de desempeño como herramientas que permita definir sus criterios de diseño, instalación, y mantenimiento.

5 Conclusiones

Considerando los diversos análisis realizados, las siguientes conclusiones pueden ser desarrolladas

- Los modos de fallo mas importantes son la operación esporádica, la desviación de procesos, y fugas externas. De estos, los eventos iniciadores relacionados a los dispositivos de alivio representan mas de la mitad de los eventos.
- La mayor proporción de incidentes ocurren durante operaciones de transferencia como carga, descarga, y llenado; Sin embargo, la contribución de operación normal, transporte y equipo fuera de uso es importante. Esto puede ser por la introducción de otros mecanismos de fallos como la fragilización por hidrogeno, sin embargo, actualmente no se encuentran datos y/o estudios que permita evaluar contribución por mecanismo.
- Existe una necesidad de mejorar las investigaciones realizadas en eventos que involucren dispositivos de alivio, y que son documentadas en bases de datos de incidentes. Estas investigaciones permitirán el desarrollo de análisis de confiabilidad y riesgos con mayor certeza alimentando así un mejor control de riesgos.
- Los controles de factores de degradación del Bowtie genérico muestra una importancia significativa en la gestión de la integridad y sistemas de control de calidad durante diseño e instalación. En tal sentido, el desarrollo e implementación de estándares de desempeño es una herramienta crucial para asegurar que estos sistemas sean diseñados, instalados, y mantenidos adecuadamente.

6 Referencias

1. Chemical Center for Process Safety (CCPS). *Guidelines for Risk Based Process Safety*. American Institute of Chemical Engineers (AIChE) and John Wiley & Sons, Inc; 2007.

2. Pacific Northwest National Laboratory, US Department of Energy. Lessons Learned. Hydrogen Tools. https://h2tools.org/

3. European Commission. Hydrogen Incidents and Accidents Database (HIAD).

4. Pipeline and Hazardous Materials Safety Administration (PHMSA). Hazmat Incident Report Search Tool. Incident Statistics. https://portal.phmsa.dot.gov/analytics/saw.dll?Portalpages&PortalPath=%2Fshared%2FPublic%20Website%20Pages%2F_portal%2FHazmat%20Incident%20Report%20Search

5. High Pressure Gas Safety Institute of Japan (KHK). High Pressure Gas Safety Institute of Japan (KHK) Record Statistics. Published online 2022.

6. Center for Chemical Process Safety of the American Institute of Chemical Engineers, Energy Institute. *Bowties in Risk Management, A Conceptual Book for Process Safety*. John Wiley & Sons, Inc.; 2018.

Ammonia Mitigation System for Large Release Scenario in Storage Tanks - A Case Study

Hormando Leocadio
USIMINAS STEEL
Ipatinga, MG, Brazil
hormando.leocadio@usiminas.com

Tiago Gomes
USIMINAS STEEL
Ipatinga, MG, Brazil
tiago.gomes@usiminas.com

Keywords: Anhydrous ammonia; Accidental ammonia release; Automatic system; Neutralization of ammonia; Mitigation system.

Abstract

Abstract here. Anhydrous ammonia has high toxicity (tolerability ≤ 20 ppm) and flammability, and is generally transported by ships, trains, tank trucks, and pipelines. It is a raw material to produce fertilizers and an effective coolant in industry. Supplied in liquefied form, vaporizes immediately at atmospheric pressure, forming a floating cloud. At Usiminas steelworks plant, anhydrous ammonia is generated as a byproduct of the metallurgical coke manufacturing process and stored in pressurized liquefied form (15 bar) in seven tanks of 123 m^3 each. An automatic ammonia neutralization system, with a 5-second response time, was developed and installed in the storage tank and truck loading area to neutralize 100% of the worst-case scenario of ammonia leakage (7,668 kg/h – 12 m^3/h), thus ensuring safety and minimizing risk for the industrial plant. Ammonia leak neutralization tests were carried out to demonstrate the spray nozzle efficiency.

1 Introduction

Anhydrous ammonia is highly toxic and flammable, with human tolerability ≤ 0.002% and IDLH (immediately dangerous to life) index of 0.03% (300 ppm), being transported by ships, trains, tanker trucks, and pipelines. It is a raw material for the production of agricultural fertilizers and an effective refrigerant used in industry. Usually supplied pressurized and in liquefied form, vaporizes immediately at atmospheric pressure forming a floating cloud. Figure 1 shows the leakage of liquid ammonia after the rupture of a transport pipe in Kansas, United States in 2004, where the cloud reached a distance of 2.5 km.

Figure1. Ammonia cloud of 2.5 km formed after pipeline rupture [1].

At Usiminas steelwork plant, anhydrous ammonia is generated as a by-product of the metallurgical coke manufacturing process. This ammonia is stored in seven tanks with a capacity of 123 m³ each, in liquefied form pressurized to 15 bar and kept below 50°C. The partial view of the storage tanks and truck-tank loading area is shown in Figure 2.

Figure 2. Partial view of the anhydrous ammonia storage tanks and truck loading area at Usiminas.

This study presents the development and implementation of an automatic ammonia neutralization system, with a fast 5-second response time. Situated in the tanks and tank-truck loading area, this system is designed to neutralize 100% of the worst-case scenario of estimated ammonia leakage (7,668 kg/h - 12 m³/h), enhancing safety and reducing risk for the industrial plant.

2 Development

2.1 Characteristics of Anhydrous Ammonia

Anhydrous ammonia (NH_3) is a cryogenic liquid with a boiling point of -33°C 1 atm. It floats in water and rapidly reacts resulting in ammonium hydroxide (NH_4OH), while simultaneously boiling and migrating into the atmosphere as a gas. The portion of liquid ammonia that dissolves in water, in cases of spills on the surface of the water, is 85%. If the spill is underwater, the portion is greater than 90% [2]. The properties of anhydrous ammonia are shown in Table 1.

Table 1. Characteristics of anhydrous ammonia.

Human Exposure	Human tolerability of 20 ppm for exposure of 48 h/week [3].
	Eye irritation at 70 ppm [4]
	Immediately Hazardous to Life or Health (IDLH) at 300 ppm [4].
Specific Mass	0.77 kg/m³ (gas) and 639 kg/m³ (liquid) at 0°C and 1 bar [4]
Purity Grade	99,99% [4]
Water Solubility	Vol/Vol: 0.85 [2]
Vapor Pressure	– 33°C at 1 bar and +35°C at 13 bar [4]

2.2 Storage Tank Data

There are seven carbon steel storage tanks with a design pressure of 16 bar, operating at 15 bar. Each tank has a useful volume of 123 m³, a length of 10 m and a diameter of 3.6 m. Figure 3 shows the tank location plant and the containment basin.

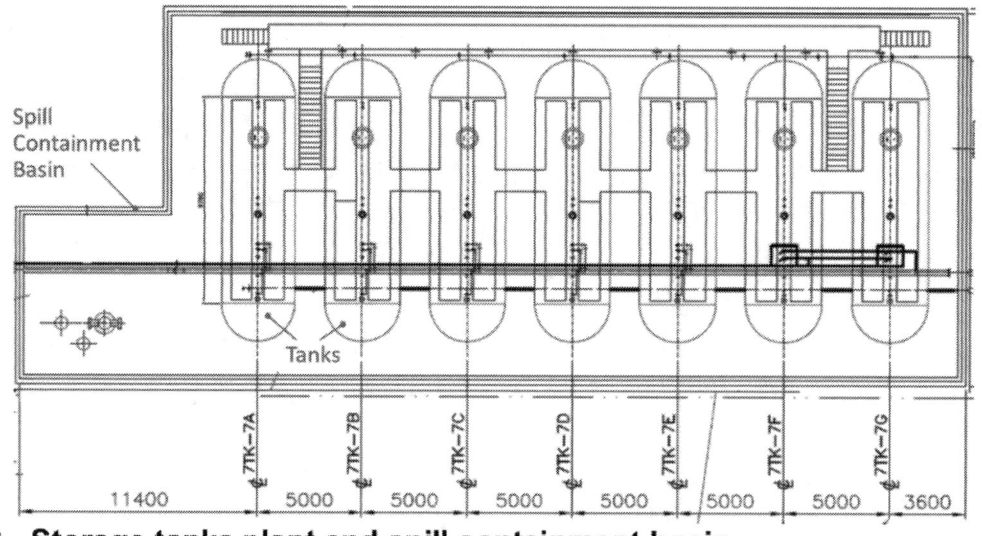

Figure 3 - Storage tanks plant and spill containment basin.

2.3 Experimental apparatus

Due to the strong affinity of anhydrous ammonia with water, the neutralization system uses water spray nozzles. Figure 4 shows the outline of the experimental apparatus used to evaluate the efficiency of the water spray nozzles in neutralizing ammonia. The spray nozzle was positioned 3.5 m above the ammonia leak point producing a full cone water jet with a base circle of 3.8 m in diameter.

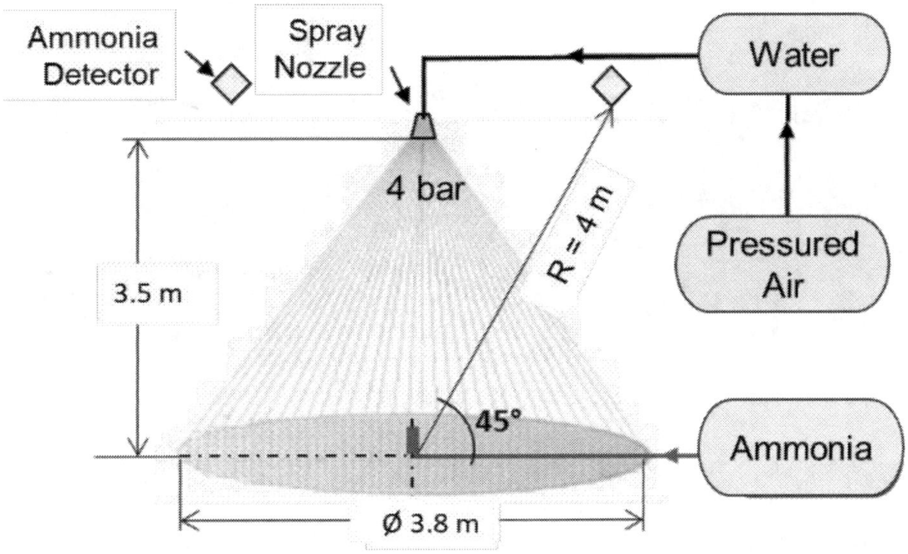

Figure 4- Experimental apparatus to evaluate the neutralization efficiency of spray nozzles.

As the tests were carried out in an open area where ammonia vapor could be influenced by wind, two ammonia detectors were used in order to compare the measured concentrations. They were positioned equidistant, radially at 4 meters from the ammonia leak point, forming an angle of 45° to the central vertical axis of impact. The water spray nozzle was pressurized by a compressed air vessel. The anhydrous ammonia used came from tanks.

Two types of spray nozzles were developed (B1 and B2) and tested. The B1 model with a flow capacity of 17 ℓ/min at 6 bar. The B2 model with a flow capacity of 32 ℓ/min at 4 bar. Droplets produced by nozzle B2 were twice the size of the droplets of nozzle B1. The original design of these two nozzles has been altered to better suit the conditions of the neutralization system. During water spraying, a smaller droplet diameter increases the contact area between ammonia and water, thereby enhancing the ammonia neutralization efficiency.

3 Results

3.1 Testing with Spray Nozzles

Two types of spray nozzles, named B1 and B2, were tested. The mitigating capacity of the B1 model was below 50%. This was attributed to the small size of the B1 droplets, which made them susceptible to be carried away by the wind, resulting in low efficiency. Nozzle B1 was discarded. Figure 5 shows nozzle B2 in operation.

Figure 5- Spray nozzle model B2 operating at 4 bar.

Model B2 demonstrated better mitigation results. Tests with ammonia leak was simulated with water and without water spray. The detectors recorded the concentration of ammonia in order to evaluate the efficiency of the spray nozzle in neutralizing ammonia. Table 2 presents the results of the efficiency tests of the B2 model for neutralization of ammonia leakage, in terms of concentration (ppm) of ammonia in the environment.

Table 2 – Results of ammonia neutralization tests for B2 spray nozzle.

Ammonia Concentration in the Environment after Simulated Leak					
Detector 1			Detector 2		
NO spray (ppm)	WITH spray (ppm)	Efficiency (%)	NO spray (ppm)	WITH spray (ppm)	Efficiency (%)
356	31	91	240	11	95
325	26	92	301	27	91
320	16	95	310	25	92

Liquefied ammonia has high solubility in water. In the gaseous state, there is a need for a larger quantity of water to neutralize it. The average efficiency of the B2 spray nozzle was 93%. The ratio (R) between the water mass flow (mw) and ammonia mass flow (ma) was 15 (Equation 1), i.e., 15 kg of water to neutralize 1 kg of gaseous ammonia, with efficiency between 91% and 95%.

$$\boldsymbol{R = m_w/m_a} \qquad (1)$$

The concentration values observed after spraying with water are close to the human tolerability and well below 300 ppm that poses risk to human life, as shown in table 1. Therefore, the 93% efficiency of the B2 nozzle meets the design requirements. Performance testing is crucial to validate the nozzle spray efficiency in neutralizing ammonia leaks.

The efficiency of a water spray system, for the mitigating ammonia leaks in storage areas, depends directly on the water-to-ammonia ratio and how the sprinkling of this water is carried out. Figure 6 shows an efficiency graph of a neutralization system using spray nozzles used in the design of the ammonia leak neutralization system in the storage tank area of the Chevron refinery, United States. The characteristics of the spray nozzles and the operating conditions of this system were not presented.

The results found in the tests performed with nozzle B2 showed efficiency ranging between 91% and 95% for a water-to-ammonia ratio R = 15. This value plotted on the graph, shown in Figure 6, finds an efficiency of 96%, which is close to the value measured in the tests carried out.

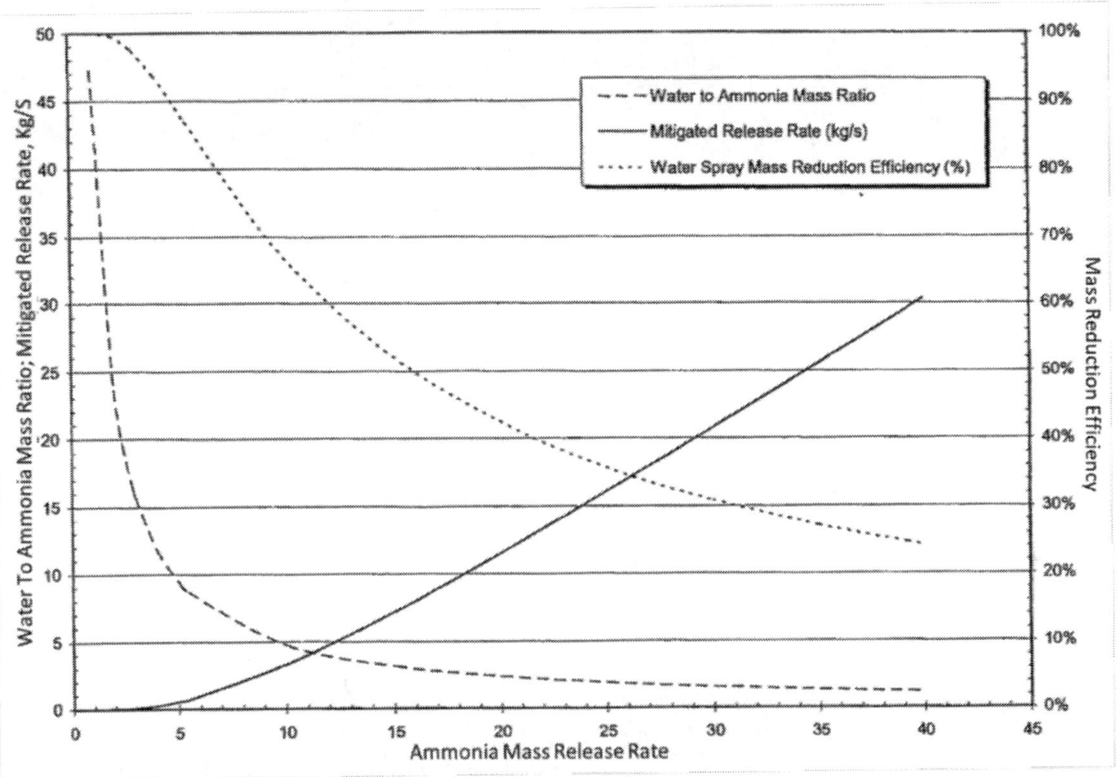

Figure 6- Efficiency of neutralization spray system for ammonia storage tanks [5].

3.2 Ammonia Neutralization System

To ensure complete coverage of both the tanks and truck loading area, a total of 154 spray nozzles were distributed. The nozzles were installed with interference to increase the density of water per square meter. Figure 7 shows the isometric view of the loading area and the tanks, while Figure 8 shows the isometric view of the spray system covering both the tanks and loading area.

The total capacity of the system is 378,840 kg/h of water spray covering all seven tanks and loading area, representing an area of 476 m². In tanks area, the specific water flow rate is 749 kg/(h.m²). The loading area has a specific water flow rate of 963 kg/(h.m²), i.e. a water density 29% higher than the tanks area.

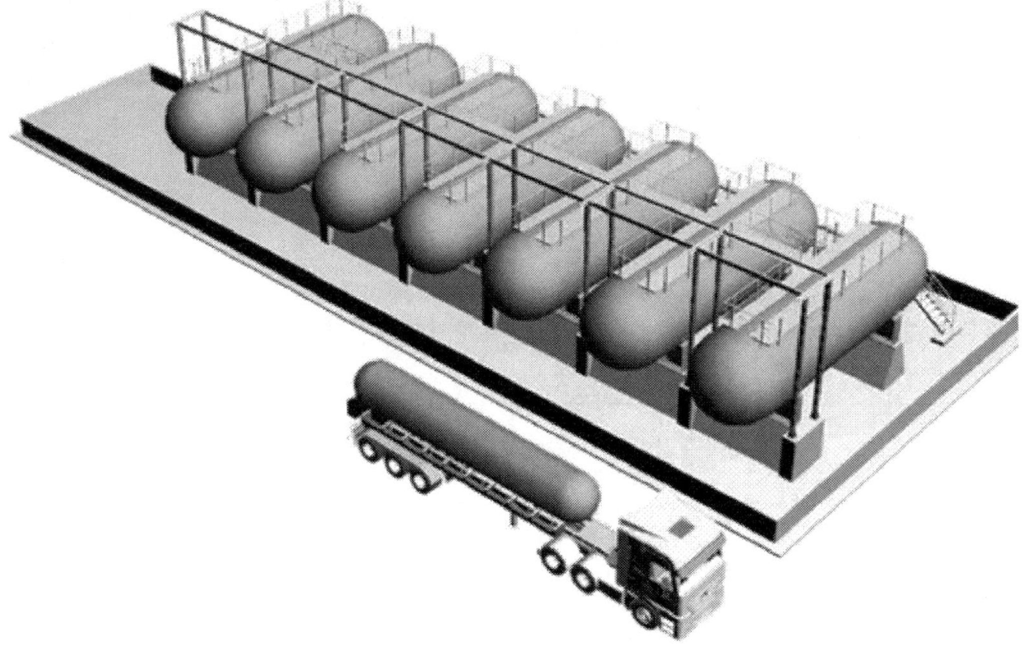

Figure 7- Isometric view of tanks and loading area.

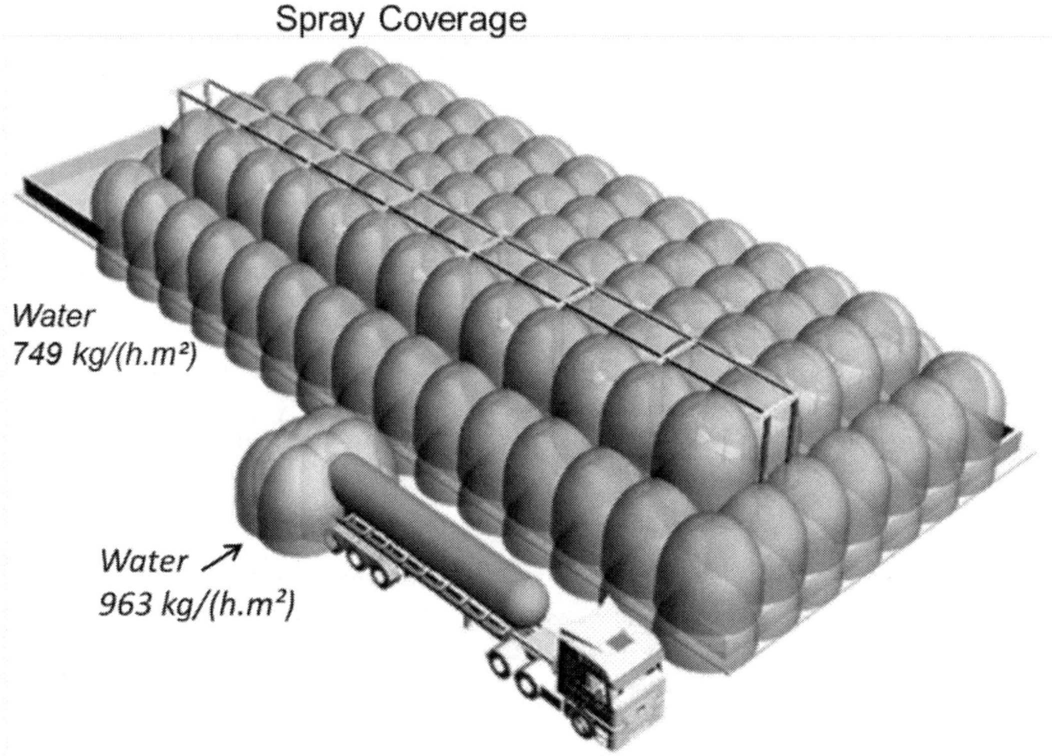

Figure 8- Isometric view of the spray system covering on tanks and loading area.

Figure 9- View of the seven tanks with neutralization spray system installed.

Figure 10- Photograph of the performance test of the water spray system, after 5 seconds of operation.

For a conservative analysis of the efficiency of the spray system, it was assumed that all ammonia leakage occurs within the tank area. In the worst-case scenario, a net ammonia flow rate of 7,668 kg/h (12 m^3/h) was estimated. Within the tank area, the water flow rate is 313,000 kg/h. Therefore, a water-ammonia ratio (R) of 41. Inserting the value R = 41 in the graph of the Figure 6, a 100% ammonia neutralization efficiency is found.

In the event of tank rupture, with a spill of 123 m^3 of ammonia inside the containment basin, the liquid ammonia will be exposed to atmospheric pressure initiating the evaporation process. The evaporation rate estimated is 24 kg/(h.m^2). Dividing the specific water flow rate of 749 kg/(h.m^2) by the ammonia evaporation rate, a ratio R = 31 is found. Inserting this value in the graph of the Figure 6, 100% of the gases from evaporation will be neutralized.

The pipes are permanently filled with water to ensure a system response time of less than 5 seconds. Moreover, the system is designed with pump redundancy. If the first pump fails, the second will automatically go into operation. The system is automatically triggered when a leak is detected, and the flow of ammonia is blocked. Figure 9 shows the view of the seven tanks with a neutralization spray system installed. Figure 10 shows the performance test of the spray system, 5 seconds after its activation, where a dense cloud of water is formed by the action of the spray nozzles.

4 Conclusion

The anhydrous ammonia neutralization system has been successfully implemented in the storage tanks and loading area of tanker trucks at Usiminas steelworks plant. Tests have shown that the spray nozzle developed has good performance in neutralizing ammonia leakage.
The system cover an area of 476 m^2 (truck loading and storage tanks area) with a total capacity of 378,840 kg/h of spray water, the system achieves water densities of 963 kg/(h.m^2) in the truck loading area and 749 kg/(h.m^2) in the storage tanks.
The system is capable of neutralizing 100% of the worst-case scenario of ammonia leakage (7,668 kg/h or12 m^3/h) and flooding of the containment basin due to the rupture of a storage tank. The system is activated automatically when a leak is detected, and within 5 seconds it reaches full operation, ensuring the safety of the industrial plant.
It´s worth noting that performance testing remains essential to validate whether the spray nozzle selected will promote the necessary neutralization of ammonia leakage.

5 References

[1] Rosenker, M. V. Hersman, D. A., National Transportation Safety Board, Washington, Pipeline Accident Brief, Anhydrous Ammonia Pipeline Rupture, Report No. NTSB/PAB-07/02, 2004.

[2] Becker, S. M., Pitt, R., Clark, S. E., Environmental Health, Public Safety, and Social Impacts Associated with Transportation Accidents Involving Hazardous Substances, UTCA Report No. 214, 2001.

[3] NR 15 - Unhealthy Activities and Operations, Ministry of Labor, 2014.

[4] Fingas M., Handbook of hazardous materials spills technology, 1st edition, McGraw-Hill, 800 pg., 2001.

[5] Evaluation of an accidental anhydrous ammonia release from the Chevron Refinery, Anhydrous Ammonia Consequence Modeling Report 510-620-6841, 2007.

Aplicação De Segurança De Processos Em Projetos De Capital - Características Aplicáveis a Indústria De Mineração

Rodrigo Guimarães Horta
Gerente Área Segurança de Processos e Riscos Operacionais em Projetos
Vale SA
Brasil
rodrigo.horta@vale.com

Ronaldo Vieira de Faria
Engenheiro Master
Vale SA
Brasil

Keywords: Segurança de Processos em Projetos, Análises de Riscos de Processos, RBPS, Mineração, Projetos

Abstract

A Vale tem como objetivo tornar seus ativos e projetos cada vez mais seguros. Logo, foi decidido incluir a disciplina de Segurança de Processo no seu sistema de gestão de projetos, tendo como um dos principais pilares do RBPS (*Risk Based Process Safety*) a Identificação de Perigos e Análise de Riscos ao longo do ciclo de vida do projeto, durante as fases de desenvolvimento de projeto (FEL1, FEL2 e FEL3) e Execução, permitindo assim, a busca cada vez mais de projetos inerentemente seguros. Como resultado, foi desenvolvido um manual de segurança de processos em projetos com objetivo de introduzir a disciplina de Segurança de Projetos através da orientação e padronização da aplicação de análises de riscos de processos, como também, definir critérios que devem ser observados no momento da aprovação dos projetos para garantir que eles estejam cada dia mais seguros para serem operados.

1 Introdução

Conforme a 7ª Edição do Guia PMbok [1], considera-se como projeto um esforço temporário empreendido para criar um produto, serviço ou resultado único e seus riscos como um evento ou condição incerta que, se ocorrer, provocará um efeito positivo ou negativo em um ou mais objetivos do projeto.

Porém, nos tempo atuais, as organizações não estão somente preocupadas em entregar os projetos dentro das metas de prazo e custos estabelecidas, sendo que a qualidade e operabilidade são fatores crucias para que o projeto seja considerado como um sucesso dentro das organizações.

Logo, se o projeto possui níveis de segurança operacionais elevados, as taxas de incidentes durante as operações tendem a ser menores, tornando as operações das empresas cada vez mais seguras e confiáveis.

Pensando na operabilidade cada vez mais segura nos projetos após suas entregas, o tema de segurança de processos na mineração tem se tornado cada vez mais importante para que os projetos sejam entregues e operados da melhor forma possível.

Da mesma forma, segundo a AIChE [2], nem todos os perigos e riscos são iguais ou podem causar as mesmas consequências. Perigos/riscos pessoais ou ocupacionais, tais como escorregões, quedas, cortes e acidentes com veículos geralmente produzem efeitos sobre um único trabalhador. Por outro lado, perigos/riscos de processo podem ocasionar acidentes maiores envolvendo o vazamento de materiais potencialmente perigosos, incêndios e explosões, ou ambos. Incidentes e acidentes envolvendo segurança de processo podem ter efeitos catastróficos e podem resultar em múltiplas mortes e feridos, assim como danos substanciais à economia, à propriedade e ao meio-ambiente. Incidentes e acidentes desse tipo podem ferir tanto os trabalhadores no interior das indústrias quanto o público que reside nas vizinhanças. Essa é a razão pela qual o gerenciamento da segurança de processo está focado no projeto e engenharia de instalações, análises de perigos e riscos, análise e investigação de incidentes e acidentes, gestão da mudança, inspeção, testes e manutenção de equipamentos, alarmes e controles de processo efetivos, procedimentos de operação e manutenção, treinamento de pessoal e fatores humanos.

Tendo em vista que a Vale tem como um dos seus comportamentos-chave a Obsessão por Segurança e Gestão de Riscos para atingir a ambição de Ser uma Empresa Referência em Segurança, se fez necessário melhorar a aplicação da Segurança de Processos nos projetos do seu portfólio.

Como resultado, a Vale decidiu incluir a disciplina de Segurança de Processo no seu sistema de padronização de documentos de gestão de projetos, tendo como um dos principais pilares do RBPS a Identificação de Perigos e Análises de Riscos durante todo o ciclo de vida do projeto, durante as fases de desenvolvimento de projeto (FEL1, FEL2 e FEL3) e Execução, permitindo assim a busca cada vez mais de projetos inerentemente seguros.

Assim, foi desenvolvido em 2022 um manual de Segurança de Processos em projetos, chamado de MA-E-240 [3], o qual estabelece diretrizes e orientações para a aplicação do tema de segurança de processos e análises de riscos de processos nos projetos do portfólio da Vale.

2 Manual de Segurança de Processos em Projetos

O desenvolvimento do manual de Segurança de Processos para a aplicação nos projetos do portfólio da Vale contou com a participação de vários setores da empresa, além de uma consultoria externa com experiência na aplicação de Segurança de Processos em projetos e operações químicas.

Abaixo é mostrado os setores que contribuíram para o desenvolvimento desse manual:

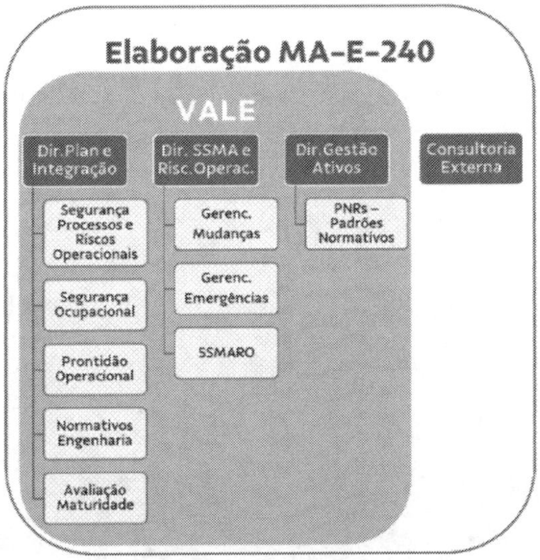

Figure 1. Áreas participantes na elaboração do manual de segurança de processos em projetos

2.1 RBPS como base conceitual

O VPS (Vale Production System), atual sistema de gerenciamento da Vale, cobre os elementos do PSM propostos pelo CCPS. Portanto, o PSM adotado pela Vale, baseia-se nos 20 elementos do PSM distribuídos em 4 pilares descritos a seguir na figura abaixo [4].

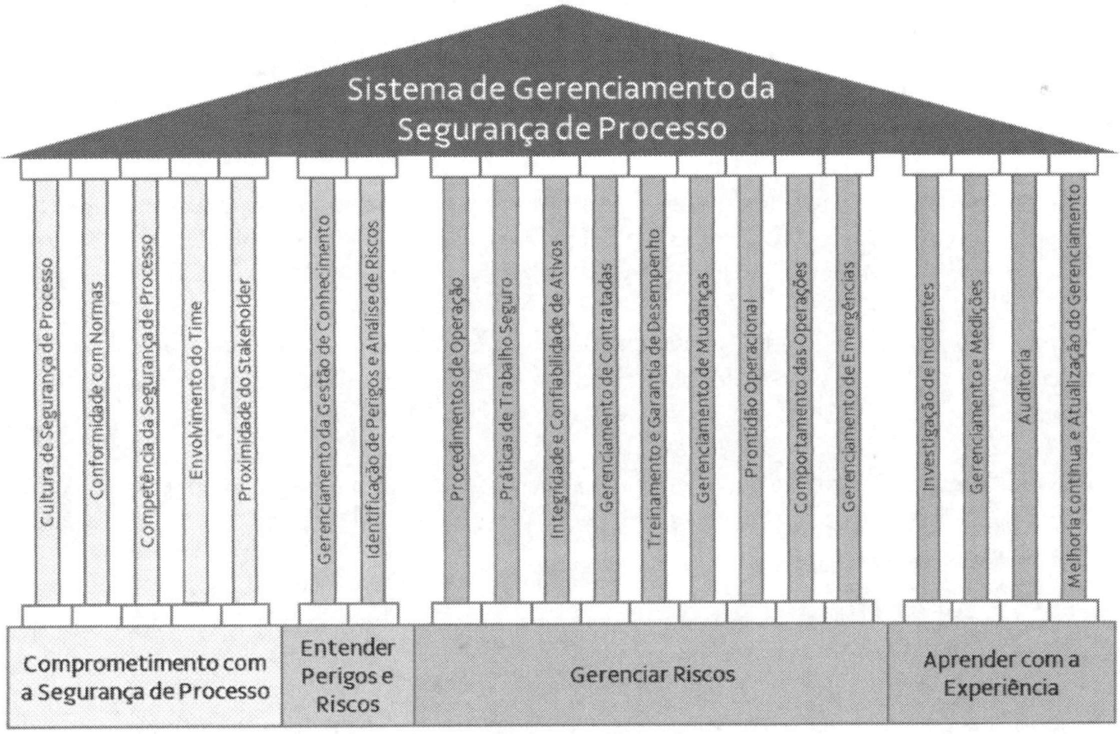

Figura 2.1 – Pilares e elementos do RBPS associados que constituem um sistema de gestão robusto

Dentre os 20 pilares de Segurança de Processos, destaca-se o de Identificação de Perigos e Análise de Riscos que será objeto do próximo capítulo e será detalhado como foi desenhada a estratégia de aplicação nos projetos da Vale.

3 Análise de Riscos de Processos em Projetos

Os riscos de processos na Vale são definidos e classificados conforme a matriz de riscos [6] mostrada na figura abaixo:

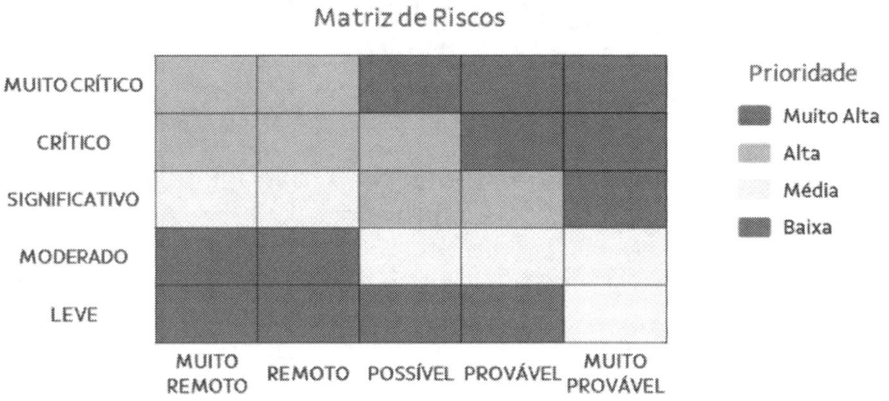

Figura 3.1 – Matriz de Riscos da Vale

Para a classificação dos riscos, os cenários de riscos de processos são divididos em 3 estágios conforme descrito abaixo:

- **Risco Inerente**: é a medida de risco presente no pior cenário plausível, desconsiderando a ação de controles. É o "risco na sua origem" no qual ainda não se aplicou nenhum controle ou estratégia de resposta ao risco;

- **Risco Residual**: é a medida de risco presente atenuada pelo ambiente de controles existentes na Vale. É a condição atual do risco, após as medidas de prevenção e mitigação estarem em vigor;

- **Risco Residual Projetado**: a projeção da medida do risco presente considerando os controles eficientes e a mudança esperada na classificação de risco devido à implementação de planos de ação.

Destaca-se que a tolerância de cada cenário de risco é estabelecida conforme limites internos da companhia.

Para a identificação de perigos e análise de riscos na Vale são utilizadas metodologias de riscos que são classificadas em três categorias: qualitativa, semiquantitativa e quantitativa. A figura 3.1

destaca como as metodologias devem ser aplicadas tanto para projetos quanto para as operações da Vale, conforme estabelecido no procedimento de Diretrizes Gerais para Análise de Riscos de Processo [5] da Vale.

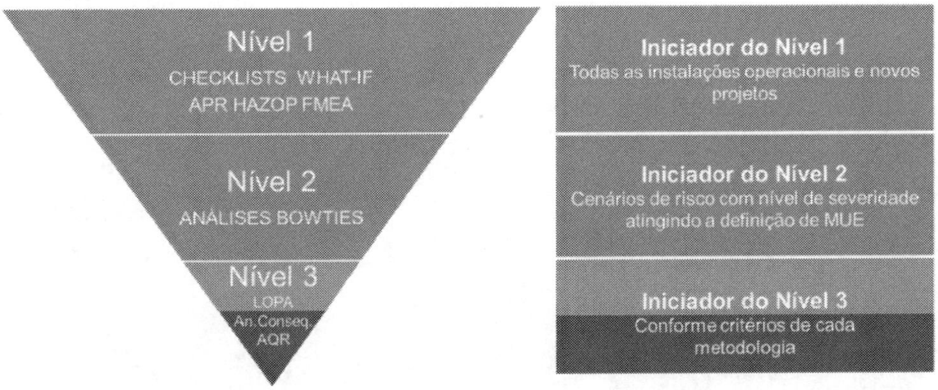

Figura 3.2 – Aplicação Metodologias de Riscos conforme Níveis

Abaixo é mostrado na tabela 3.1 de forma resumida como são definidos os níveis para aplicação de cada metodologia:

Tabela 3.1. Níveis para definição das metodologias

Níveis	Detalhamento
Nivel 1	A primeira etapa do processo é aplicar um método de identificação de cenário sistemático e estruturado que garantirá que eventos indesejados serão devidamente identificados, bem como seus iniciadores (causas) e potenciais consequências. Uma primeira classificação do nível de risco (probabilidade x severidade) servirá de base para a escalada da análise para os próximos níveis
Nivel 2	Obrigatório para todos os cenários de risco que atinjam a definição de MUEs - severidade muito crítica e/ou crítica para a dimensão pessoas e meio ambiente e a partir de significativa para a dimensão financeira
Nivel 3	Elegível para riscos que requerem avaliações de risco semi ou totalmente quantitativas

Devido a necessidade de definição de maneira mais clara e objetiva sobre quais metodologias seriam aplicadas para o portfólio de projetos, foi desenvolvido no Manual de Segurança de Processo para Projetos uma árvore de decisão o qual é apresentado na figura abaixo qual metodologia deve ser aplicada conforme fase e características dos projetos.

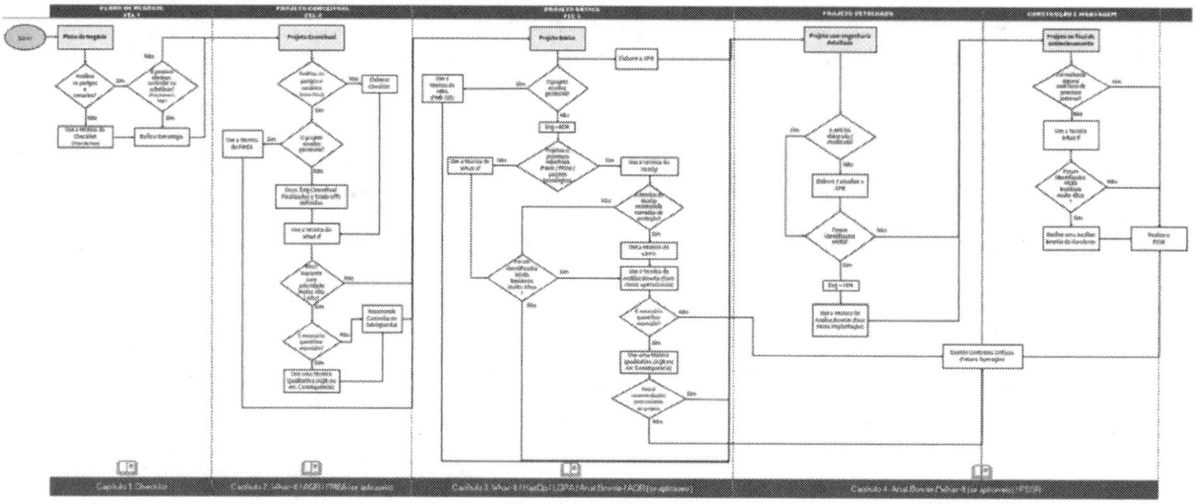

Figura 3.3 – Aplicação Metodologias de Riscos conforme Níveis

Com essa definição, é possível elaborar um book de riscos de processos para projetos, tendo em vista que as metodologias são complementares e são aplicadas a cada avanço de fase do projeto (FEL1, FEL2, FEL3 e Execução) para que a futura operação da Vale receba projetos com todos os riscos de processos mapeados e classificados conforme matriz de riscos da Vale [6] e com seus controles preventivos e mitigatórios monitorados.

Nos próximos capítulos serão detalhadas as metodologias aplicadas nos projetos.

3.1 Checklist

O Checklist é uma metodologia para a identificação de perigos e tem como objetivo identificar perigos ou riscos que podem ocorrer ao longo do desenvolvimento e operação do projeto. Ela consiste em uma série de itens para serem confirmados ou não que possam resultar em cenários de riscos separados por tópicos.

A aplicação dessa metodologia é recomendada ao final de FEL 1, ou seja, com o plano de negócio já finalizado ou no início do FEL 2, antes do início do desenvolvimento da engenharia conceitual.

A aplicação dessa metodologia nessa fase já permite o projeto avaliar a possível eliminação de perigos buscando o projeto inerentemente seguro desde as fases iniciais do projeto. Recomenda-se que a cada fase de desenvolvimento o checklist seja atualizado buscando a eliminação dos principais perigos da futura operação.

3.2 What-If

Após a finalização da engenharia conceitual com a definição das melhores alternativas de engenharia, uma nova análise de riscos de processos deve ser aplicada com objetivo de identificar os cenários de riscos existentes para o projeto, além de avaliar se os perigos e riscos mapeados anteriormente no Checklist podem se tornar cenários de riscos ao longo da implantação e operação.

Portanto, com a participação de um grupo multidisciplinar de trabalho, contanto com pessoas com devida experiência em operações similares à desenvolvida pelo projeto, é utilizado uma dinâmica com a avaliação de riscos utilizando a metodologia de pergunta "E se?" o qual o analista de riscos discute com o grupo de trabalho (representantes da operação, manutenção e equipe do projeto) a possibilidade de ocorrência de cenários de riscos para a futura operação de um determinado ativo.

Tendo em vista que em FEL 2 os projetos ainda não detalham os controles de instrumentação (P&Ids) para os ativos, a identificação é voltada para a identificação de Cenários de Riscos Inerentes, ou seja, sem controles para verificação dos piores cenários de riscos que podem ocorrer no projeto ou na futura operação do ativo.

Portanto para essa fase de FEL 2, somente são elaboradas recomendações de salvaguardas e controles que devem ser estudados e analisados em FEL 3 para os cenários de riscos identificados.

Destaca-se que a metodologia de What-If também pode ser aplicada em FEL 3, com a inclusão de controles nos cenários de riscos residuais para projetos de menor complexidade, quando estiverem com em torno de 70 a 80% de desenvolvimento da engenharia básica na fase.

3.3 HazOp

A metodologia do HazOp visa a identificação de riscos que possam comprometer a segurança, tanto de pessoas quanto das instalações industriais e suas vizinhanças, o meio ambiente, e o desempenho operacional das instalações, e propor ações de mitigação ou eliminação desses riscos.

O HazOp deve ser realizado em FEL 3 para projetos com processos industriais (com base em documentos já emitidos, como por exemplo: P&Ids / PFDs / pacotes tecnológicos) sendo aplicado em torno de 70~80% de avanço da engenharia básica.

A metodologia utiliza uma lista de possíveis desvios operacionais que podem ocorrer durante a operação dos projetos e através da aplicação dos controles previstos, são identificados os 03 tipos de cenários de riscos previstos na Vale, conforme descrito abaixo:

- Cenário de riscos inerente (sem a adoção de controles), geralmente com base nos resultados do What-If executado em FEL 2;
- Cenário de risco residual, com a inclusão dos controles previstos no desenvolvimento da engenharia básica;
- Cenário de risco projetado, caso seja necessária a inclusão de novos controles para que ocorra a redução da prioridade do cenário de riscos (caso o cenário residual identificado posteriormente esteja classificado como muito alto ou alto)

Destaca-se que após a finalização do HazOp pode ocorrer a recomendação de execução de uma análise de riscos de processo complementar para a avaliação das camadas de proteção dos controles de instrumentação, também conhecida como LOPA caso os riscos sejam classificados como altos ou muito altos com a necessidade de confirmação da efetividade da instrumentação planejada.

3.4 FMEA

Atualmente o FMEA - Análise de Modos de Falha e Efeitos, é utilizado pela Vale para avaliação dos modos de falhas que podem ocorrer em projetos geotécnicos e em projetos de descaracterização de barragens.

A metodologia tem como objetivo identificar previamente de maneira sistemática as fragilidades de um sistema, subsistemas e/ou componentes a partir de uma análise criteriosa de possíveis modos de falha, causas e efeitos, com o objetivo de mitigar, eliminar ou controlar os riscos que podem afetar a implantação e futura operação de projetos tanto industriais quanto geotécnicos.

Dessa forma, modos de falha podem ser priorizados para apoiar decisões sobre controles de prevenção ou mitigação.

3.5 LOPA

A metodologia LOPA (Layer of Protection Analysis) ou Análise de Camadas de Proteção é uma ferramenta de avaliação de risco semiquantitativa simplificada para avaliar o valor das camadas de proteção independentes (IPLs) para um cenário de incidente bem definido. Situa-se entre técnicas puramente qualitativas, como a APR (Análise Preliminar de Risco) ou o HazOp (Análise de Perigos e Operabilidade), e técnicas totalmente quantitativas, como a AQR - Avaliação Quantitativa de Risco.

O objetivo do método LOPA é determinar se há camadas de proteção independentes (IPLs) suficientes contra um cenário de incidente (o risco é baixo o suficiente?) e quanta redução de risco cada camada fornece ou precisa fornecer. Para um determinado cenário, apenas uma camada deve funcionar com sucesso para que a consequência seja evitada. No entanto, considerando que nenhuma camada é perfeitamente eficaz, camadas de proteção suficientes devem estar disponíveis para reduzir o risco ao nível desejado.

Atualmente a LOPA é aplicada de maneira complementar ao HazOp realizado caso seja necessária a confirmação de camadas de proteção de controles para cenários de riscos que possam levar a uma ou mais fatalidades ((ou seja, categorias de severidade Grave, Crítica e Muito Crítica, de acordo com a Matriz de Risco da Vale.

3.6 Análise Bowtie

Caso as análises anteriores tenham identificado cenários de riscos inerentes com severidades crítica ou muito crítica nas dimensões pessoas ou meio ambiente ou significativa, crítica ou muito crítica na dimensão financeira, conforme a matriz de riscos da Vale é necessário que os projetos executem a Análise Bowtie para melhor entendimento dos controles preventivos e mitigatórios previstos.

A essência da Análise Bowtie consiste em cenários de riscos possíveis em torno de certos perigos, e maneiras de como a organização se previne de suas ocorrências. O nome da metodologia advém da forma de seu diagrama, que parece uma gravata borboleta.

A análise Bowtie é algumas vezes descrita como a combinação de duas ferramentas de análises de riscos, uma Árvore de Falhas (analisando as causas de um evento) e uma Árvore de Eventos (analisando as suas consequências).

3.7 Análise de Consequências e Vulnerabilidade

A metodologia é aplicada em projetos caso seja necessária a confirmação e avaliação das consequências e severidades dos Cenários de Riscos Inerentes Altos e Muito Altos (FEL 2) ou Eventos Materiais Indesejados – MUEs (FEL 3) e consequentemente auxiliar na qualificação desses cenários de riscos conforme análises realizadas.

Ela tem como objetivo auxiliar na definição de parâmetros de projeto de equipamento, ou seja, parede resistente a fogo e explosão, altura da pilha, exigência de dilúvio de água, colocação de detectores de gás/fumaça, dentre outros.

3.8 Análise Quantitativa de Riscos

Por fim, caso a análises de consequência e vulnerabilidade necessite de confirmação quantitativa dos impactos dos Cenários de Riscos Inerentes e MUEs analisados é recomendado que os projetos executem a AQR.

A AQR é uma abordagem formal e sistemática para estimar a probabilidade e as consequências de eventos perigosos e expressar os resultados quantitativamente em uma medida geral de risco (risco cumulativo) para as pessoas, o meio ambiente e os negócios. Ela também avalia a robustez e a validade dos resultados quantitativos, identificando premissas críticas e maiores contribuidores do risco.

3.9 Monitoramento e Gestão das Recomendações das Análises de Riscos de Processo

Com objetivo de executar uma gestão de riscos de processos eficaz para que os projetos sejam cada vez mais seguros, é de vital importância que a equipe do projeto execute o monitoramento das recomendações elaboradas em cada análise de risco realizada.

Dessa forma, com o avanço do desenvolvimento da engenharia, as recomendações podem se materializar como salvaguardas ou controles que podem reduzir os níveis de prioridade dos cenários de riscos mapeados, tornando a respectiva operação mais segura e robusta.

Assim, ocorre a redução gradativa dos níveis de prioridade de cenários de riscos conforme o avanço das fases até que os controles sejam efetivamente testados e avaliados na operação dos novos ativos entregues pelos projetos.

Na figura abaixo, é mostrado um resumo de quais metodologias devem ser aplicadas conforme as fases do projeto / ativo e como é esperado que ocorra a redução de prioridades dos cenários de riscos conforme vai ocorrendo a implantação e monitoramento das recomendações que são elaboradas pelas análises de riscos:

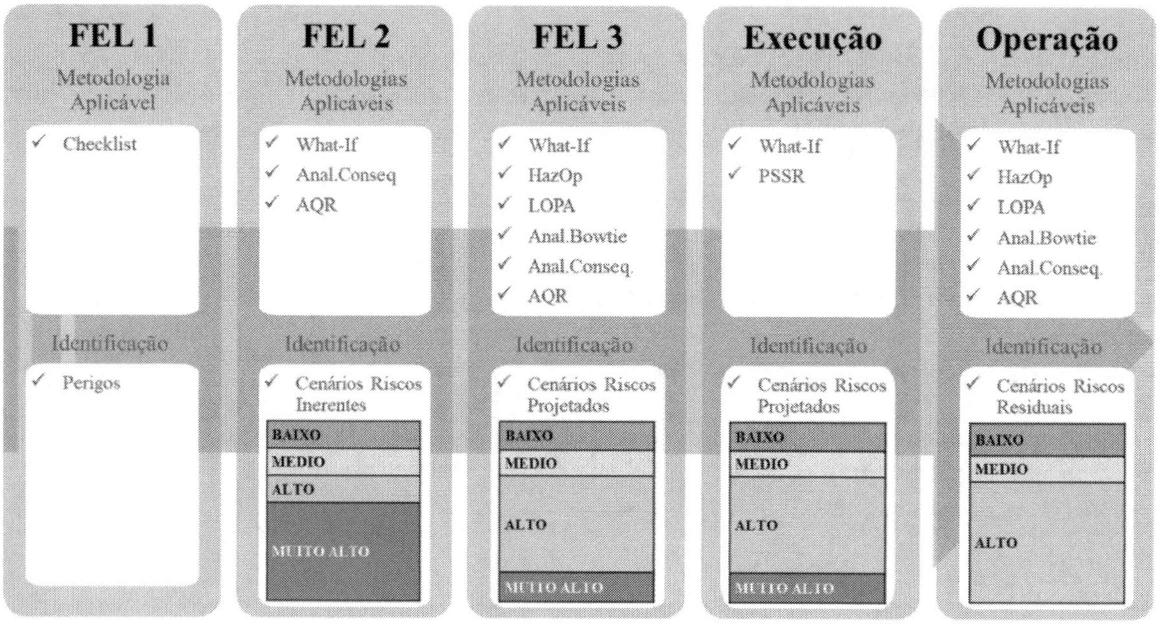

Figura 3.4 – Análises de Riscos de Processos e Cenários de Riscos

4 Conclusão

As análises de riscos de processos apresentadas são complementares e caso sejam executadas nos momentos adequados, podem contribuir para a identificação e monitoramento dos riscos de processos para tornar os projetos cada vez mais seguros para operar.

Elas devem ser executadas conforme necessidades identificadas em análises anteriores com objetivo de evitar o desperdício de energia da equipe de projetos e de custos para realização de análises de riscos de processos caso não sejam necessárias conforme nível de informação existente em projetos.

Dessa forma, a execução sistemática das análises pode proporcionar a elaboração de um livro de riscos que deverá ser avaliado pela futura operação no momento da entrega do projeto com objetivo de avaliar se os cenários de riscos de processos foram adequadamente identificados e se os controles planejados estão eficientes e eficazes para evitar a ocorrência desses riscos e seus impactos durante a operação.

De forma resumida a figura abaixo mostra como as análises de riscos de processos podem ser executadas para a formação desse livro de riscos de processos:

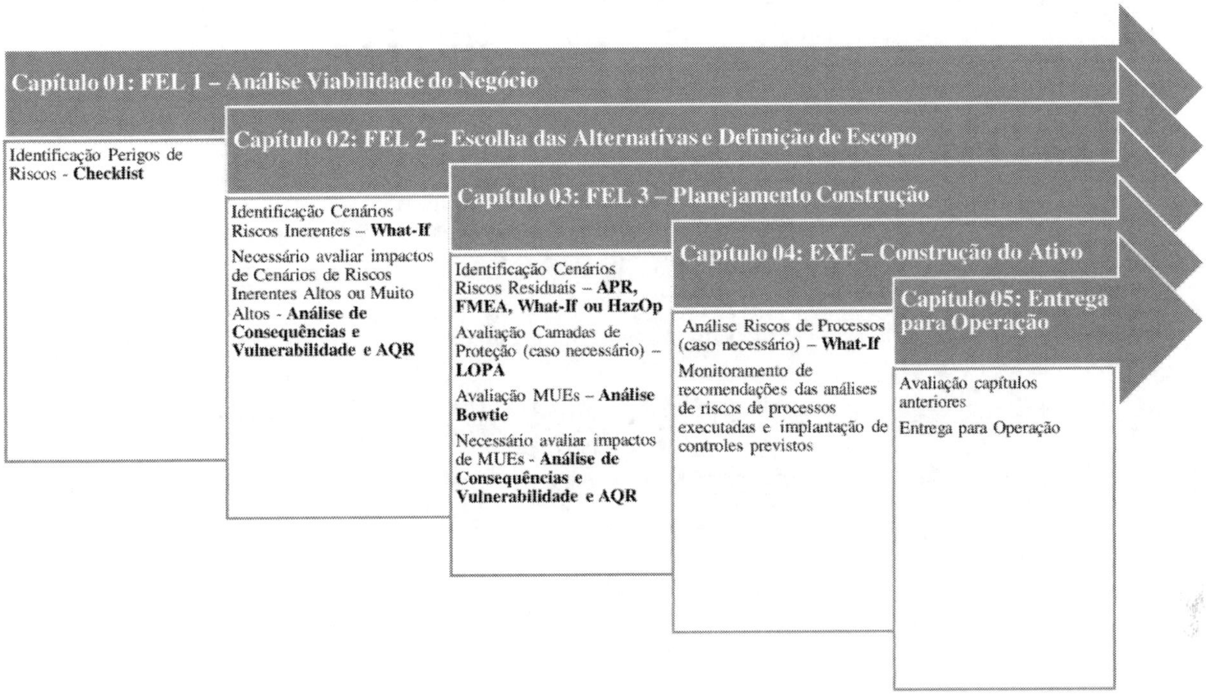

Figura 4.1 – Book de Riscos de Processos para Projetos

5 Referências

[1] PROJECT MANAGEMENT INSTITUTE (PMI). Guia do Conhecimento em Gerenciamento de Projetos (Guia PMBOK). 7 ed. Estados Unidos: PMI, 2021.

[2] AIChE. 2008-07 - Beacon-Portuguese (Brazil). https://www.aiche.org/sites/default/files/2008-07-Beacon-Portuguese%20%28Brazil%29.pdf. Acessed December 29, 2023

[3] Vale, MA-E-240, Manual de Segurança de Processos em Projetos, novembro de 2023

[4] CCPS / AIChE – Guidelines for Risk Based Process Safety, 1 ed, november-2011

[5] Vale, PNR-000192, Diretrizes Gerais para Análise de Riscos de Processo, novembro de 2022

[6] Vale, NOR-0003G, Norma de Gestão de Riscos, dezembro de 2022

Dynamic Process Safety Barriers Management at Petrobras: Challenges for Data Integration

Presenter/Principal Author
Sandro da Cruz Malvão
Petrobras S.A.
smalvao@petrobras.com.br

André Rodrigues Dantas
Petrobras S.A.
andrerd@petrobras.com.br

Other Author(s)
Fabio Leandro Rossi
Petrobras S.A
flrossi@petrobras.com.br

Keywords: Dynamic Process Safety Barriers Management, Process Safety, Bow Tie, Risk Management, Data Integration, Cyber security

Abstract

Dynamic Process Safety Barriers Management is a systematic approach to monitoring the integrity and reliability of preventive and mitigating barriers in industrial facilities. It involves defining responsibilities and consolidating information across all levels of the organization to actively prevent accidental scenarios.

Since 2022, Petrobras started implementing a software to support this process across various facilities in the upstream and downstream segments.

This paper details the main challenges related to data integration with existing informational data sources within the company, explaining the strategy used for building each end-to-end integration and using some use cases to illustrate the difficulties evolved.

Furthermore, the article explains how Petrobras addresses cybersecurity concerns like access control, secure data storage and communication. It also highlights the main challenges in maintaining the software, not only its availability and user's support but specially with more updated and accurate data, returning a more powerful tool for risk management and decision making.

1 Introduction

Dynamic process safety barriers management is a systematic monitoring of the integrity and reliability of preventive and mitigating barriers, defining responsibilities, and consolidating information across all levels of the corporation – operational, tactical, and strategic. This proactive process actively and dynamically contributes to the prevention of possible accidental scenarios.

Monitoring the status of process safety barriers in industrial facilities enables proactive measures to maintain their integrity and manage the risks of the facility. Implementing a software and keeping it up to date with accurate data from internal systems are crucial to correct support this process.

This article presents the challenges of implementing a solution for automatic and dynamic update of process safety barriers, aiming to provide an integrated and near real-time view of the state of these barriers in Petrobras facilities.

2 Business Context

2.1 Project History

In 2019, Petrobras began specifying requirements for implementing Dynamic Process Safety Management using the Bow Tie methodology. Throughout 2020, Petrobras carried out a market consultation (Request For Information) to identify possible suppliers of commercial software solutions to support this process. Tests were carried out to ensure the effectiveness of these software. In 2021, Petrobras carried out a competitive bidding process to contract the software tool with implementation services. Throughout the period 2022-2023, the project has been implemented in several industrial facilities.

Some of these project implementation phases were previously presented at events such as (AICHE/CCPS [1]. The project involved approximately 100 professionals, both internal and external, organized into diverse multidisciplinary teams. These teams were responsible for tasks ranging from organizing technical documents and converting them into bow ties to software loading, consolidation, and displaying data in the software. Over approximately two years, 32 installations were integrated into the system, with an additional 14 installations planned for incorporation throughout 2024.

2.2 Barriers Management x Dynamic Barriers Management

Safety barrier management consists of identifying, assessing, and controlling the risks associated with industrial activities that can cause serious accidents, such as explosions, fires, or leaks. Security barriers are technical, organizational, or human measures aimed at preventing, detecting, or mitigating these undesirable scenarios. Safety barrier management involves defining the roles and responsibilities of each barrier, monitoring its performance, and implementing corrective actions when necessary.

The dynamic management of safety barriers is an evolution of traditional management, which seeks to make the process more agile, efficient and integrated, using information and communication technologies. Using data integration mechanisms, it is possible to automatically update the status of barriers in real time, from diverse and reliable data sources, such as sensors, alarms, control systems, reports and inspections. In this way, it is possible to obtain an analytical view of data, through dashboards, indicators and alerts, which help in decision-making and prioritization of actions.

3 IT Context

3.1 Data Integration Architecture

The data integration architecture, shown in figure 1, follows standards commonly adopted by the market and adheres to the state-of-the-art technologies adopted by Petrobras for this type of solution.

Considering the mapped internal data sources (A) and their technologies – Oracle for databases, and REST APIs, for services, it was defined that the model would follow a first data flow, where the information goes from the Petrobras environment to the Supplier's environment using the GLUE tool[2] – AWS solution for ETL activities (Extraction, Transformation and Load)[3] which stores the captured data in the Staging Area (B).

The staging environment is an intermediate database that enables processing and cleaning before final consumption by the application. In it, business rules are applied through SQL scripts [4] (Universal Language for Querying/Updating in Databases) for subsequent transfer to the application database (C).

All this processing is orchestrated by a complex structure of Jobs(D) or programs that run daily automatically and culminate in the updating of data and the preparation for consumption of users in the cloud-based web application – Dynamic Barrier Management (DBM) Solution (E)

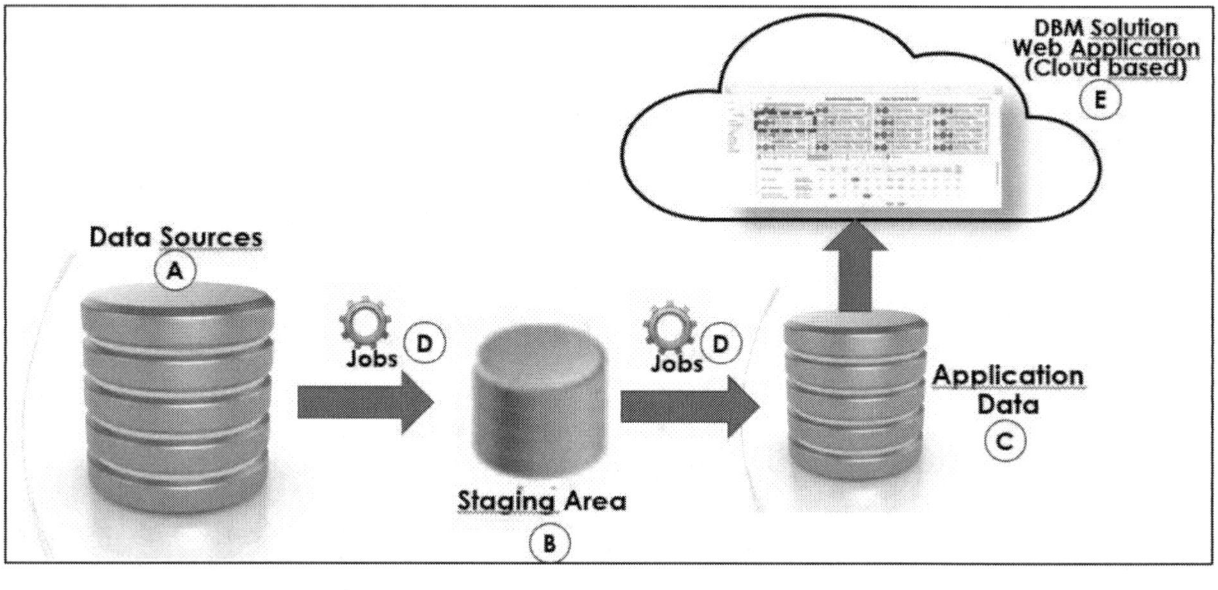

Figure 1. Data Integration Architecture

3.2 Integration Steps

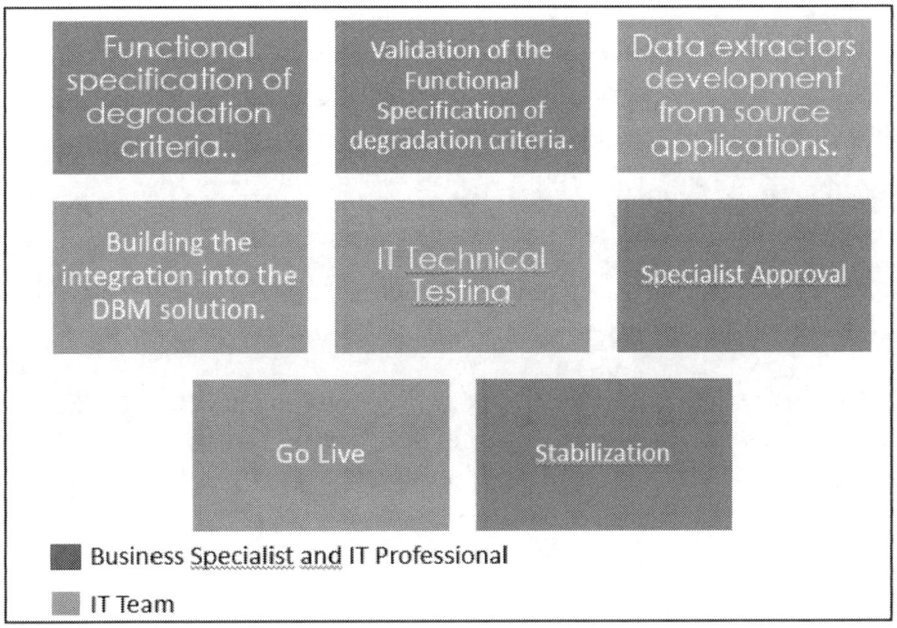

Figure 2. Integration Steps

The data integration process, as shown in figure 2, was partitioned into 8 steps, with their respective responsible (IT and Business) in each of them:

3.2.1 Functional specification of degradation criteria

In this step, the main objective is to define what the degradation criterion is, that is, what information will be evaluated to define the degradation or not of a barrier element. Along with this definition, it is evaluated in which data source this information is contained and how we can obtain it more correctly and quickly.

3.2.2 Validation of the Functional Specification of degradation criteria

At this stage, the main goal is to record all the knowledge obtained in the previous step, adding all the IT details that are necessary for the correct capture in order to meet the business definition for the degradation criterion. This record is formalized in a document that describes all the details so that the business specialist is sure of what has been defined as the degradation criterion and the IT team is clear about where and how to obtain the data at the time of construction.

3.2.3 Data extractors development from source applications

The 3rd stage aims to build the data extraction programs, as specified in the previous step, considering security issues, such as formalization/approval of data access, data quality issues, and execution performance so that the integration is as efficient as possible. As a good practice in the search for a reduction in the volume of data and greater agility in loads, the strategy of searching only for "degradation" data was adopted, that is, records that do not impact the status of the elements/barriers are not captured, thus avoiding unnecessary data traffic between the Petrobras and supplier networks.

3.2.4 Building the integration into the DBM solution

In this stage, the objective is to create routines in the DBM solution environment that organize the execution of the programs built in the previous stage, ensuring the effective transfer of Petrobras data to the solution, in a secure way and within a processing time appropriate to the updating needs expected by the business.

3.2.5 IT Technical Testing

The next step in the process includes the technical IT tests. They encompass the comparison of the source data predicted in the integration with that obtained at the input layer in the DBM solution environment. As an example, the number of records communicated, data samples, and processing time are common tests during this step.

3.2.6 Specialist Approval

In addition to the previous step, in this phase of the process, the business specialist analyzes the result of the integration, that is, through the data captured from the source of origin, it is evaluated whether the degradation criteria were correctly applied, reflecting on the status of the barrier (degraded or intact). A good practice at this stage was the use of real data, even though the validation takes place in a separate environment from the Productive one.

3.2.7 Go Live

Once the previous step is completed, the next step is the application of all the programs built and adjustments identified in the previous steps in the Production environment so that the new integration starts to reflect correctly in the relevant analyses and scenarios. As a result, the solution gains greater reliability with the automatic acquisition of data and improves the quality of analysis of software users.

3.2.8 Stabilization

The last step is the stabilization of the integration, where the quality of the data, runtime and, eventually, small corrections not found in the previous steps are evaluated. This step is carried out through integrated daily monitoring of the IT and business teams that effectively use the solution and lasts, on average, 30 days after integration's go live. Errors found after this time are addressed with tickets to the support team.

3.3 Use case – Integrating with Oil Platform Control System (APLAT)

One of the management elements evaluated as a barrier is the control of the training of the operation teams in the facilities in the respective operating standards. For some offshore units, this control is carried out in the Platform Control System (APLAT) application – Petrobras internal development. The following details all the steps followed for the construction of this degradation criterion, identified as "Training in System Operating Procedures".

3.3.1 Functional specification of degradation criteria

As previously described, this step defines the degradation criterion and how to search for it in internal data sources. In the example in question, we have the following criterion:

Training in System Operating Procedures?

As a result of the analysis and interactions between business experts and the IT team, we have the following result:

Comprehensiveness of the Criterion
- E&P – 4 offshore units.

Data Source
- Data on whether the onboard professional is trained according to the operating procedures of the systems will be obtained from the APLAT system.

Monitored Element
- The monitored element will be the onboard professional verifying that it is trained according to the operating procedures of the systems.
- It will be verified if the professional has performed the most up-to-date version of the training.
- The list of criterion trainings by platform is defined in Excel spreadsheets

Element Degradation Rule
- If the onboard professional has completed the training, if it is approved and within the validity period, then the element will be considered "available".
- If the onboard professional has any training planned for him in arrears, then the element will be considered "unavailable".

- If the onboard professional has completed the training provided for him/her, but is not approved in the last one, then the element will be considered "unavailable".
- If the onboard professional has not completed one of the trainings, then the element will be considered "unavailable".

Barrier Degradation Rule

- The scope of this Degradation Criterion for the relevant barriers.
- The barrier is considered "available" (green) if all elements are available.
- The barrier will be considered "Unavailable" (red) if at least one of the elements is unavailable.

Data made available in the DBM Solution for users to consult

- Installation (platform)
- Professional Registration (SAP)
- Name of the professional
- Training Name
- Training validity (days)
- Code of the PAdrão
- Version of Pattern
- Date of completion of the practitioner's last training
- Have you passed the training? (YES/NO)
- Practitioner Training Status (CANCELED/COMPLETED/SCHEDULED/PROGRESSED)
- Professional Training Cycle Number
- Embarkation date
- Expected disembarkation

With this data, the final document begins, which fully specifies the criteria and how the IT team should build it.

3.3.2 Validation of the Functional Specification of degradation criteria

In this phase, the teams work on a new document enriching the definitions obtained in the previous phase. In it, fundamental details are added for the correct understanding of the definition of the criterion, business rules, data sources and fields to be considered by the IT construction team. Figure 3 shows a graphical and more intuitive view of the degradation criterion decision flow. Figure 4, on the other hand, exemplifies a part of the consultationwith the database that allows the extraction of the information necessary for the criterion. Finally, figure 5 illustrates the mapping of the data extracted from the APLAT system to the correct load in the target system (DBM Application).

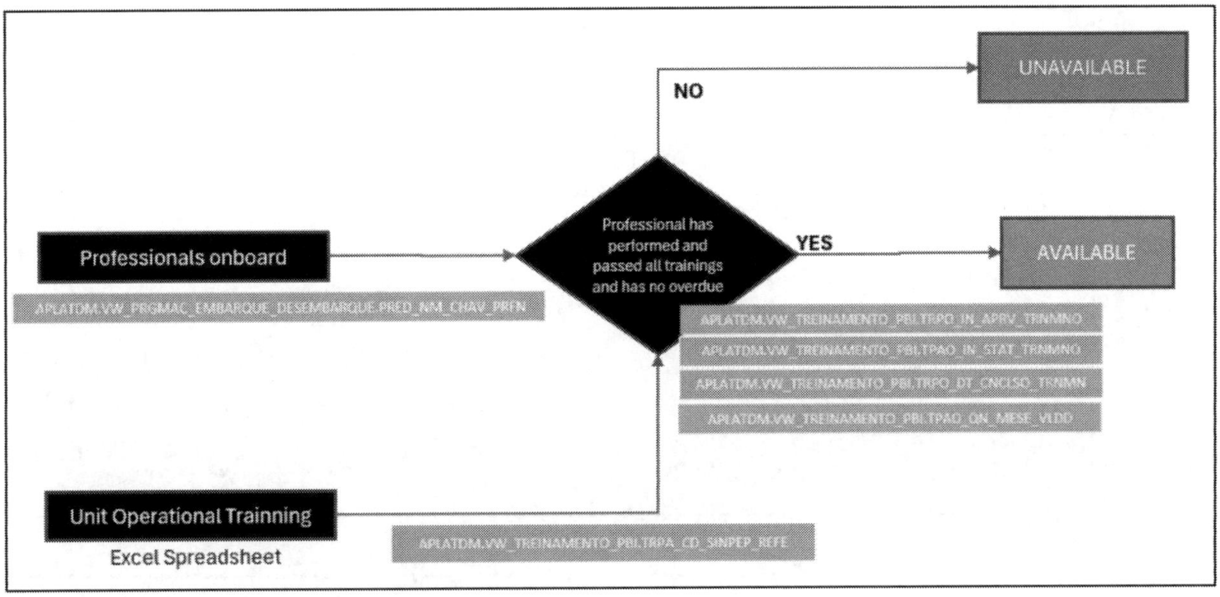

Figure 3. Graphic representation of degradation criteria

```
SELECT
e.pred_sg_area_oprl, -- PLATAFORMA
e.pred_nm_chav_prfn, -- CHAVE DO PROFISSIONAL
e.pred_nm_prfn,  -- NOME DO PROFISSIONAL
e.pred_nm_empr, -- NOME DA EMPRESA
e.PRED_CD_EMPR, -- CODIGO DA EMPRESA
e.PRED_IN_PRGMAC_EMBRQE_DSMBRE, -- PREVISAO DE EMBARQUE/DESEMBARQUE
e.PRED_DT_EMBRQE, -- DATA EMBARQUE
e.PRED_DT_DSMBRE, -- PREVISAO DESEMBARQUE
t.trpa_cd_sinpep_refe, -- PADRAO
t.trpo_in_revs_padr, -- REVISAO DO PADRAO
t.TRPA_NM_TRNMNO_PAD, -- NOME DO TREINAMENTO
t.tpao_in_stat_trnmno, -- STATUS DO TREINAMENTO
t.avpr_in_stat_agndmo, -- STATUS AGENDAMENTO
t.trpo_dt_cnclso_trnmno, -- DATA DE CONCLUSÃO DO TREINAMENTO
t.trpo_in_stat_trnmno_prfn, -- STATUS DO PROFISSIONAL NO TREINAMENTO
t.trpo_in_aprv_trnmno, -- APROVAÇÃO NO TREINAMENTO
t.tpao_qn_mese_vldd, -- VALIDADE EM DIAS
t.vcaq_nr_ciclo -- CICLO
FROM APLATDM.vw_prgmac_embarque_desembarque E
INNER JOIN  APLATDM.VW_TREINAMENTO_PBI T ON TRIM(E.PRED_NM_CHAV_PRFN) =
TRIM(t.PRFN_NM_CHAV_PRFN)
WHERE E.PRED_IN_PRGMAC_EMBRQE_DSMBRE IN ('EMBARCADO', 'PRÉVIA DE EMBARQUE')
AND TRIM(E.PRED_CD_EMPR) = 1771 -- EMPRESA PETROBRAS
AND TRIM(E.pred_sg_area_oprl) = 'P-43' --- PARAMETRIZAR CONFORME PLATAFORMA
AND (t.tpao_in_stat_trnmno = 'ATIVO')
AND (t.avpr_in_stat_agndmo <> 'CANCELADO')
AND (((t.tpao_qn_mese_vldd is not null  AND  (t.trpo_dt_cnclso_trnmno +
t.tpao_qn_mese_vldd) < trunc(sysdate)) - vencido
     OR(t.tpao_qn_mese_vldd is null AND t.trpo_dt_cnclso_trnmno IS NULL) --
```

Figure 4. SQL Query for data extraction

APLAT Field	DBM Solution Field	Description
APLATDM.VW_PRGMAC_EMBARQUE_DESEMBARQUE.PRED_SG_AREA_OPRL	SITE_CODE	Instalação
APLATDM.VW_TREINAMENTO_PBI.TRPA_NM_TRNMNO_PAD	COMPETENCY_ROLE	
APLATDM.VW_PRGMAC_EMBARQUE_DESEMBARQUE.PRED_NM_CHAV_PRFN +	PERSON_NAME	Chave + Nome
APLATDM.VW_PRGMAC_EMBARQUE_DESEMBARQUE.PRED_NM_PRFN		
APLATDM.VW_TREINAMENTO_PBI.TRPO_IN_STAT_TRNMNO_PRFN	PERCENTAGE_COMPLETE	Status do Profissional no Treinamento

Figure 5. Data mapping (from data source to target system)

3.3.3 Building the integration into the DBM solution & IT Technical Testing

In this phase, the teams work on the construction of data capture and technical validation programs in the IT vision. The DBM solution has reports that allow you to verify the data that arrives in your environment before any processing. This allows you to compare with the source data source and evaluate the quality of the solution. Figure 6 shows a sample of data captured by DBM application side. Figure 7, on the other hand, illustrates the sample of the source data, which is essential for the performance of the technical tests.

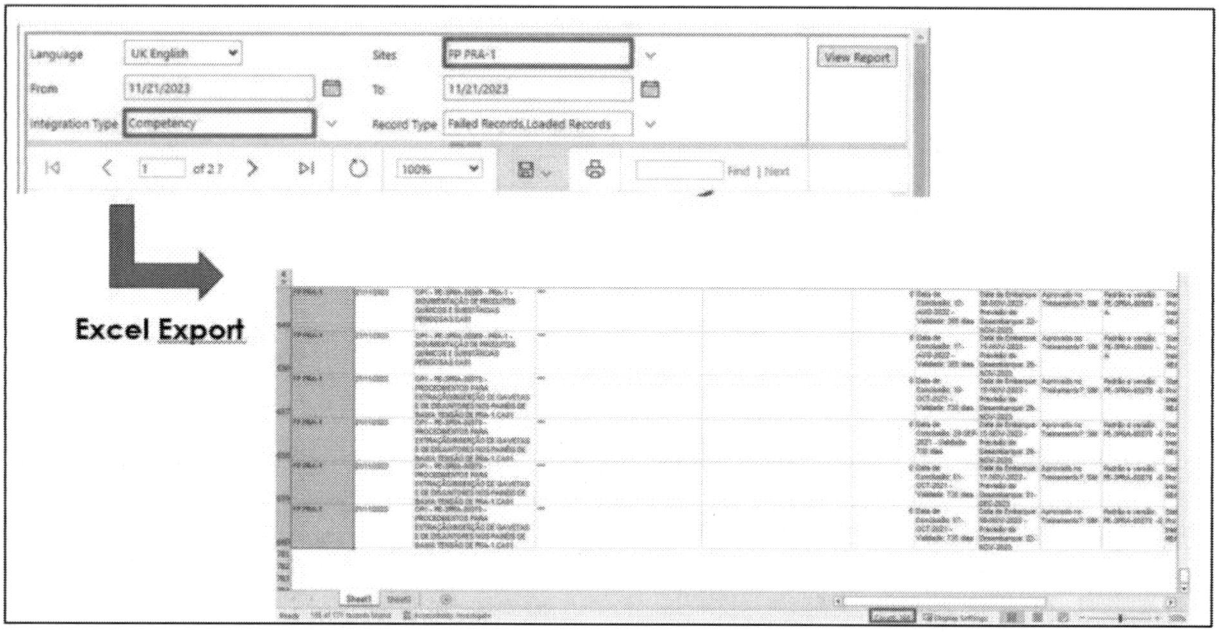

Figure 6. Data acquisition (DBM - Target system)

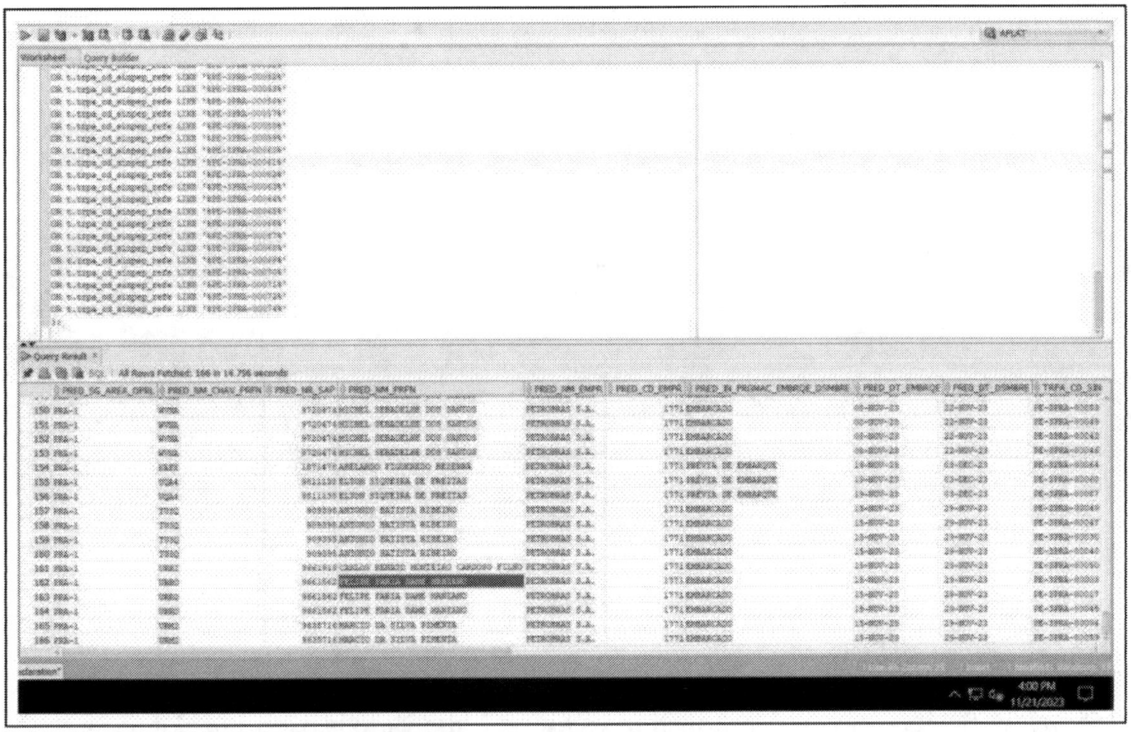

Figure 7. Data Source (APLAT)

In this example, the teams compare the results of data output from Petrobras with those arriving in the DBM solution.

3.3.4 Approval by the customer, Go Live, Stabilization

After the construction and IT testing, the final steps are the validation of the business experts and delivering to the Production environment. The validation of the experts consists of the complement of the technical validation of IT, that is, the results are evaluated directly in the DBM solution, where it is evaluated whether, considering the data set of the data source, the elements and barriers behave as expected. Figure 8 shows an example of DBM Application report used for customer review and approval.

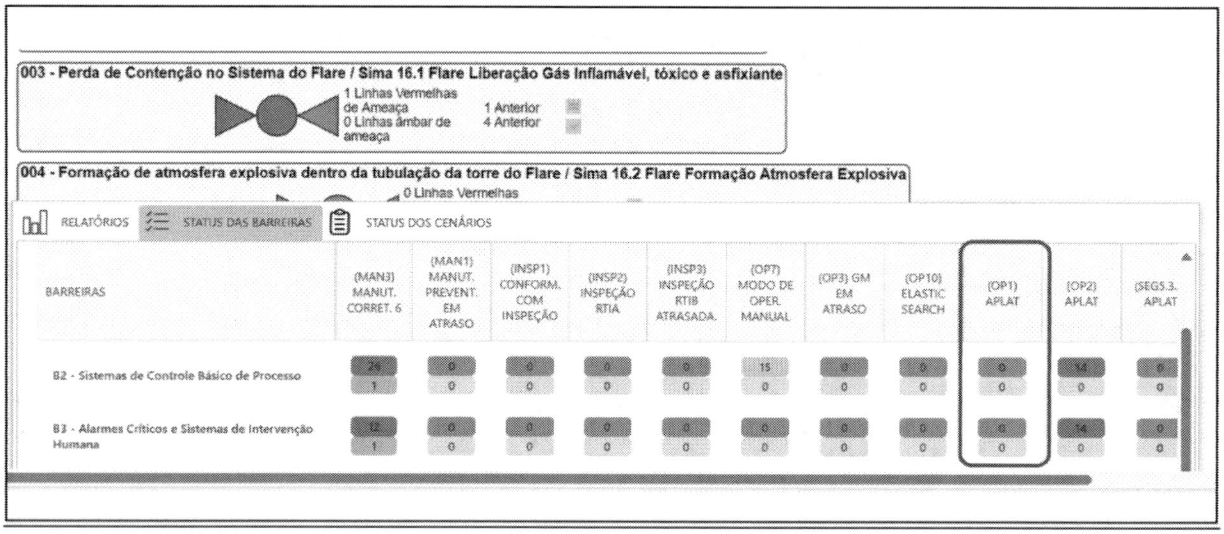

Figure 8. Specialist Validation example

3.4 Data Integration Challenges

Despite the well-defined process, as described in section 3.2, the various executions of these steps for each integration showed a variety of challenges listed below:

- Definition of the degradation criterion – aligning concepts and criteria for multiple installations can be challenging.

- Business process assessment – the same process may vary between business areas and/or facilities.

- Technical mapping of tables/fields – in several cases, a degradation criterion required a query on multiple tables, eventually combining more than one data source, bringing complexity to the integration.

- Construction of programs for data exposure – some data mappings generated complex programs, involving queries to several data sources, and with low efficiency in the execution time.

- Data transfer to the DBM solution – the multiple data loading and processing processes bring the challenge of ensuring execution in the correct order, within the time frame available for update and with the appropriate performance so as not to cause delays in the availability of data to end users.

- Calculation and consolidation of element/barrier statuses – the business rules that define the status of elements and barriers can be highly complex and require several tests to ensure the integrity of the results.

3.5 Integrations big numbers: one day at DBM application operation

Considering the current deployment in 32 facilities, one day in operation of the DBM solution represents a great computational effort and a robust end-to-end process monitoring scheme. The following are the key figures:

- 30 systems consulted.
- About 100 tables queried.
- About 1,000 columns extracted.
- About 30,000 Maintenance Orders/Maintenance Plans.
- Nearly 4,000 records of Alarms / PI(Aveva) [5].
- About 10,000 people-related records (up-to-date process compliance check and training performed).
- About 300,000 pieces of equipment monitored.
- About 100 integration jobs.
- Daily load processing of 1h:30.
- Daily update of Dashboards and reports in the DBM solution in 40 minutes.
- 80 business rules/calculations applied.

3.6 Data Integration Status Monitoring

With all the complexity and high volume of processing, varied data sources and components such as network connectivity or server unavailability, it was necessary to ensure constant monitoring, not only to identify and recover from any failures, but also to ensure timely reporting to end users in case of failure in any planned integration. This is essential to avoid wrong decisions based on outdated or non-existent data.

As part of this monitoring solution, a report was deployed within the DBM software itself – it is available for consultation by end users, bringing status information for each load, whether from the current day or from a period in the past.

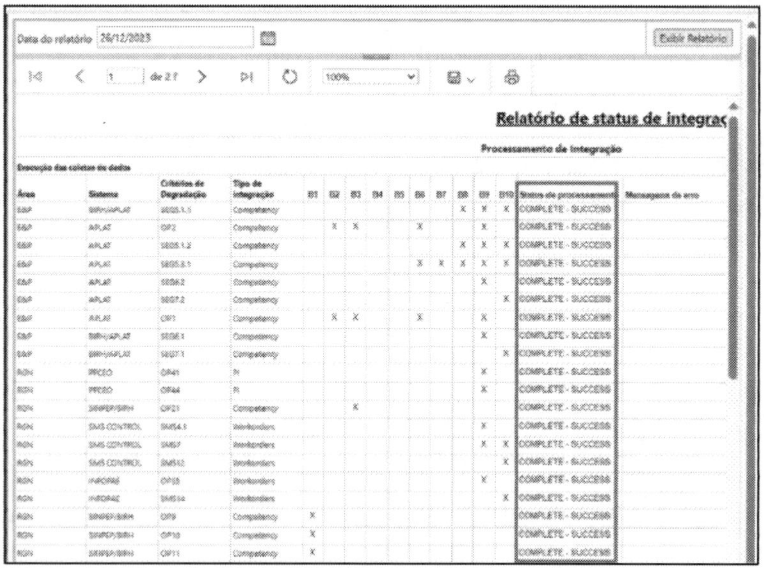

Figure 9. Data Integration Status Report

The report in figure 9 can be exported to Excel, with multiple tabs of varied information that allow a more complete analysis of the complete process, from the output of Petrobras data to receipt and processing on the supplier side. More consolidated dashboards and views for management analysis are planned for future releases.

4 Cyber security

Within the complex context of integrations, diversity of data and forms of access, the discipline of cyber security has been a concern since the conception of the integration architecture. Ensuring security in the connectivity between Petrobras' network and the supplier's network, using secure and standardized protocols during data transmission and establishing access control for end users in an integrated manner with Petrobras standards were the main challenges addressed throughout the implementation. As a result, a specific security architecture was designed considering all the characteristics of this scenario.

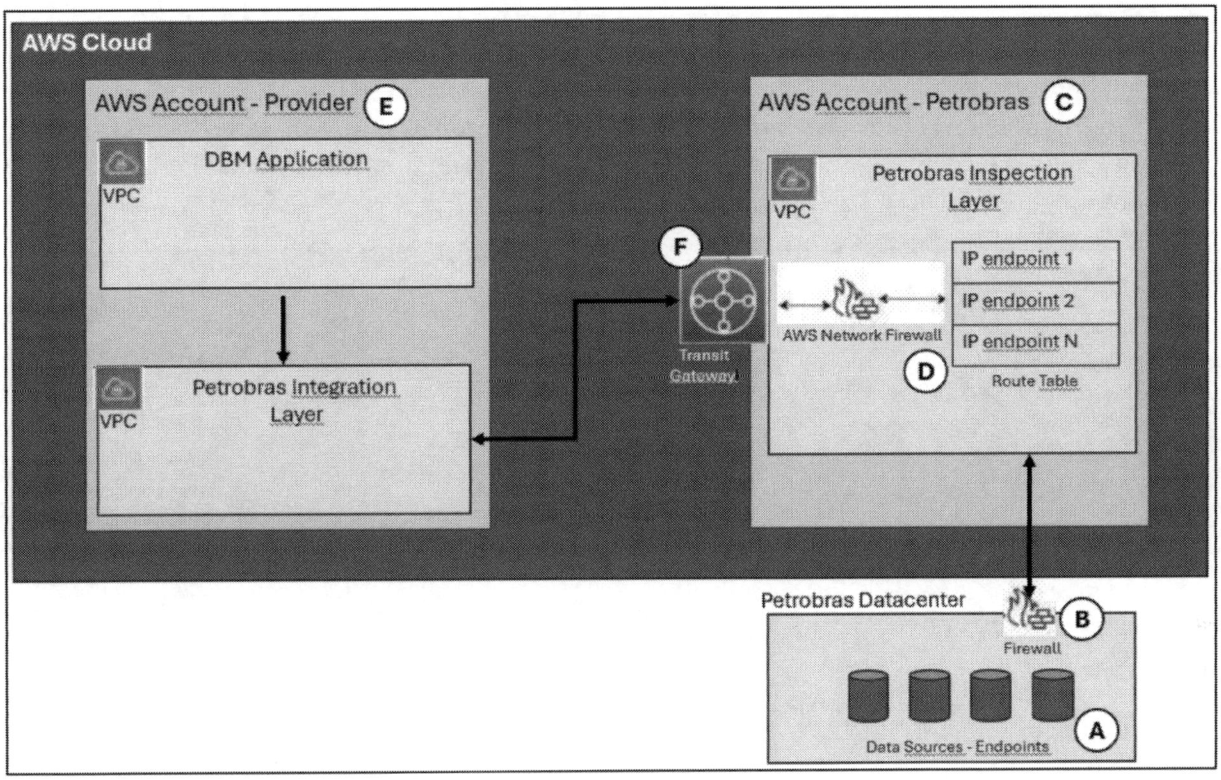

Figure 10. Data Integration with Cyber Security components

4.1 Secure Data Communication

Within the overall cyber security architecture, some components have been designed and organized in such a way as to ensure secure end-to-end communication. Following the diagram in figure 10, the secure communication process is as follows:

- Petrobras data sources are mapped individually – endpoints **(A)**
- Petrobras network firewalls **(B)** are configured to enable communication from each endpoint with Petrobras account in the AWS**(C)** cloud using a private network (VPC)
- In the Petrobras cloud, another set of firewalls receives the configuration of each endpoint **(D)** to enable sharing with the vendor's AWS cloud where the DBM application runs **(E)**
- To establish communication between the two AWS accounts (Petrobras and Supplier), a standard component is used - Transit Gateway **(F)** - which ensures the secure connection between the Petrobras private network and the Supplier's private network.

4.2 Secure Access Control

Another part of the Cyber Security architecture has the mission of ensuring secure access for end users and integrated with authentication mechanisms already implemented at Petrobras. The diagram in Figure 11 demonstrates the components:

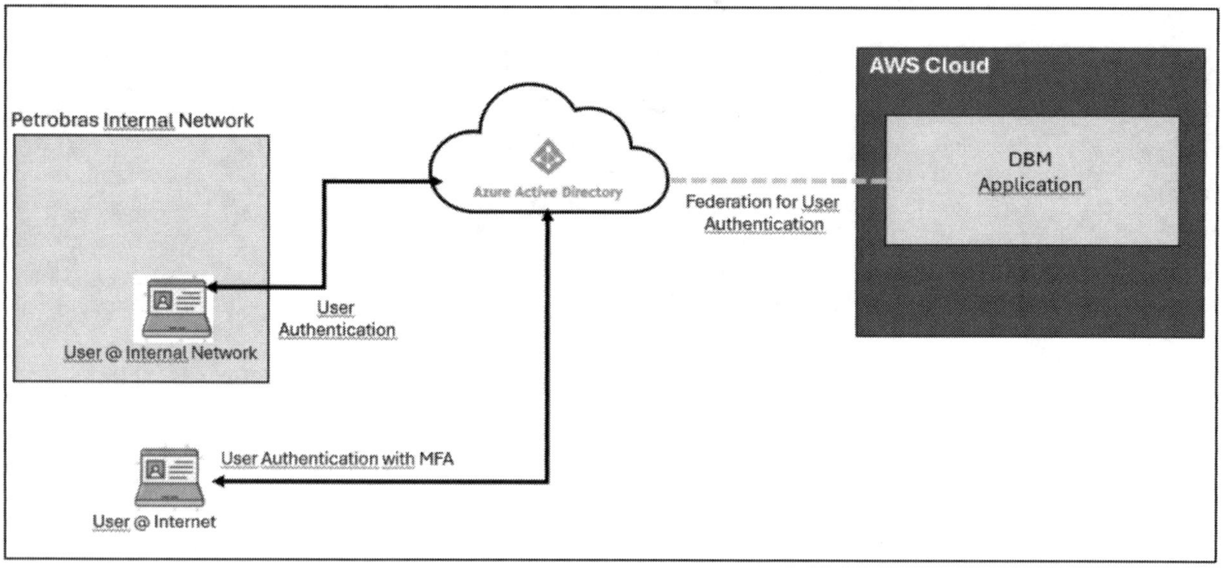

Figure 11. User Authentication diagram

The strategy adopted was to use a standard authentication solution already tried in similar solutions – Through the federation of the supplier's cloud with Petrobras' MS Azure Active Directory cloud service, it was possible to ensure the correct authentication of users using the same credentials used to access Petrobras' internal resources and systems. The solution of access directly from the internet, using the multi-factor authentication (MFA) method, was also planned.

5 Error Support and New Features

5.1 Error Support

With the advancement of DBM application deployments in dozens of installations and the greater complexity involved with the various data integrations, it was essential to adopt a complementary solution to offer the necessary support to system users in case of errors or questions. Therefore, the standard ITSM (IT Service Management) platform was adopted, where users are already accustomed to requesting support in solutions provided by IT.

The procedure consists of opening a ticket in some of the available channels (phone, web interface, whatsapp or virtual assistant) reporting the error or doubt. The generated ticket is then followed up by the team at level 1 of service and forwarded to specialized teams, including the supplier, depending on the type of problem found. Through the ticket registration system itself, the user is updated on the progress, elapsed time and resolution of their problem.

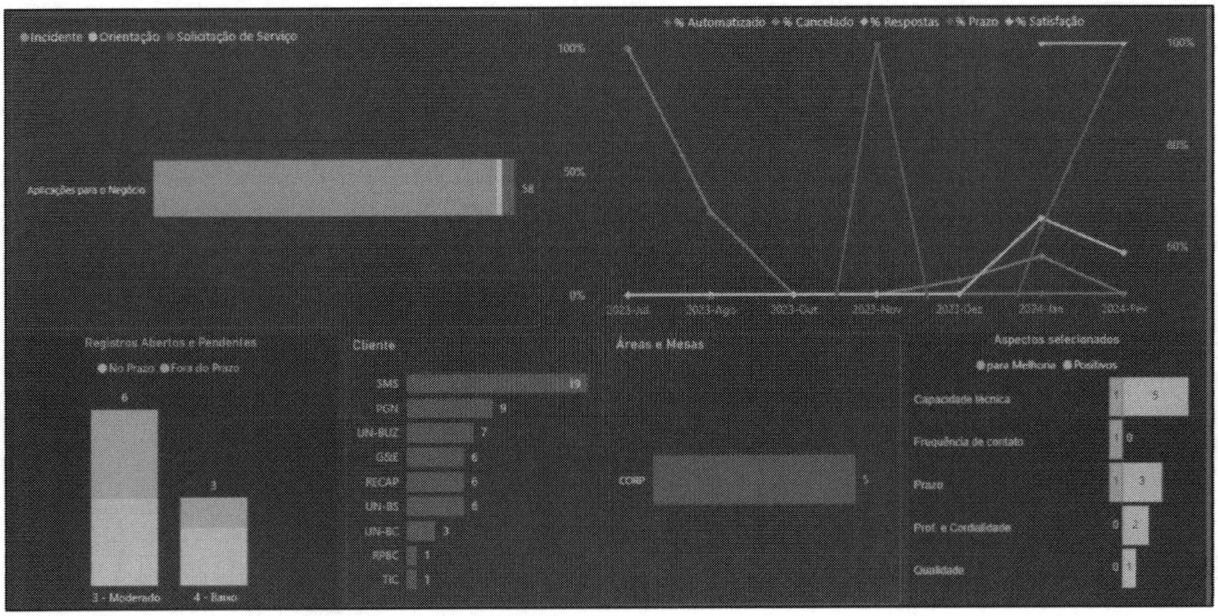

Figure 12. User Support Dashboard

In addition, the ticket registration system is linked to dashboards that allow various analyses, such as average ticket resolution time, number of tickets solved or pending resolution, and the volume of open tickets over time. This data allows for better management of the support process and adjustments of procedures with internal and supplier teams. Figure 12 demonstrates examples of analysis available in dashboards.

5.2 New Features

As well as supporting errors and doubts, it was necessary to structure an IT governance and solution to capture requests for improvements or new features requested by users and go through the stages of evaluating each request and deciding whether to forward it to IT team to provide the development of the approved improvement.

The solution adopted was the corporate system for requesting services, which is widely used at Petrobras. Service offers have been created, where the user can choose the type of improvement/functionality he wants and fill out a form with more details and attached documents that bring as much information as possible to facilitate the correct understanding and importance of the suggested improvement. Figure 13 illustrates an example of a new feature registration form.

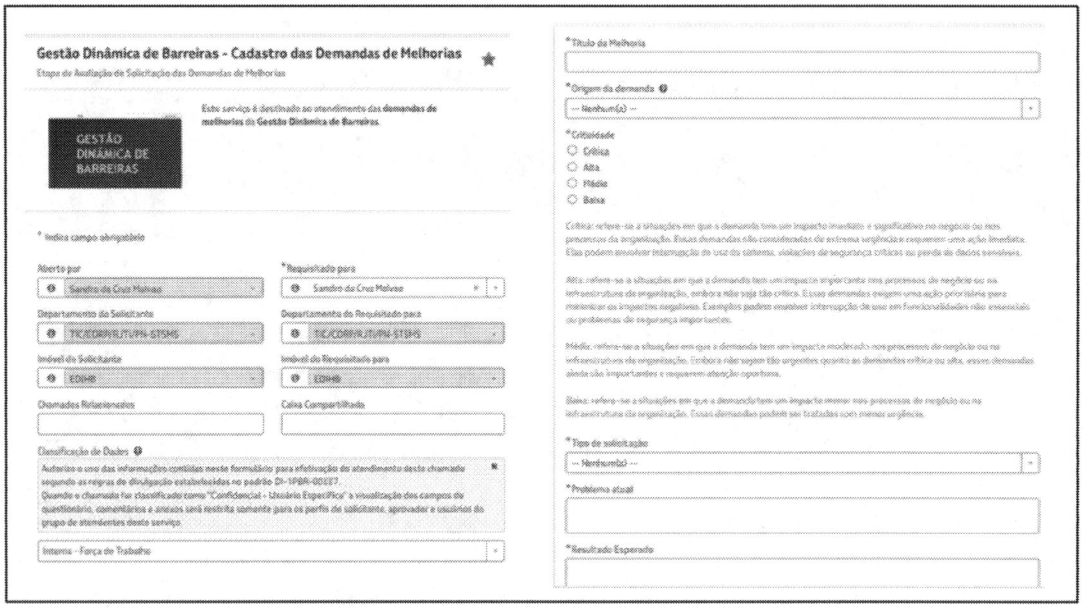

Figure 13. Example of New Feature Register Form

From the opening of the service, the application follows a programmed workflow so that the evaluation procedures provided for in the improvement governance are followed, that is, all the areas involved (locally and globally) make their evaluations and register the approval or disapproval of the improvement suggestion. All steps and advances are recorded in the service, giving visibility to the requesting user and transparency to the process. Approved and prioritized enhancements/features are sent to the vendor for IT review. After defining the effort and service expectation, the implementation planning of the new improvement is informed to the user, who can follow the process until the final delivery in the DBM application.

6 Next Challenges

With the BDM application increasingly widespread in the company and with the high volume of integrated data, it is expected that the main challenge is to ensure high availability and minimal failures in data integrations so that there is no loss of accuracy in the analysis of the scenarios available in the system. The control of availability and tracking of scheduled maintenance of the source systems also represent challengesso that the integrations are prepared for shutdowns or adaptations that the systems may have.

Another important challenge will be the continuous review of the integration architecture, always seeking optimization and simplification in data capture and processing, ensuring the use of state-of-the-art technology, with high performance and compliance with cyber security requirements.

Finally, the evaluation of new features that add value to the current DBM solution is highlighted. As an example, the creation of management dashboards, at tactical/strategic levels, which allow broader analyses of the company's risk situation. Another example would be the use of AI to enrich the data generated by the DBM application or to interpret and capture changes in bow ties from risk studies.

7 Conclusion

The results of this work show that the definition of a robust and flexible integration architecture is essential to ensure the coverage of the various types of data sources mapped for consumption in the DBM application. In the same way, it is a success factor to establish a complete process of specifying data integration, from the capture of business requirements, through the definition of the data source and how to obtain the information, to the final verification of users as to the expected results.

On the cyber security side, the adoption of market standards and cloud solutions from the main players in the IT market proved to be the best option to meet the security challenges in communication and data hosting and in user access to the DBM application.

Finally, structuring a solution that is easy to adopt and in compliance with the governance of the tool's deployment was crucial to ensure effective support to the system's users and to capture opportunities for improvement pointed out by the community that adopted the DBM application.

8 References

[1] Dantas, André Rodrigues. Et al.: "Dynamic Process Safety Barriers Management at Petrobras", 18 Th Global Congress on Process Safety, San Antonio 2022.

[2] AWS Glue (2023). https://aws.amazon.com/pt/glue/. Accessed February, 16, 2024.

[3] What is ETL (Extract Transform Load)? (2024). https://aws.amazon.com/what-is/etl/?nc1=h_ls. Accessed February, 16, 2024.

[4] ISO/IEC 9075-1:2023 Information technology Database languages SQL (2023) https://www.iso.org/standard/76583.html. Accessed February, 16, 2024.

[5] AVEVA PI System (2024). https://www.aveva.com/en/products/aveva-pi-system/. Accessed February, 16, 2024.

[6] What is: Multifactor Authentication (2024) https://support.microsoft.com/en-gb/topic/what-is-multifactor-authentication-e5e39437-121c-be60-d123-eda06bddf661. Accessed February, 16, 2024.

[7] ISO/IEC 20000:2019 Information technology Service management (2019) https://www.iso.org/standard/72120.html. Accessed February, 16, 2024.

Reliability in MASS ALERT System for DAM Emergencies: Case Study of Vale S/A

Flávio Thimótio
Vale S.A.
Address
flavio.thimotio@vale.com

Cilene Faria
Flávio Morais
Germano Ferraz
Roberto Delamora
Priscila Guedes
Daniela de Miranda Souza
Vale S.A.

Keywords: mass alert, dam, reliability, mining

Abstract

In its iron ore operations, Vale, the largest mining company in Brazil and the fourth largest mining company in the world by market value, has 77 dams considered with High Potential Damage in case of a failure in the stability controls. This classification follows the criteria of the sector regulatory agency in Brazil and determines that, for this class, a mass alert system is necessary for the resident population and workers immediately downstream of the dam within a region pre-set by legislation, considering a simulated flood area.

For Vale, the mass alert system for dam emergencies is part of a critical mitigating control unfolding a business risk analysis considering the hypothetical dam break as an Unwanted Material Event. Therefore, both the probability of failure on demand, spurious failure, and availability of the assets of this protection barrier are important elements for this purpose, as well as the emergency response plan and the evacuation simulation of flood areas which are part of a risk management system that require continuous verification and continuous improvement.

Currently, Vale has 375 sirens, and it is planned to reach 420 sirens by the end of 2024. The purpose of this article is to present how the safety mesh concepts applied to these systems were developed, the good practices applied from this study and the lessons learned from the dam emergency management arising from the break of the Brumadinho dam in 2019. The reliability analyses of the systems involved and the main contributors to the Safety Critical Unavailability are presented, as well as the results of the reduction of the reparation time and the system availability.

1 Introduction

Brazil has great geological and mineral diversity in its territory. According to the Brazilian Mining Institute – IBRAM [1], the country ranks first in the world in the production of Niobium, second in the production of Iron and third in the production of Bauxite.

The mining industry plays a crucial role in the Brazilian economy, and it is one of the main sources of revenues and employment, present in 48% of Brazilian municipalities, gathering 7,300 companies generating USD$24.9 billion in the trade balance (40% of the Brazilian balance) and 240 thousand direct jobs in 2022 [1]. One of the biggest challenges of the sector is the disposal of the mining tailings in a socially and environmentally safe, as well as economical, way, especially after the accidents involving breaks of dams, the most used technique for tailings disposal.

Tailings disposal facilities are the set of engineering structures and their components involved in the management of solid tailings or sediments, including tailings distribution and water recovery systems, dams, walls, dikes, pits with built dams and stacks.

Azam, S. and Li, Q. analyzed the failure events of dam structures in 100 years between 1910 and 2010 [2] and built an impact classification. Specifically, regarding the impact of loss of life, mass alert systems, or outdoor Public Alert System (PAS), are mitigating measures recommended and provided about by law to warn the population and the workers within the flood area about the risks of these events. In Brazil, the entrepreneurs are responsible for the maintenance and the operation of these assets related to the risk of dam break.

In the last 9 years, two major dam break events occurred, which led several countries around the world, including Brazil, to discuss and take stricter actions about the safety of mining tailings dams. Among these actions, there is the creation and the update of standards and laws linked to this topic [3][4][5][6][7]. Specifically, regarding the break of the Brumadinho dam which occurred in 2019, information about the impacts of the event can be checked in the reparation reports published by Vale [8].

For Vale, the risk and control management systems are based on the ICMM Critical Controls Management (GCM) process. According to the bow-tie analysis methodology, the Material Unwanted Event (MUE) is the scenario of a hypothetical dam break. One of the Critical Mitigation Controls is the effectiveness of the Emergency Action Plan for Mining Dams (PAEBM). The mass alert system is part of this control. Given the complexity of the control, a second bow-tie analysis was made (Table 1) and we obtained the reliability and availability requirements of the mass alert system as critical success factors (Table 2)

Table 1. Critical control bow-tie analysis

Material Unwanted Event	Failure to Activate and Effectiveness of the Dam Emergency Action Plan
Critical control name	Inspection and Preventive/Predictive Maintenance of the Siren System
Critical control objectives	Establish the minimum requirements and resources necessary to maintain the availability and reliability of the siren system.

Table 2. Alert system reliability and availability requirements

INDICATOR	REQUIREMENT	RESULT 2023
Siren system availability	99.51%	99.82%
Siren system performance (Test failures)	0	7
Siren systems in maintenance plans	100%	100%
Maintenance backlog	10 to 30 days	6.10 days
Adherence to Systematic Maintenance	90%	98.30%

Considering the 2023 results, we can conclude that the indicators related to asset availability and maintenance strategy are aligned with the defined goals; however, the events related to performance failure are below the desired. These events consider spurious failures, demand failures and failures in the monthly audible tests, including failures in emergency simulations or undue activations.

In the context of the mass alert systems, it is crucial to consider strategies and technologies that enable scheduled tests to simulate and prepare the potentially affected population and validate the system itself. The challenge these systems face is the maintenance of their reliability. It is imperative that the alerts are triggered only in situations of actual demand, avoiding false positive and false negative. These errors can impair the company's image, the credibility of the systems and result in fines and sanctions for the mining company.

2 Vale's Mass Alert System

112 of the 927 dams cataloged in Brazil belong to Vale. Ninety-one of these are included in the National Dam Safety Policy, the purpose of which is to guarantee the effective adoption of safety standards, in addition to regulating the safety actions and standards [9]. The safety classifications and standards are the basis for the selection of equipment and technologies installed at the dam and along the flood area.

Two concepts used in this classification are: (i) Risk Category (CRI), which takes into account aspects of the dam, and (ii) Associated Potential Damage (DPA), the concept and the grading of which is related to the damages that may occur in the dam, causing loss of human life and social, economic and environmental impacts, according to requirements set forth in legislation [10].

The human occupation in areas located downstream of dams over the years constitutes an aggravating factor in the risk scenario. Given the construction of this scenario, the public authorities, together with the private sector, prepare a contingency plan for assistance in case of disasters. This plan relates to the Emergency Action Plan for Mining Dams (PAEBM) to assure the safety of the communities in case of a dam containment failure.

2.1 Emergency Procedures and Actions

The purpose of the protocols outlined in the plan is to facilitate the decision-making process, seeking to speed up the effective response, considering that it is crucial to rely on the experience of the people responsible for the emergency management. The coordinated evacuation emerges as one of the most effective strategies to safeguard the lives of the people located in potentially threatened areas.

According to Balbi [11], the evacuation process can be divided into five phases according to the schematic representation of the process (Figure 1). These aspects must be prepared within contingency plans for general disasters developed by the municipal civil defense in a broader scope [11].

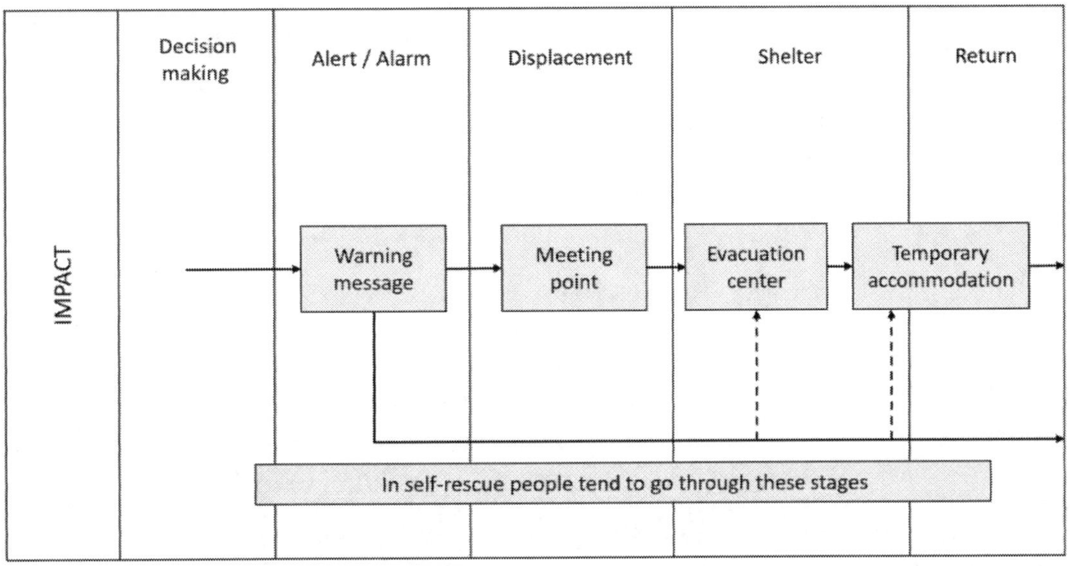

Figure 1. Evacuation process [11]

The alerts emitted by sirens represent a direct and immediate form of warning, although their effectiveness may be limited by the ability of the population to understand the signals. They are recognized as effective means of communication to warn densely populated areas; however, not the only one [12]. Figure 2 illustrates sirens used by Vale.

In these sirens, solar panels, and a set of redundant batteries to assure their continuous operation stand out. The evacuation plan highlights that the sound emitted in case of a dam break is different from that used for other purposes, such as simulations and tests.

2.2 *Monitoring of the facilities*

According to Martins [13], the possibility to protect human lives in case of a dam break depends on three factors: the type of the dam (earth or concrete), the distance between the dam and the inhabited areas and the existence of alert systems.

Figure 2. Example of sirens used by Vale.
Source: Vale's internal archive

The Global Industry Standard for Tailings Management (GISTM) was launched on August 5, 2020, from an initiative of the United Nations Environment Program (UNEP), the Principles for Responsible Investment (PRI) and the International Council on Mining and Metals (ICMM), becoming the first global standard for the mineral sector and a global benchmark for the safety of tailings dams [14].

In October 2020, Vale's Administration Board approved a new Policy on Dam and Geotechnical Mining Structure Safety, which contains GISTM as one of its references. Among other guidelines, the policy determined that all components of its current Tailings and Dam Management System (TDMS) were designed with elements of continuous improvement, using, and applying the best technologies and practices available in accordance with international institutions, including ICMM [2].

One of these elements is the monitoring system for dams, which are equipped with specified technologies to assure the monitoring and the control of variables that may indicate the state of conservation and safety in their structure. The function of these instruments and equipment are divided into preventive monitoring and the mass alert system (Table 3), operated, and controlled uninterruptedly by means of an integrated room called Geotechnical Monitoring Center (CMG), part of the dam safety, risk, and emergency process.

2.3 Automated siren activation system

In the Brazilian legislation, the Self-Rescue Zone (SRZ) is the region where the entrepreneur is responsible for providing warning notices to the population in the section of the valley downstream of the dam by means of a distance metrics that corresponds to a flood wave arrival time equal to 30 (thirty) minutes or 10 km (ten kilometers).

Table 3. Types of Monitoring and their main elements

Types of monitoring	Elements	Function
Preventive	Visual inspections and through instruments such as Water Level Indicator, Piezometers (automated or conventional), Flow Meters, Rain Gauges, Robotic Stations, Prisms, Cameras, Interferometric Radars, Seismographs, Automated Linimetric Rulers and Orbital Radars.	Monitoring the state of conservation and stability of dam structures
Emergency	Tiltmeters, Doppler radars, PLCs, and Sirens	Quick detection of abnormalities and automatic activation (Automated siren activation system - AAS) or manual activation of sirens

Brazilian legislation ANM Resolution No. 95/2022, Article 8, determines that dams with high potential damage must be provided with automated systems for siren activation installed outside the flood area and other mechanisms adequate for the efficient warning in the affected zone, installed in a safe place and equipped with a failure mode in case of break of the facility, complementing the manual activation systems [10].

Table 4 shows the distribution of signaling structures and sirens under Vale's responsibility [15].

Table 4. Distribution of dams and sirens

State – Region	Dams	High Potential Damage Dams	Sirens
Minas Gerais – Southeast	99	73	337
Pará – North	13	4	79

Source: SIGBM [15]

As for the mass alert and signaling methods, the mining companies generally use specific technologies to assure the safety of their operations. This includes alarm systems, sirens, and real-time communication methods to warn the workers and the public in case of emergency. Figure 3 shows the sirens installed in the region of Minas Gerais.

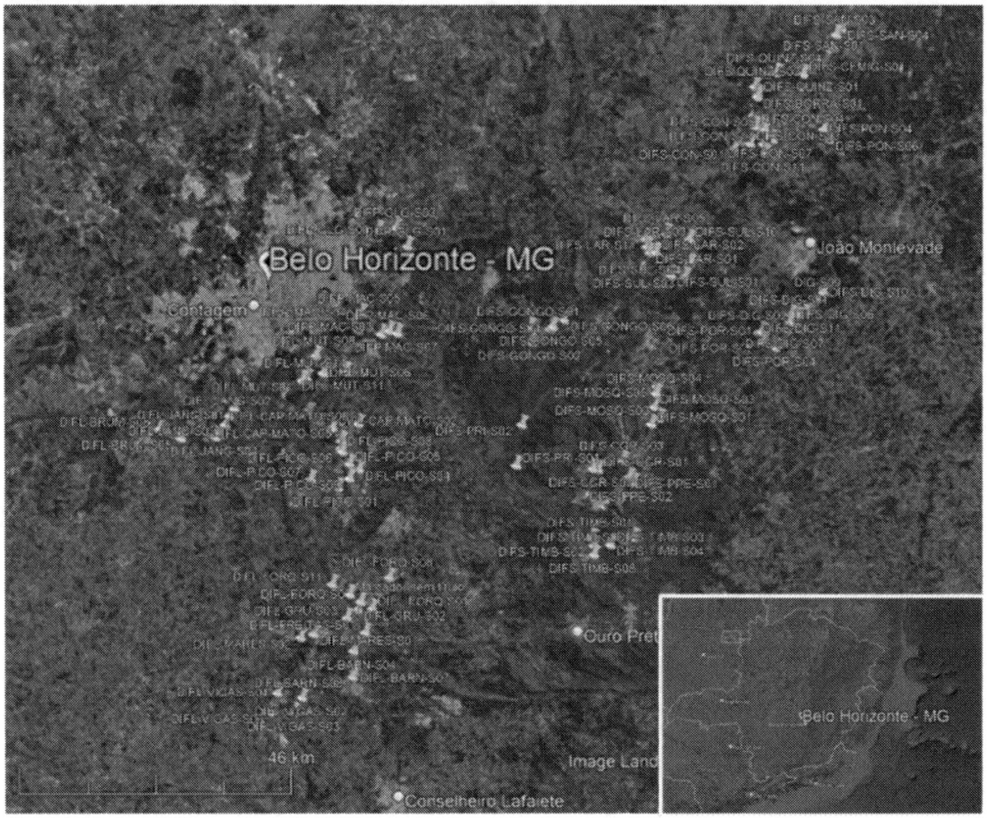

Figure 3. Location of sirens and dams – Minas Gerais
Source: Vale's internal archive

The automated dam system consists of a Break Detection System and an Alert System. Together, these systems can be interpreted as an instrumented safety barrier starting from the sensor element (Break Detection System), logic solver and actuator (Alert System). This barrier is called Automatic Siren Activation (AAS). Its components and functions are presented in Table 5.

Table 5. Elements of the Automatic Siren Activation System

Technology	Function	Elements	Objective
Tiltmeters	Tilt sensors that allow you to identify displacements in the structure.	**Sacrificial sensors:** installed in the structure's mass which, in the event of rupture, will be hit and permanently damaged.	On-site slope detection of dam slopes.

		Integrity sensors: they are outside the structure and their basic function is to validate the correct functioning of the system. **Logic Solver:** analysis and activation of individual or group movement triggers.	
Doppler radar	Detects, tracks and alerts in real time movements and geotechnical risks.	**Low frequency transmitting and receiving antenna:** spatial detection of movement using the Doppler principle. **Signal processor and logic solver:** analysis and activation of motion triggers.	Quick detection of movements (displacement, speed, and direction) from a distance.
Mass Alert System	Alert with an audible and guiding tone, within the range, of possible emergency processes, simulations, or system tests.	**Redundant amplifiers and digital sound system:** send pre-recorded sound tone depending on the specific situation. **Redundant communication system:** connect the siren to the control and monitoring system. **Onboard Monitoring and Command System:** monitor the asset and process remote commands. **Electrical Power System:** provide redundant electrical power for the sirens. **Control System:** monitor, operate and trigger the siren system manually or automatically.	Mass emergency alert.

Source: Vale's internal archive

The Figure 4 shows the simplified diagram of the Break Detection and alert systems.

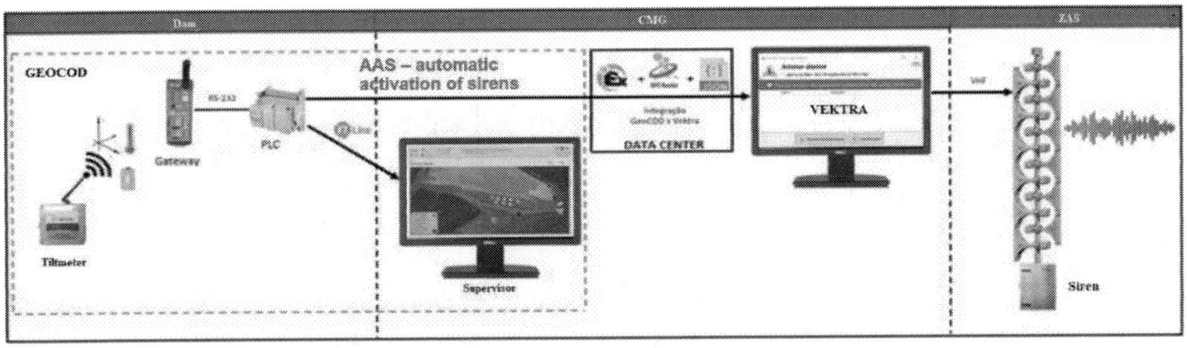

Figure 4. Break Detection and alert systems diagram
Source: Vale's internal archive

The physical availability of all system components is measured daily by means of tests, as for the sirens, or online in case of the break detection system. The unavailability of a siren or detection system is a priority over all scheduled maintenance and is handled as an emergency. The test processes for the elements occur daily, weekly, and monthly, as shown in Table 6, and guide the maintenance and design actions of the alert system. The alert system availability history is shown in Figure 5.

The maintenance strategy for these critical assets is based on maintenance plans registered in the maintenance management system (SAP) grouped into routes for optimization, reducing the service and displacement time. The preventive strategy considers concentrating the highest number of maintenance operation during dry seasons. Anomalies detected in the preventive maintenance generate conditional maintenance for repair. Events diagnosed in the periodic tests lead to corrective or emergency corrective maintenance for restoration of the system functions.

Table 6. Test frequency

Test	Frequency	Target	Actions
Status and timing test	Daily	Evaluates 9 technological parameters – ability to activate sirens	Identify possible problems and drive corrective or emergency maintenance to restore electrical and technological function
Silent Test	Weekly	Evaluates 9 parameters and tests sound system – horn system capacity	Low frequency test (inaudible) to evaluate power system. Siren diagnosis to drive corrective or emergency maintenance to restore sound function
Sound Test and scheduled drills	Monthly and on demand	Evaluates 10 parameters and tests the sound system – Effective sound of the sirens	Complete system test, with pre-configured music or simulated shout. Diagnostics of the siren and acoustic coverage. Design guidance for the acoustic coverage of the warning system, in addition to maintenance actions if necessary

Source: Vale's internal archive

To guarantee the critical control management requirements, two complementary aspects can be followed: identifying failure elements derived from the event tree, in a continuous improvement and maintenance management action, or analyzing the technical information of the components with the highest probability of failure. Starting from the second approach, here are the results of a reliability study using SINTEF (2013) approach and the relationships with the loss profile consolidated in 2013.

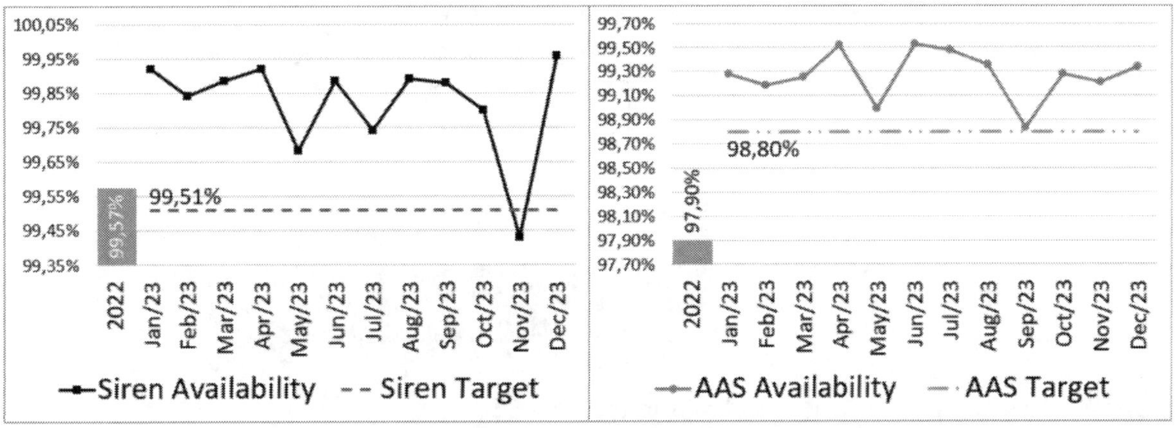

Figure 5. Availability Indicator of sirens and AAS
Source: Vale's internal archive

2.4 Reliability Studies of the Automatic Siren Activation (AAS)

The main purpose of the reliability study was to calculate the probability of failure, and therefore, the reliability of the system configuration, and direct possible investments to improve the system.

The specific goals of the stage included:

- Quantification of the Critical Safety Unavailability (CSU) per siren point in the facility, including the photovoltaic electrical energy supply system,
- Definition of the contribution of each of operational discipline involved in the operation (detection, telecommunications, servers, sirens, and electrical supply) to the obtained value of CSU,
- Quantification of the system Spurious Trip Rate – STR per siren point; and
- Presentation of recommendations for improvement of the system reliability to be included in the system improvement investments.

The methodology applied in this study considered the good industry practices, such as IEC 61.508: 2010, IEC 61.511: 2018 and SINTEF guidelines (2013), especially, the "Method for Reliability Prediction of Instrumented Safety Systems (PDS) [8].

The method is used to quantify unavailability and loss of production due to safety of a Safety System by calculating the Critical Safety Unavailability (CSU), which generates the probability the component or the system not to be able to carry out a safety action automatically upon occurrence of a dangerous or accidental event.

Critical Safety Unavailability (CSU)

The critical safety unavailability, in PDS method, is composed of three main categories, according to Figure 6.

Undetected danger	1a) Random hardware failures 1b) Systematic failures	PFD
Unavailability due to system inactivity	2a) Unavailability due to repair	DTU
	2b) Unavailability due to test	
Test Independent Failure (TIF)	3) Failures not covered by functional testing	P_{TIF}

Figure 6. Contributors to loss of security in PDS.
Source: Adapted from SINTEF [8].

Calculation of the Critical Safety Unavailability (CSU)

According to [8], CSU is defined as the probability of failure of the component/system when performing a safety action during the occurrence of a dangerous/accidental event. To calculate CSU, the equations of low-demand systems were considered, and according to guidelines from [8], the Probability of Failure on Demand (PFD) was used to estimate the reliability of this kind of systems. CSU was obtained by Equation 1:

$$CSU = PFD + DTU + P_{TIF} \tag{1}$$

Probability of Failure on Demand (PFD)

The term PFD in the CSU calculation equation is associated with the unavailability of the safety system due to failures which are detectable only during the functional test or in real demand situations. Thus, PFD quantifies the loss of safety resulting from undetectable dangerous failures during the period in which the safety function is unavailable, and this condition is not known. The formulas used to calculate PFD, for low demand systems, can be found in [8]

In relation to the common cause failure (FCC) modeling and when quantifying the reliability of redundant safety systems, it is crucial to differentiate independent from dependent failures. In general, random hardware failures caused by natural factors are considered independent failures. Common cause failures have the potential to reduce the effectiveness of the redundancy.

The traditional approach to account the Common Cause Failures has been the factor model β. In this model, it is assumed that a specific fraction of failures is of common cause, i.e., failures which can lead to concurrency or occur within a short period of time, affecting all redundant components. The PDS method recognizes that this approach is limited once the failure rate associated with FCC does not vary in relation to the amount of equipment in the redundant system.

To correct such limitation, the PDS method suggests an improved version of the factor model, which distinguishes the different types of voting based on the parameter C_{MooN}. In [8], it is possible to summarize the values of C_{MooN} for some typical voting configurations.

Downtime Unavailability (DTU)

Downtime Unavailability refers to the period associated with the unavailability of the safety system due to unplanned and planned maintenance activities. It was decided to consider the contribution of this factor to the system CSU as negligible. This decision is supported by the fact that the Mean Time to Repair (MTTR) of the elements of the system is significantly shorter than the interval between the functional tests ($\boldsymbol{\tau}$).

As indicated in [8], the term DTU becomes significant in mathematical models of systems with high technical repair complexity. Once the system addressed by this study does not present such characteristics, it was concluded that the term DTU was not relevant to the model in question.

Independent Test Failures (P_{TIF})

In the reliability analyses, the functional test rarely detects 100% of the failures due to the differences between the test conditions and the actual demand conditions. The PDS method incorporates the Probability of Independent Test Failures (P_{TIF}) into the Probability of Failure on Demand (PFD).

P_{TIF} represents the probability of a component/system to fail in its intended function due to a failure not detected by the functional tests. For an individual component, P_{TIF} expresses the probability of failure on demand, regardless of the functional test interval. The formulas for the calculation of P_{TIF} in different voting logics can also be obtained in [8].

Spurious Trip Rate (STR)

The IEC 61508 standard focuses on the safety loss and the calculations of PFD. Safety systems can cause triggers even without real demand, characterizing spurious triggers. These failures, categorized as safe failures, can result in unintentional shutdowns. The PDS method quantifies this rate with STR, representing the expected number of spurious activations of the safety system per unit of time. The equations to obtain STR for different voting logics are mentioned in [8].

2.5 Reliability Block Diagram

Following the guidelines of the methodology proposed by SINTEF, the system functional reliability blocks were structured. Figure 7 shows an example of one of the assessed structures, allowing the redundancy configurations to be explained and supporting the reliability calculations. It was possible to obtain contributors for each stage of the safety system, namely: detection, processing, and action.

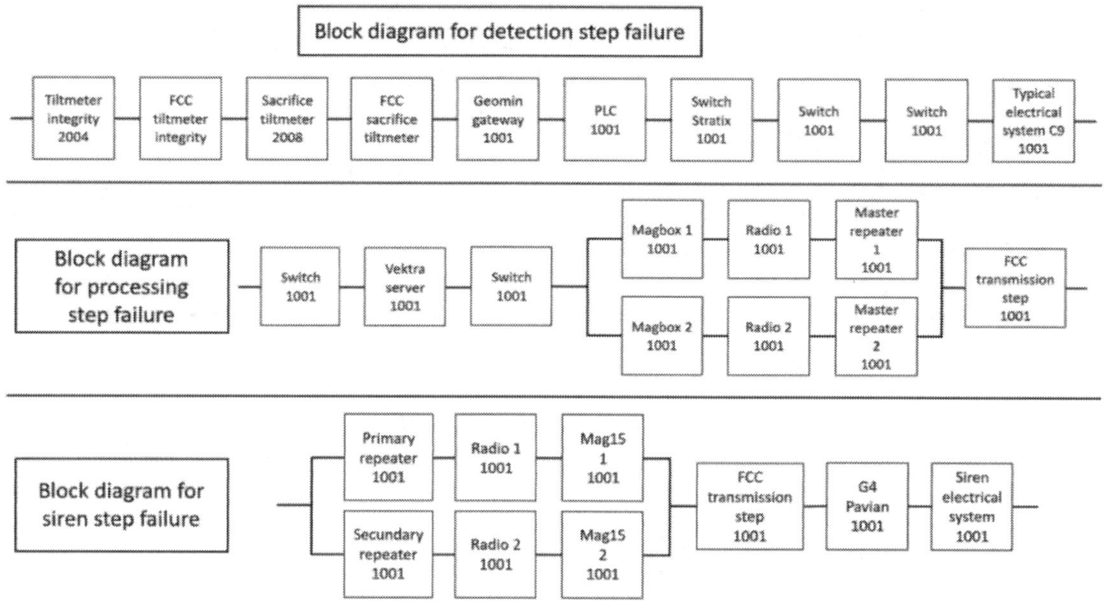

Figure 7. Forquilha V reliability block diagram
Source: Vale's internal archive

2.6 Obtaining of Failure Rate Data

To obtain the model input data referring to the equipment/components of the analyzed system, the following sources of information were taken into consideration, listed in order of priority:

1. Specific data of the equipment published by the manufacturer.
2. Data on similar equipment published by other manufacturers.
3. Information published in public databases [9][12].

Failure rates which were not made available by the manufacturer, by VALE or found in public databases, were obtained from similar equipment in terms of functionality and possible failure modes. The estimates adopted were always taken in favor of safety.

2.7 Contribution of CSU and STR

Figure 8.1 shows the percentage contribution of each subsystem to the CSU result of the system. It is possible to observe that the electrical supply section is responsible for approximately 58.5% of the total CSU of the system.

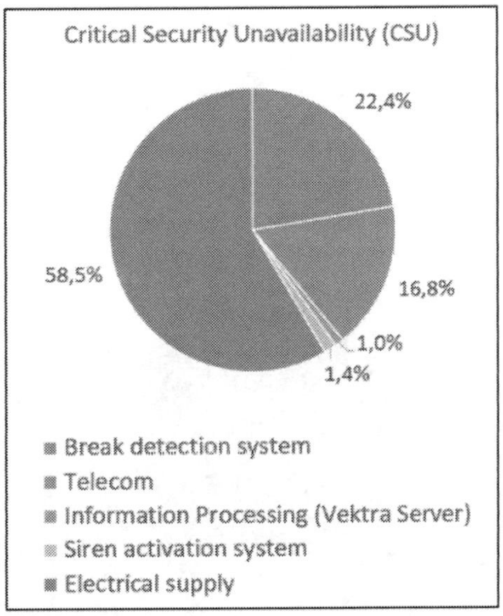

Figure 8.1. Distribution of CSU by subsystem of the AAS system.

Figures 8.2, 8.3 and 8.4 show the distribution of CSU in the sections of break detection, electrical supply of the detection system and electrical supply of the sirens, respectively. It is possible to observe, individually for each subsystem, that the gateway is responsible for 48% of the total CSU of the break detection section, the source is responsible for 36% of the total CSU of the electrical supply, and the voltage converter is responsible for 65% of CSU of the electrical supply of the sirens, the latter being the equipment most likely to fail in each subsystem.

In relation to STR (Figure 8.5), it is observed that the detection section of the system is responsible for 40% of the total STR. Figure 8.6 shows STR distribution in the detection section. It is possible to observe that PLC is responsible for 92% of the total STR of the break detection section.

As a proposal for improvement, redundancy, simplification, and equipment replacement techniques were sought. With the implementation of these measures, it is expected to increase the system reliability and assure safe and efficient operation. Among the main measures, there are:

- Optimize the current photovoltaic system and/or installation of conventional electrical systems,
- Replace Tiltmeters with Tiltmeters/inclinometers with higher reliability,
- Install the Fiber Optic system as redundancy for Tiltmeters for the detection of dam break,
- Replace some switches with switches with higher reliability,
- Improve redundancy of telecommunication components,
- Replace PLC of the detection systems with PLCs with λ_{DU} of at least $1.15e^{-06}$ /hour, and λ_{SU} of at least $1.24e^{-06}$ /hour,

- Optimize the logic solver redundancy architecture,
- Optimize functional testing interval,

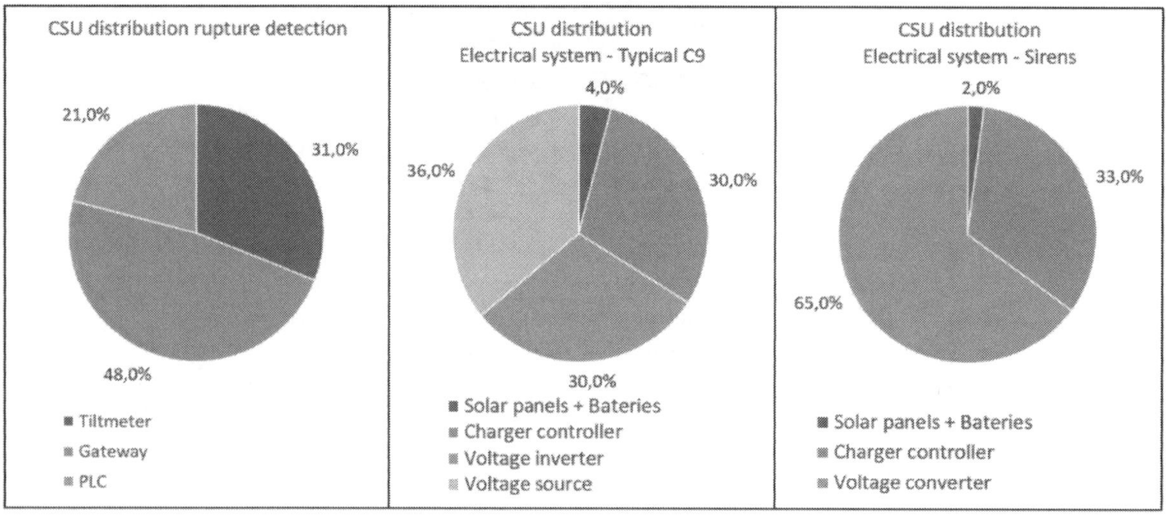

Figure 8.2. Distribution of the CSU of the rupture detection system per equipment.
Figure 8.3. Distribution of the CSU of the typical C9 electrical supply system by equipment.
Figure 8.4. Distribution of the CSU of the sirens' electrical supply system per equipment.
Source: Vale's internal archive

3 Results

In a mineral processing plant, the automation process began several decades ago and continues to evolve and generate historical milestones in terms of use of new technologies, therefore, their degree of maturity regarding the process automation is high.

The Geotechnical Automation faces particularities regarding the degree of maturity in its process and the maturity of the implemented technological solutions, namely: geotechnical particularities of each dam making it difficult to standardize the use of new instruments, SRZ that extends outside the perimeter of the company and involving the community in the process of installing and testing new instruments; difficulty for installation of network infrastructure that allows communication with good latency and without data loss; sensitivity of the affected population, making it difficult to execute end-to-end emergency drills. Another factor that impacts the reliability of the systems correlates with the novelty of the application of emergency monitoring solutions and automatic activation of sirens. Figure 9 represents the main improvements applied to Vale's mass alert system since 2018.

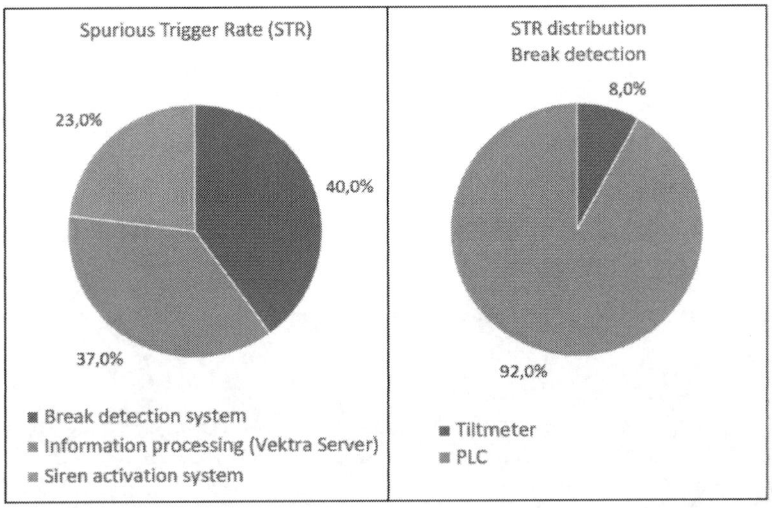

Figure 8.5. Distribution of STR by subsystem of the AAS system.
Figure 8.6. STR distribution of the rupture detection system per equipment.
Source: Vale's internal archive

The reliability study of the alert system, combined with the failure analysis and comprehensive maintenance management, set improvements that can be applied to the process. The implementation of break detection redundancy, definition and daily monitoring of availability indicators, replacement of components with more modern ones, creation of working groups to act on problems are actions which have contributed to increasing the reliability and the availability of the system, as shown in Figure 10.

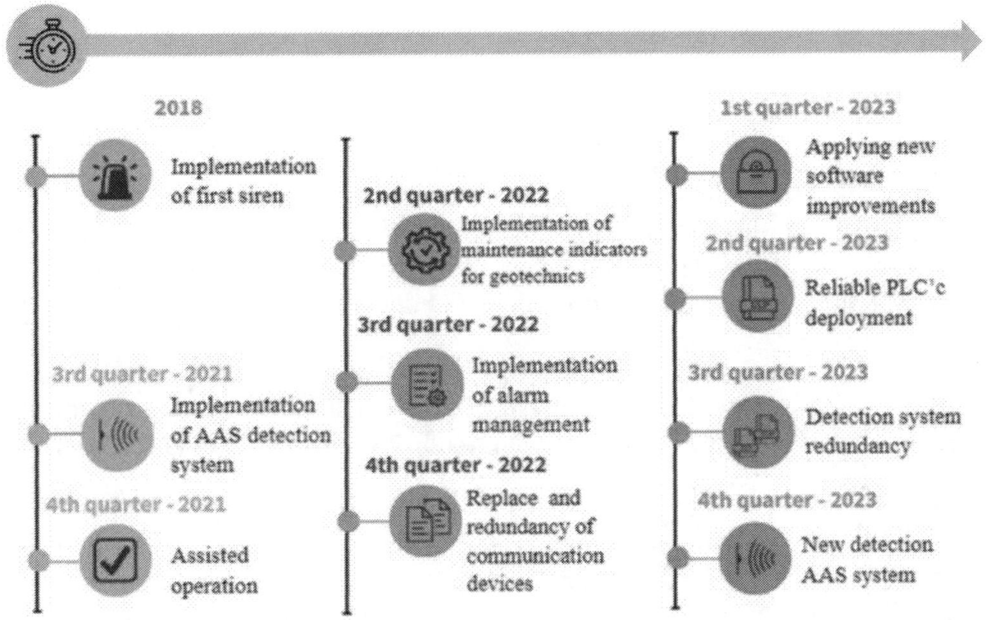

Figure 9. Evolution of the maintenance process of the AAS system
Source: Vale's internal archive

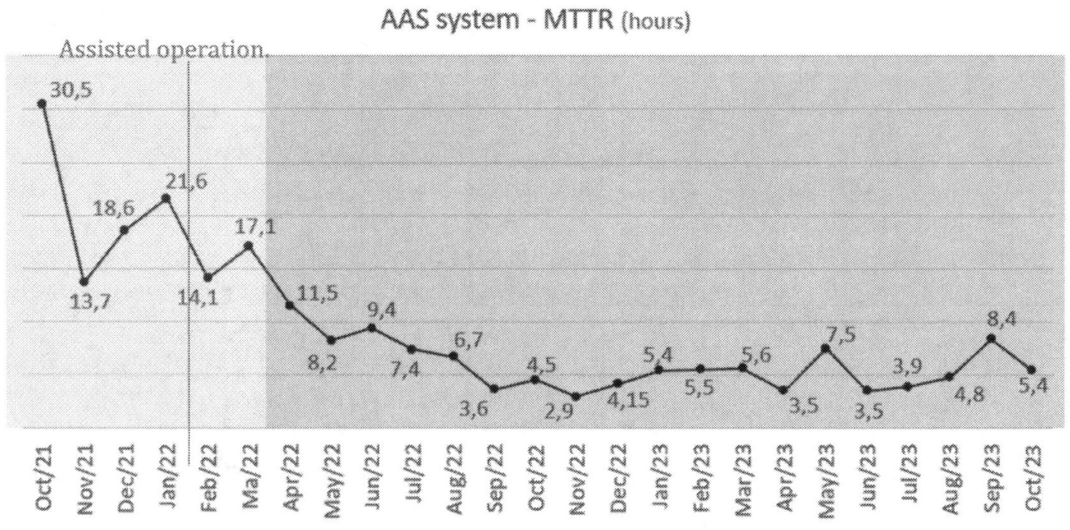

Figure 10. Evolution of the MTTR of the AAS system
Source: Vale's internal archive

4 Conclusion

Vale developed its production management program, called Vale Production System (VPS), using world-renowned process management and control methodologies based on those developed by Toyota in its Toyota Production System (TPS) [16]. The use of tools, such as MASP [17] (Figure 10), applied to the alert system environment showed to be quite efficient in the improvement and the stabilization of the maintenance processes.

The presented data enables identifying offenders of the alert system and seek improvement. Regarding the break detection, the improvement in MTTR of AAS with actions carried out over the last 2 (two) years is remarkable.

To maintain the system reliable for use only when demanded is the challenge faced, since as the result of the increased reliability of the alert system, there are trust and higher engagement of the population in the alert system. A siren can operate only when demanded and as specified in the project.

The improvement actions, associated with the maintenance projects and execution, based on the reliability studies, address the priorities of the system adaptation actions to increase reliability and mitigate false positive or false negative.

The electrical power supply systems using solar panels and batteries are a sensitive point in the solutions and have been the target of improvement projects for capacity strengthening and adaptation, in addition to replacement of components with low reliability. Simple telecommunications systems or those mapped in the reliability studies have been restructured and redundancies are being implemented.

5 References

Online Reference:

[1] Panorama Mineração do Brasil 2023. (2023). Instituto Brasileiro de Mineração, IBRAM. https://panoramamineracao.com.br/pmb2023/. Accessed December 28, 2023.

[2] Health and Safety Critical Control Management - Good Practice Guide. International Council on Mining and Metals (ICMM). https://www.icmm.com/website/publications/pdfs/health-and-safety/2015/guidance_ccm-good-practice.pdf. Accessed January 8, 2024.

[3] Law N° 12.334, dated September 20, 2010. (2010). https://www.planalto.gov.br/ccivil_03/_Ato2007-2010/2010/Lei/L12334.htm. Accessed December 22, 2023.

[4] Ordinance No. 70,389, dated May 17, 2017. (2017). https://www.defesacivil.pr.gov.br/sites/defesa-civil/arquivos_restritos/files/documento/2018-12/PortariaDNPMn70389de17demaiode2017SegurancadeBarragensdeMineracao.pdf. Accessed December 22, 2023.

[5] Law N° 14.066, dated September 30, 2020. (2020). https://www.planalto.gov.br/ccivil_03/_ato2019-2022/2020/lei/L14066.htm#:~:text=Fica%20proibida%20a%20constru%C3%A7%C3%A3o%20ou,minera%C3%A7%C3%A3o%20pelo%20m%C3%A9todo%20a%20montante. Accessed December 22, 2023.

[6] 95 ANM Resolution No. 2002, dated February 07, 2022. (2022). https://www.gov.br/anm/pt-br/assuntos/barragens/legislacao/resolucao-no-95-2022.pdf. Accessed December 22, 2023.

[7] Ordinance No. 237, dated October 18, 2001. (2001). https://anmlegis.datalegis.inf.br/action/ActionDatalegis.php?acao=abrirTextoAto&link=S&tipo=POR&numeroAto=00000237&seqAto=000&valorAno=2001&orgao=DNPM/MME&cod_modulo=414&cod_menu=7901. Accessed December 28, 2023.

[8] Reparation Day by Day. Vale. Published 01/07/2023. https://vale.com/documents/44618/414053/revista-dia-a-dia-da-reparacao-2023.pdf/00d224b7-b582-e4be-024e-ddc81c12eaa6?version=1.0&t=1706042922477&download=false. Accessed January 8, 2024

[9] Agência Nacional de Águas e Saneamento Básico, SNISB. https://www.snisb.gov.br/perguntas-frequentes . Accessed January 29, 2024.

[12] Outdoor Warning Systems Technical Bulletin, Federal Emergency Management Agency's (FEMA). http://www.midstatecomm.com/PDF/FEMA_guide.pdf. Accessed January 8, 2024

[13] Martins, R. *Segurança hidráulico-operacional de barragens à luz da legislação internacional*. In: CONGRESSO DA ÁGUA, 5, 2000, Lisbon. https://www.aprh.pt/congressoagua2000/COMUNIC/3.pdf. Accessed January 8, 2024.

[14] Global Industry Standard for Tailings Management, GISTM. (2020). https://globaltailingsreview.org/wp-content/uploads/2020/08/global-tailings-standard_spreads_PT.pdf. Accessed January 6, 2024.

[15] Mining Dam Safety Management System, SIGBM. National Mining Agency, ANM. (2024). https://app.anm.gov.br/SIGBM/Publico/GerenciarPublico. Accessed January 6, 2024.

[16] Lean Institute Brasil. *Sistema Toyota de Produção (Toyota Production System - TPS).* https://www.lean.org.br/conceitos/117/sistema-toyota-de-producao-(toyota-production-system---tps).aspx. Accessed February 7, 2024.

[17] De Bastiani, Jeison A. *MASP (Método de Análise e Solução de Problemas): Saiba quando utilizar.* https://blogdaqualidade.com.br/masp-metodo-de-analise-e-solucao-de-problemas-parte-1/#:~:text=O%20MASP%20(M%C3%A9todo%20de%20An%C3%A1lise,%2Dlos%20ou%20minimiz%C3%A1%2Dlos. Accessed February 7, 2024.

Thesis:

[10] Souza, Thainá S. A., Evolução Histórica da Legislação Brasileiera e do Estado de Minas Gerais relacionado ao tema de disposição de rejeitos de mineração em barragens. Bacharelado, Universidade Federal de Ouro Preto, 2019. https://monografias.ufop.br/bitstream/35400000/2342/1/MONOGRAFIA_Evolu%C3%A7%C3%A3oHist%C3%B3ricaLegisla%C3%A7%C3%A3o.pdf. Accessed December 28, 2023.

[11] Balbi, Diego A. F., Metodologias para a elaboração de planos de ações emergenciais para inundações induzidas por barragens. Case Study: Peti Dam-MG. Post-Graduation Dissertation. Universidade Federal de Minas Gerais. 2008. file:///C:/Users/01187922/Downloads/diegoantonio356.pdf. Accessed January 8, 2024.

American Institute of Chemical Engineers
2024 Spring Meeting and 20th Global Congress on Process Safety
New Orleans, LA
March 24-28, 2024

Reduction of Process Accidents Based on API RP 754, by Improving the Safety Culture Measured by the Hearts & Minds Index

Eduardo Aparecido Zuin
Ultrapar
Brigadeiro Luiz Antônio Av., 1343, São Paulo
Eduardo.zuin@ipiranga.ipiranga

Livia Tavares Freitas Alves
Ultrapar
Brigadeiro Luiz Antônio Av., 1343, São Paulo
Liviaf@ipiranga.ipiranga

Vanessa Alves Pessin
Ultrapar
Brigadeiro Luiz Antônio Av., 1343, São Paulo
pessin@ipiranga.ipiranga

Vanessa Moura franco
Ultrapar
Brigadeiro Luiz Antônio Av., 1343, São Paulo
vanessa.franco@ipiranga.ipiranga

Keywords: Process safety incidents reduction; PSI reduction; Process safety Culture; API RP 754; PSM; Hearts & Minds

Abstract

In the 2022 ESPC benchmark report, it was observed that 50.4% of events are directly linked to operations, and this percentage has remained above 50% since 2018. The situation at Ipiranga is similar, with 50% of TIER 2 incidents having operations as the primary agent, and human factors in procedure execution contributing as a causative factor.

This article aims to demonstrate and validate the decrease in the accident rate year after year showed the relation the Hearts & Minds tool, which measures organizational culture. The transition to new cultural stages in the index represents an improvement in organizational culture. This is reflected in assertive actions in employee behavior, which, in turn, enhances operational discipline.

As a result, operational reliability increases, and the rate of process safety incidents decreases.

1 Introduction

1.1. The relation among the culture versus process safety incidents

Over the years, the relationship between the evolution of safety culture and the interaction between the pillars: technology, management system, and human factors is noticeable. With process safety focusing more on assets and structures, there is an increasingly evident need to improve the human-machine interface and continuous improvement in management systems, as shown in the Figure 1.

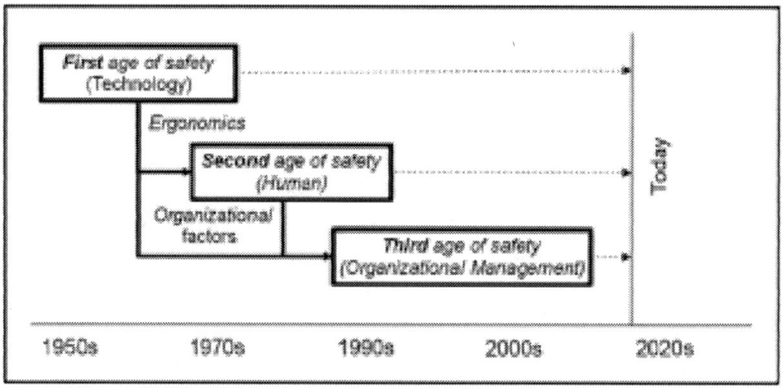

Figure 1: evolution of safety thinking[1]

When we think about the evolution of safety culture, it is possible to observe the evolution of thought also oriented towards management processes, as shown in figure 2. In the first phase, there is a greater concern involving reliability issues, which is directly related to loss prevention and the beginning of a culture focused on process safety. In the second phase, we have the start of safety management systems. Some maturity is expected, and management systems begin to gain greater visibility. The third wave is more focused on human interfaces, where concern for human factors becomes more prominent and is understood as one of the priority points when it comes to safety management like below in the Figure 2.

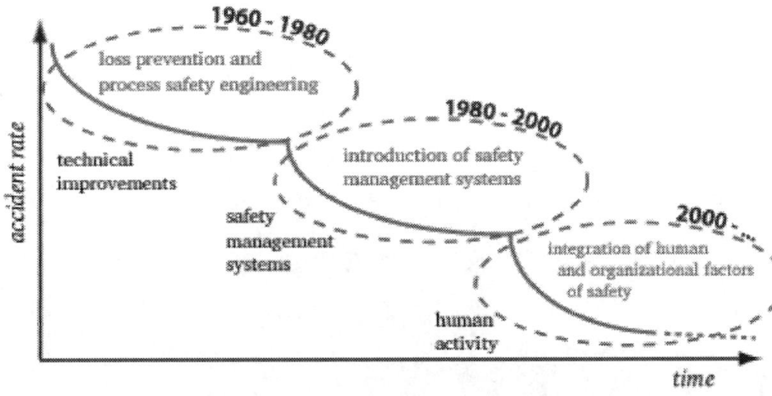

Figure 2: Successive approaches to industry[2]

Considering this, our company recognized the need to measure the maturity of its organizational culture to integrate human and organizational factors of safety. After some experiences, the chosen tool was the Heart and Minds (H&M) Curve.

1.2. Curva Hearts & Minds

The Heart and Minds (H&M) Curve is a concept that plays a pivotal role in shaping organizational culture and significantly influencing incident reduction within a workplace. This curve represents the journey an organization undertakes in cultivating a safety culture where both the hearts (emotions and attitudes) and minds (knowledge and awareness) of individuals align with the commitment to safety. Establishing a positive safety culture is not only about compliance with rules and procedures but involves fostering a genuine belief in the importance of safety at all levels of an organization.

In this context, the Heart and Minds Curve becomes a powerful tool for organizations aiming to reduce incidents and enhance overall safety performance. By understanding and addressing both the emotional and cognitive aspects of individuals, organizations can create an environment where safety is not just a set of rules, but a shared value ingrained in the fabric of the organizational culture.

The H&M methodology works with 23 dimensions that assess different aspects of organizational safety culture, ranging from communication processes to operational procedures and behavioral. Measurements are conducted across five levels: pathological, reactive, calculative, proactive, and generative, where each level expects a specific type of behavior and process implementation as figure 3.

Despite being widely used in measuring Occupational Safety and often understood as independent disciplines, they are extremely correlated. Approximately 80% of the Hearts and Minds dimensions have some correlation with Process Safety.

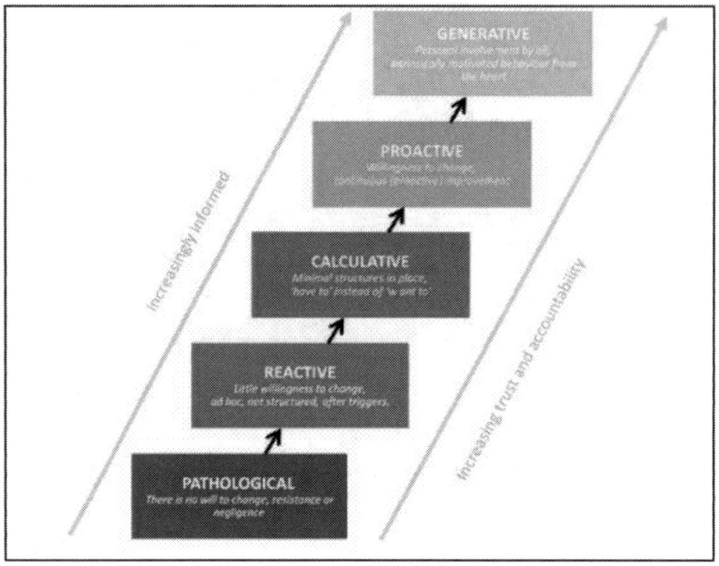

Figure 3: Culture Ladder[3]

2 Case Study

In the period between 2015 and 2020, another methodology was used within the company to measure the maturity of safety culture. In 2021, the decision was made to replace the previous methodology with Hearts and Minds. The results obtained from the evaluation of Ipiranga's employees using Hearts and Minds indicated a reactive ladder in terms of cultural aspects and a calculative ladder for the operations. The study also highlighted the main points of concern: incident communication and root cause analysis, integrity and reliability and new technology's, contractor management, training and empowering.

Based on this diagnosis, the most critical items were assessed to create an action plan for their development. For this aspect a master plan was developed with the intention of elevating the classification to the calculative level by 2025. Figure 4 shows the main programs implemented so far based on technology, management system, human factors.

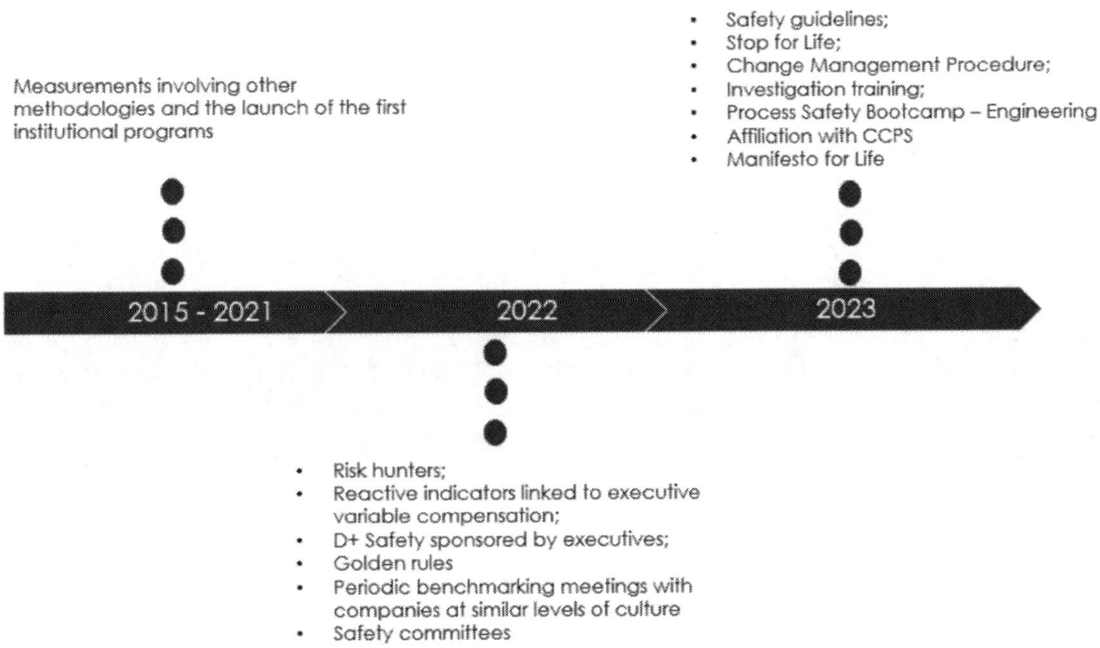

Figure 4: Implementation task and action ba

2.1. Programs based on Technology

2.1.1. Tecnology master plan

Based on the underlying causes of accidents, a technology master plan was designed to help reduce them. Among the technologies mapped, operations automation, tank level alarm, pneumatic valves to interrupt the product flow instantly, and level alarm in shipments stand out. This plan has already advanced 75% in these 3 years and new technologies are continually being studied for implementation.

2.1.2. Maintenance management system

A new system to manage corrective and preventive maintenance was implemented, as well as a corporate structure with trained professionals. With the new system, all periodicities were standardized to increase the reliability of the equipment. Furthermore, the system also talks to the management system, providing how maintenance should be carried out, as well as safety instructions for carrying out the task. The system still has a lot to evolve when compared to the best market practices.

2.2. Programs based on Management Systems

2.2.1. Seniority of Layer protection analysis and risk assessment

The maturity of our safety culture also involves a comprehensive assessment that evaluates the potential risks present in our facility, both individual and social.
As the analysis and assessment protocols evolved, there was a significant emphasis on optimizing the evaluation of items with major critical risks. Ensuring workforce involvement in creating and distributing documents, procedures, and disseminating information through committees and townhall meetings became a reliable part of this process. For the studies conducted, activities are monitored not only by local leadership but also by the HSE department.

2.2.2. Implementation of safety guidelines and Golden Rules

Safety guidelines are a guiding framework for the behaviors of Ipiranga employees, ensuring the execution of our purpose tied to recognition and consequences. On the other hand, the Golden Rules emphasize situations that are understood as "zero tolerance," meaning those that are not subject to any form of tolerance. With these guidelines, it was possible to instruct employees about what is considered safe behavior and situations that require intervention. Through this guidance, it was also possible to establish the mindset of "safety as a non-negotiable value "Transparency in the relationship with stakeholders.

To ensure involvement with third parties, the Environmental, Governance, and Sustainability reports include data related to personal and process accidents. In these reports, in addition to long-term goals, data related to primary containment losses can also be found. "Proactive Safety Measures as Leading KPIs

An advanced process safety culture goes beyond mere compliance and actively promotes proactive safety measures. Leading KPIs associated with safety audits, risk assessments, and adherence to safety procedures are explored in this section. The discussion centers on how a culture that values continuous improvement and anticipates potential risks reinforces the reliability of safety processes.

2.2.3. Learning from Experience

Based on CCPS pillar of Learning from Experience, adaptation means that themes can be better works, as they are shaped to unite the desire for culture to mature and the real level at which we find ourselves. In this way, it is possible to guarantee the involvement of senior leadership, as explained previously, and translate all the information generated by the investigation committees into lessons learned to be disseminated to the operational units.

In addition to information about accidents, they also include comprehensive actions, which are monitored in management committee meetings, which aim to prevent the recurrence of the root causes of events or reduce accidents with high potential.

2.2.4. Alignment of Process Safety Culture with Leading KPIs

A key aspect of this relationship lies in the alignment of process safety culture with leading KPIs. By fostering a culture where safety is ingrained in every aspect of operations, organizations can positively influence KPIs related to incident prevention, near-miss reporting, and proactive safety measures. This section elucidates the critical role that a strong safety culture plays in shaping these leading indicators. In addition, several initiatives to leverage the culture were planned in the strategy, as figure below

2.3. Programs based on human factors

2.3.1. Empowering Workforce Engagement for Incident Reporting

One of the cornerstones of an effective process safety culture is the empowerment of the workforce to actively engage in incident reporting. This engagement serves as a leading KPI, reflecting the organization's commitment to transparency and continuous improvement. Here, we delve into how a culture that encourages open communication and values incident reporting contributes to the reliability of safety data.

2.3.2. Enhancing leadership team involvement in the monitoring of process safety incidents life cycle.

The active engagement of top leadership played a pivotal role in driving the surge in safety reporting. Regular monitoring via monthly meetings and the direct involvement of leaders in the incident investigation process were key contributors to this notable increase, with some regions experiencing a substantial rise of up to 400%. This underscores the positive impact of leadership commitment on fostering a robust safety reporting culture within the organization.

2.3.3. Risk Hunters

With the goal to listen to employees about their work environment, a program was designed so that we could map opportunities to improve reports related to unsafe conditions and behavioral observations.

The implementation of the program initially linked to a recognition program resulted in high adherence, but the biggest gain concerns having “minds and eyes close to the task”, allowing the workforce to have a closer look. determined for the conditions that were previously part of the environment.

Nowadays, Risk Hunter is running in all operation units and have more than 10.000 reports in less then 20 months after his start.

3 Continuous learning and Adaptation for Enhanced Reliability

A dynamic process safety culture emphasizes continuous learning and adaptation. This final section examines how a culture that encourages learning from incidents, near misses, and industry best practices fosters reliability in the face of changing operational landscapes. The ability to adapt to new challenges becomes a leading KPI in assessing the resilience of both the safety culture and operational reliability.

Through these sections, we aim to provide a comprehensive understanding of how a strong process safety culture becomes the linchpin for enhancing reliability, with a keen focus on leading KPIs that drive continuous improvement.

Although certain dimensions were given priority, it is evident that others also directly impact the selected items. For example, dimensions like "how occupational safety is discussed in meetings and committees" and "what priority is given to operational safety?" play a significant role. By raising awareness among the workforce, even when focusing on specific aspects, an overall improvement can be observed.

After 3 years working with Hearths and minds methodology for measuring culture, we can identify how the evolution of the company's safety maturity leveraged engagement and resulted in an increase in Tier 3 event reports, acting at the base of the pyramid and preventing events of medium and high severity (Tier 1 and Tier 2)

4 Conclusion

Based on the graph below, we can identify a comparison between the increase in Tier 3 event reports (double 2023 vs. 2022), the reduction of Tier 2 events, and the elimination of Tier 1 events. All of this aligns with the evolution of cultural maturity on the Hearts and Minds scale.

Tracing a parallel between the process safety metrics based on API RP 754 and bird pyramid, become clear (improve clear) the impact of Tier 1 and 2 index reduction due to of Tier 3 increases.

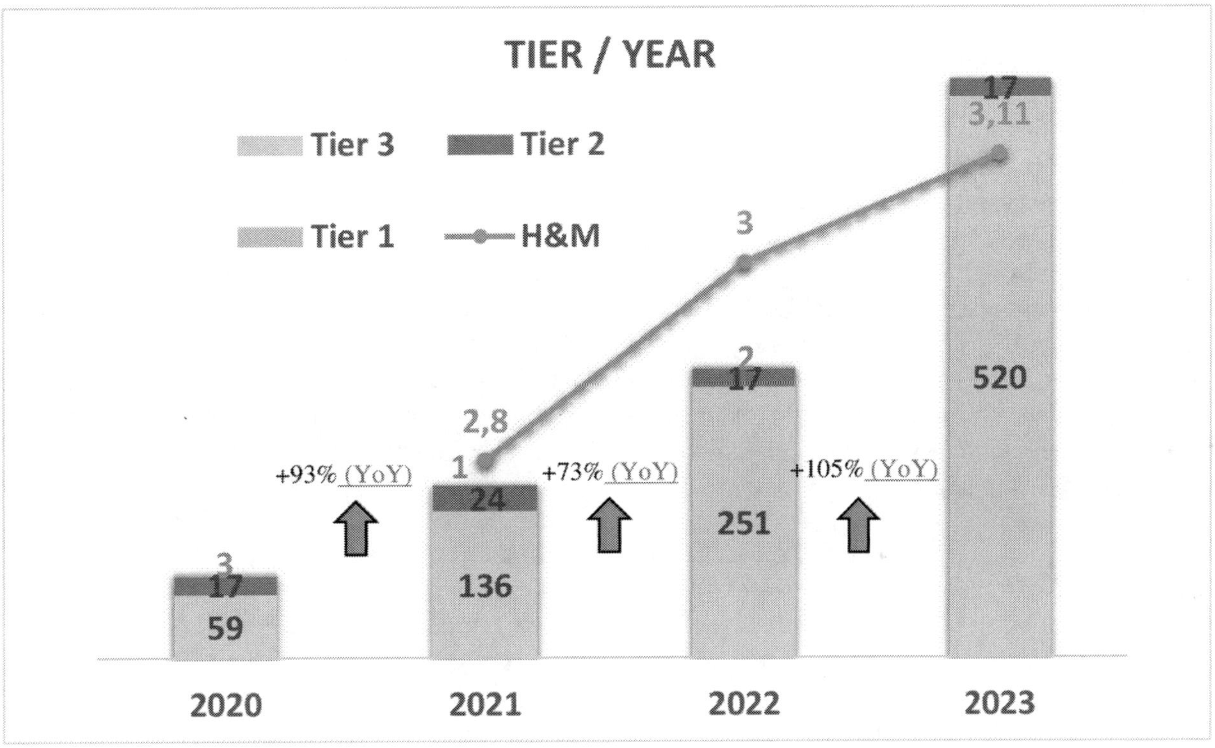

Graph 1 - Incident Tier Index x Hearts & Minds

Alternatively, if we establish a cross-analysis between the turnover volume and the amount from LOPC, we can clearly demonstrate this rate reduction, amounting to around 75% when comparing 2023 to 2022.

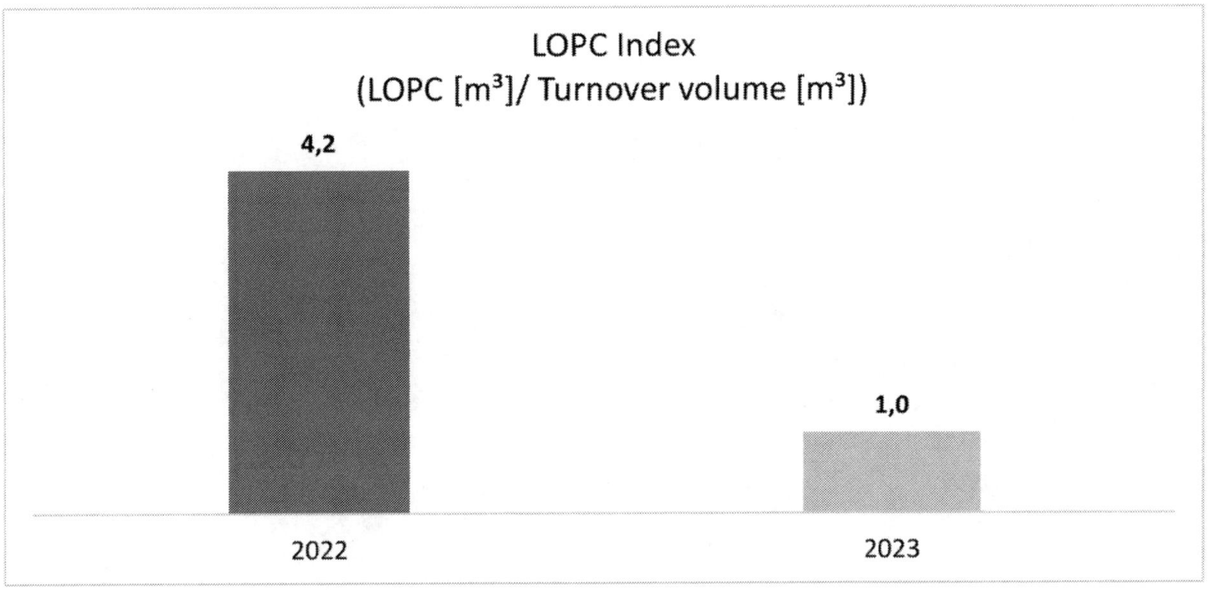

Graph 2 – LOPC volume x the turnover volume

The conclusion for Ipiranga, analyzing different Leading and Lagging KPIs, underscores the extraordinary importance of its initiative and governance strategy regarding process safety awareness, which has been bolstered by improvements in cultural understanding. The remarkable and highly motivating results demonstrate that we are making significant progress in our journey towards process safety excellence. Despite the substantial improvement measured across the 23 dimensions of H&M in our latest internal diagnosis, the implementation of effective initiatives to enhance our culture has yielded impressive outcomes. Within just one year, we have successfully transitioned from underreporting Tier 3 incidents to producing more accurate and comprehensive occurrence reports. These reports are now more robust and detailed, enabling us to establish a baseline and identify trends and systemic root causes more effectively. With the mapping of chronic and sporadic root causes, we were able to define strategies and investment plans more accurately, focusing on protective technologies to prevent top events that fall within the ranges of Tier 2 and Tier 3 incidents.

Our upcoming actions are centered on addressing the primary culprits within our structure to meet the calculated parameter within the H&M curve. We anticipate a significant boost in our reporting base and a reduction in Tier 2 incidents while striving to maintain a Tier 1-free environment. This ensures a safe work environment free of process accidents for our operational team, who are exposed to risks. Furthermore, we are committed to fulfilling our ESG objectives outlined in our 10-year plan (2020-2030), which we've undertaken alongside our shareholders and clients. This serves to showcase our position as a leader in process safety within the Brazilian market.

BOOK:

[1] Bugalia, N., Y. Maemura, and K. Ozawa. 2019. Safety Culture in High-Speed Railways and the Importance of Top Management Decisions. ADBI Working Paper 955. Tokyo: Asian Development Bank Institute. Available: https://www.adb.org/publications/safety-culture-highspeed-railways-importance-top-management-decisions.

[2] Humann and organisationa factors (HOF). https://www.era.europa.eu/domains/safety-management/human-and-organisational-factors-hof_en. Accessed 01 February

[3] Wright, M.; Opiah, S.; *Literature review: the relationship between psychological safety, human performance and HSE performance.* EI psychological safety review. 2018. Available: https://heartsandminds.energyinst.org/__data/assets/pdf_file/0007/648484/Literature-review-the-relationship-between-psychological-safety,-human-performance-and-HSE-performance.pdf

[4] Guidelines for Risk Based Process Safety, AIChE Center for Chemical Process Safety. Wiley-AIChE, New York, 2007.

[5] American Petroleum Institute. Process Safety Performance Indicators for the Refining and Petrochemical Industries (API RP 754). Washington. 2021.

[6] Guidelines for Process Safety Metrics, AIChE Center for Chemical Process Safety. Wiley-AIChE, New York, 2009.

GEXCON

Gexcon.com

Gexcon makes the world a safer place ▶▶

GEXCON

Agenda

When going beyond emperical is really needed

01 Intro to Siting Assessment

02 Hybrid Approach

03 Case Studies

04 QA

05

Gexcon.com

▶▶

GEXCON

Intro to Siting

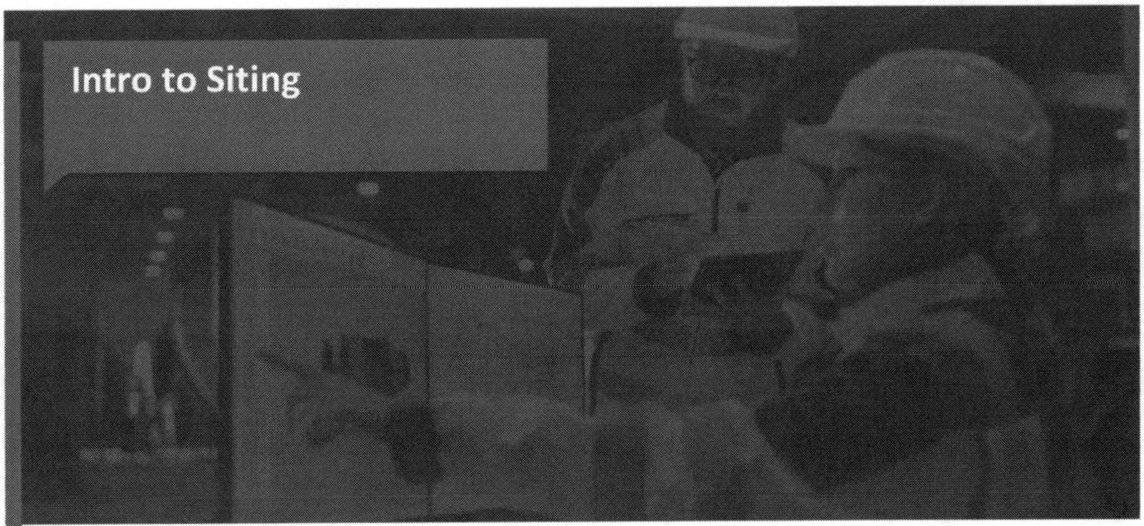

Gexcon.com

GEXCON

Introduction to Facility Siting

- Since BP Texas City Accident OSHA's 29 CFR Part 1910.119 requires a Facility Siting Study to be conducted every 5 years in order to comply with their Process Safety Management program.
- Apply process hazards consequence analysis to estimate severity.
- The overpressure, radiation and toxicity possibly can only tell part of the story.

Gexcon.com

4

GEXCON

Introduction to Facility Siting

- Scenario definition from Plotplan and P&IDs review.

- Options
  - Look-up tables.
  - Consequence Based FSS: Evaluate the results of the release scenarios, utilizing a consequence-based methodology that meet the requirements of OSHA Process Safety Management (29 CFR Part 1910.119).
  - Risk Based FSS: Inclusion of Frequency and Ignition.

Gexcon.com

5 ►►

6

GEXCON

API Usual Methodology

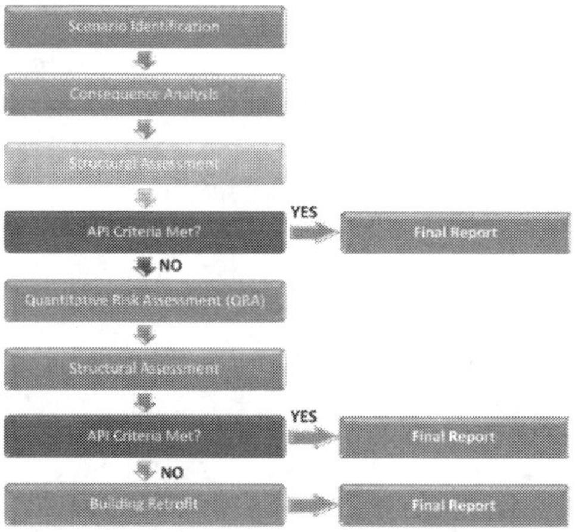

Gexcon.com

►►

GEXCON

Methodology – Consequence-Based

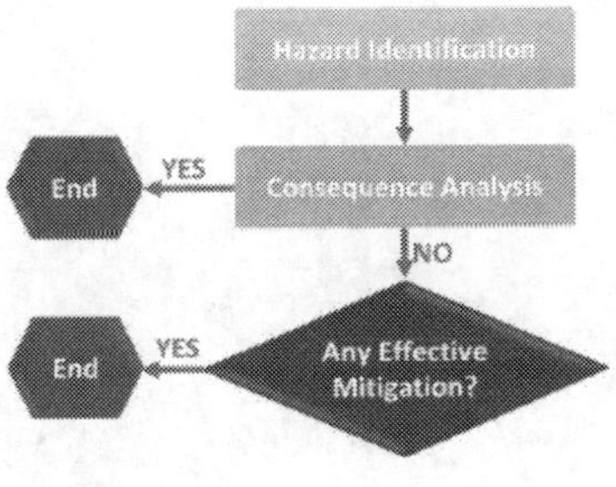

GEXCON

Methodology – Consequence-Based

- Scenario definition from Plotplan and P&IDs review
- 3 hole sizes, Maximum Credible Event (MCE) or client specific requirement
- Radiation Modeling – e.g. 4 kW/m^2
- Toxic Modeling – e.g. ERPG, Probit
- Overpressure – e.g. 0.6 psi, 0.9 psi

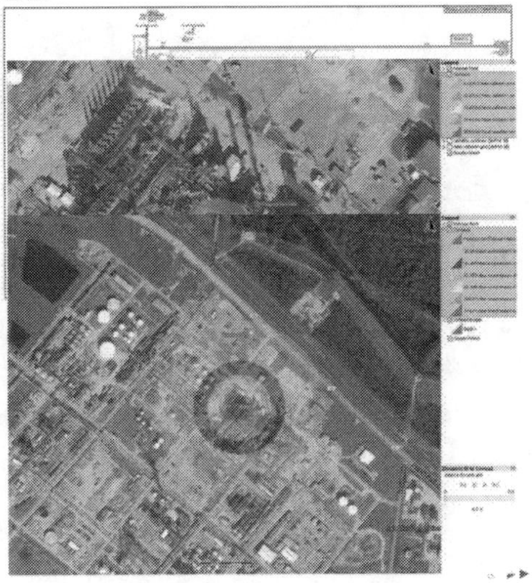

GEXCON

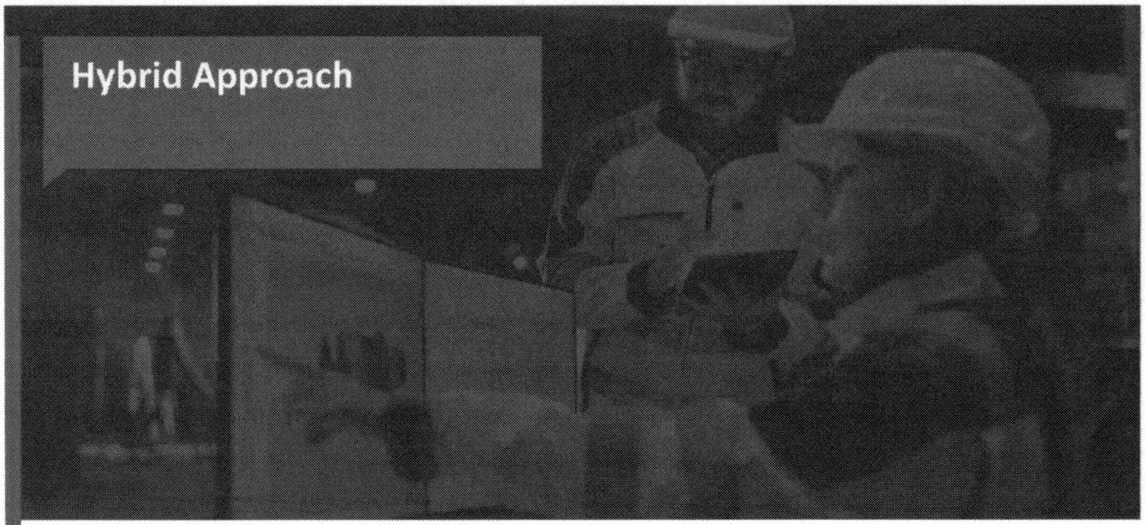

Gexcon.com

▸▸

10

GEXCON

Methodology

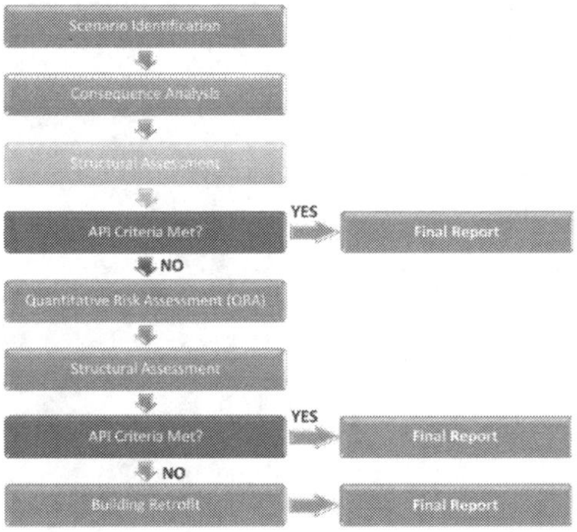

Gexcon.com

▸▸

GEXCON

Methodology – Consequence-Based (Hybrid Approach)

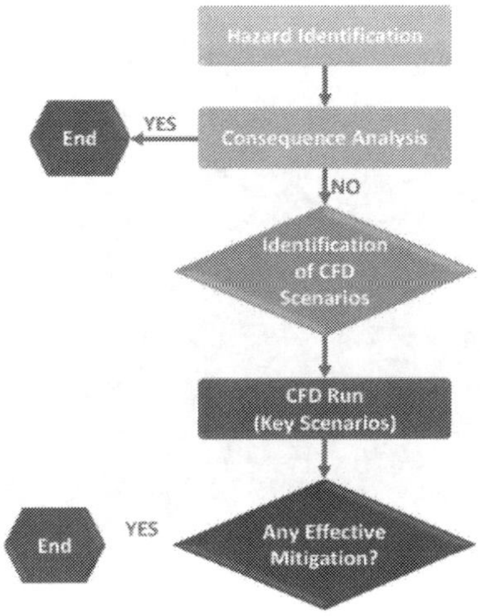

GEXCON

GEXCON

Methodology – Consequence-Based - CFD

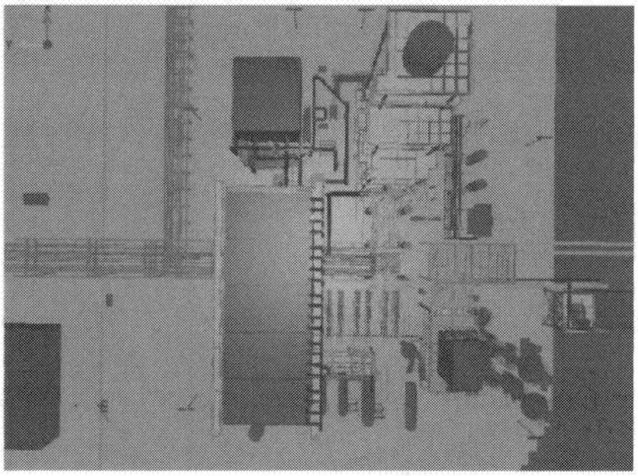

GEXCON

Methodology – Consequence-Based - CFD

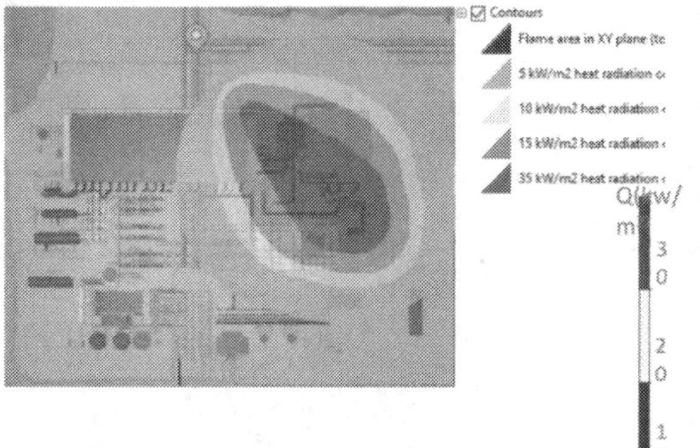

2D model Jet Fire Radiation – H2 2Kg/s Release Rate

CFD Jet Fire Radiation – H2 2 Kg/s Release Rate

GEXCON

Methodology – Consequence-Based - CFD

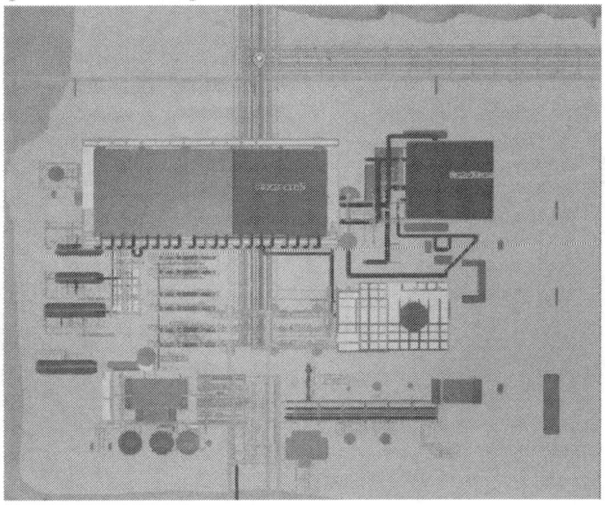

Gexcon.com

15

GEXCON

Methodology – Consequence-Based – CFD

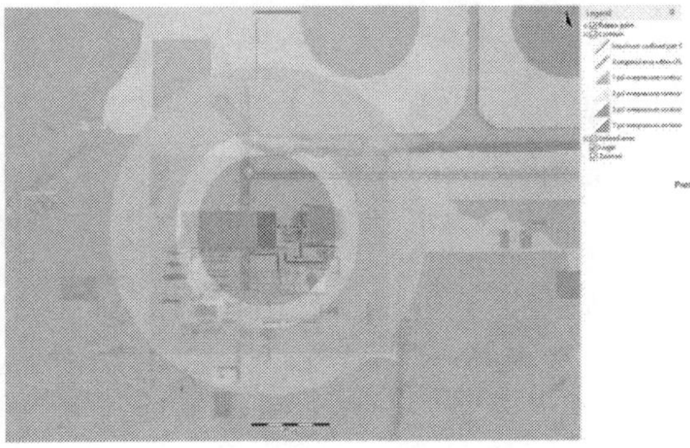

2D model Overpressure – H2 2Kg/s Release Rate

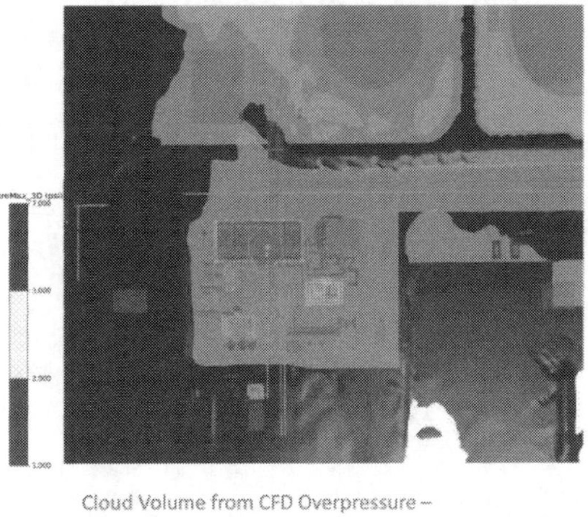

Cloud Volume from CFD Overpressure – H2 2 Kg/s Release Rate

Gexcon.com

15

GEXCON

Methodology – Consequence-Based - CFD

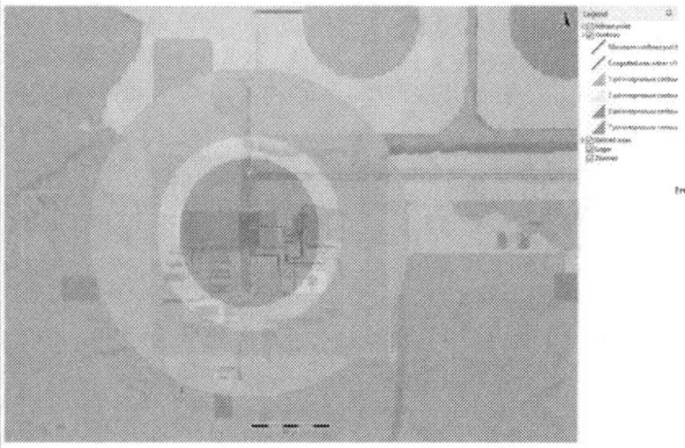

2D model Overpressure – H2 2Kg/s Release Rate

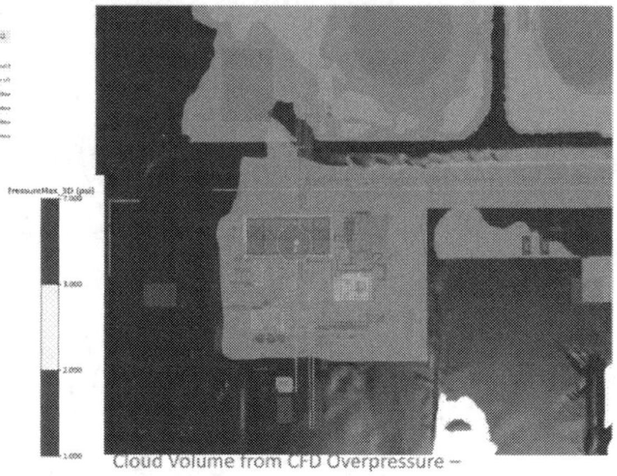

Cloud Volume from CFD Overpressure – H2 3 Kg/s Release Rate

GEXCON

Methodology – Consequence-Based - CFD

	Maximum Values					
	BDL	OV	Overpressure* (psi)	Thermal Radiation* (kW/m²)	Recommended Mitigations	ROM Cost
Conventional	4	0.838	2.0	35	- The building should be replaced by blast&fire resistant building or - The building should be retrofitted to resist the blast overpressure and thermal radiation	500K-5M
Hybrid Facility Siting Approach	2	0.00001	<1	Below 1.00	- Window glass and door and their frames should be replaced by blast resistant	5K-50K

GEXCON

Case Studies – Hydrogen 5Kg/s Release

Gexcon.com

▸▸

GEXCON

Methodology – Consequence-Based - CFD

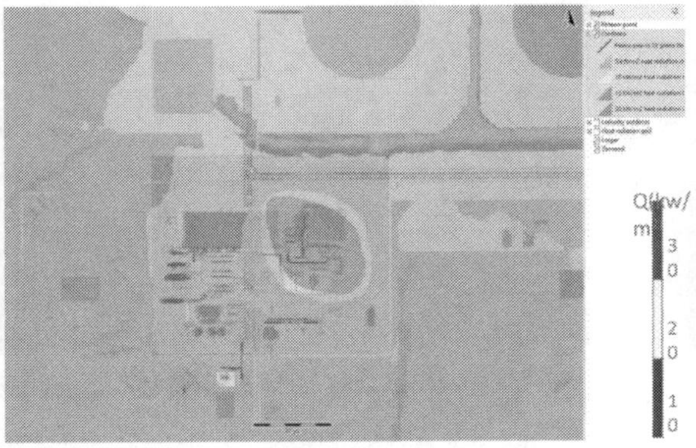

2D model Jet Fire Radiation – H2 5Kg/s Release Rate

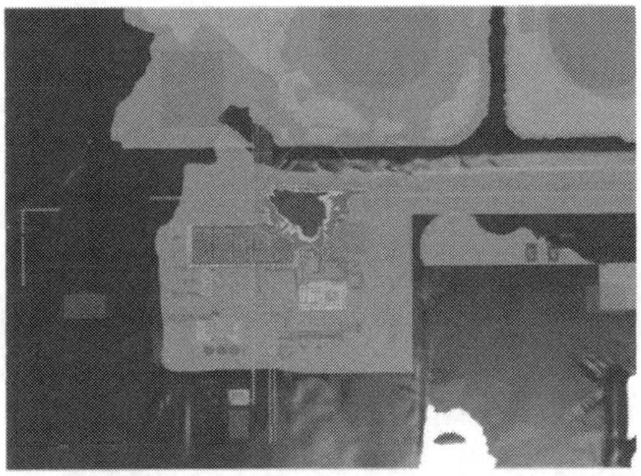

CFD Jet Fire Radiation – H2 5 Kg/s Release Rate

Gexcon.com

20 ▸▸

GEXCON

Methodology – Consequence-Based - CFD

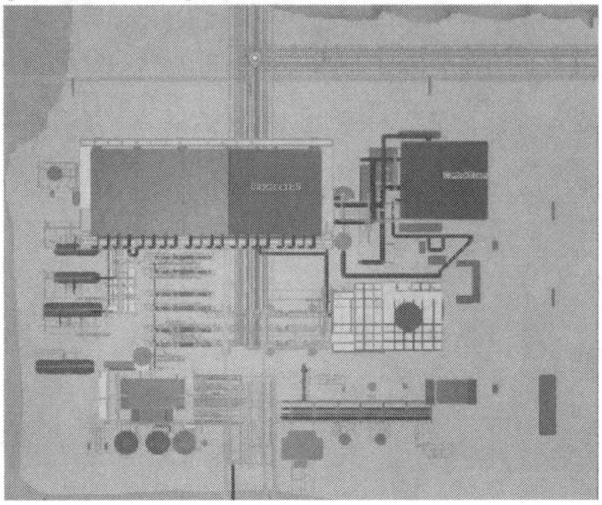

Gexcon.com

21 ▸▸

GEXCON

Methodology – Consequence-Based - CFD

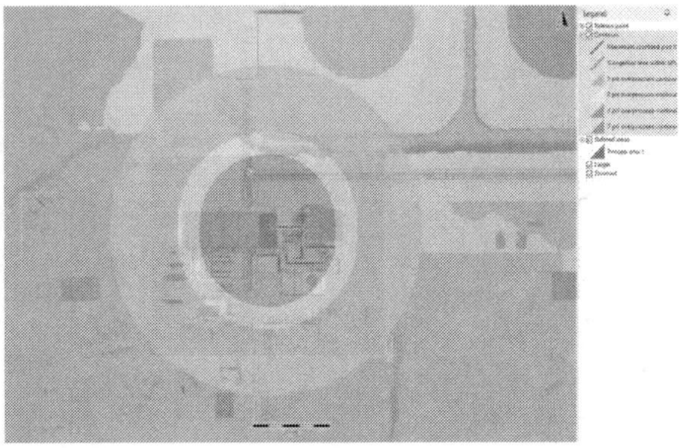

2D model Overpressure – H2 5Kg/s Release Rate

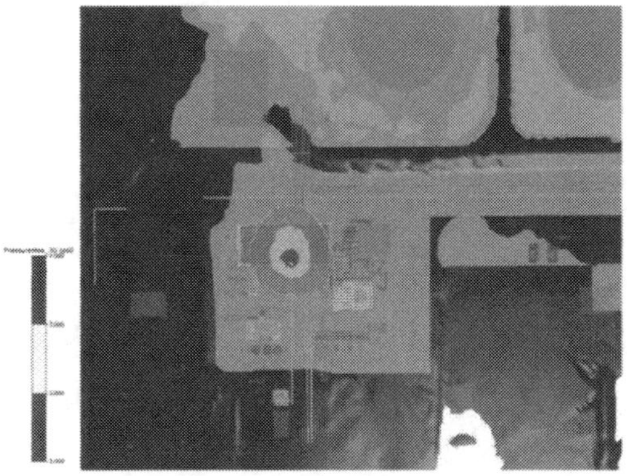

Cloud Volume from CFD Overpressure – H2 5 Kg/s Release Rate

Gexcon.com

22 ▸▸

GEXCON

Methodology – Consequence-Based - CFD

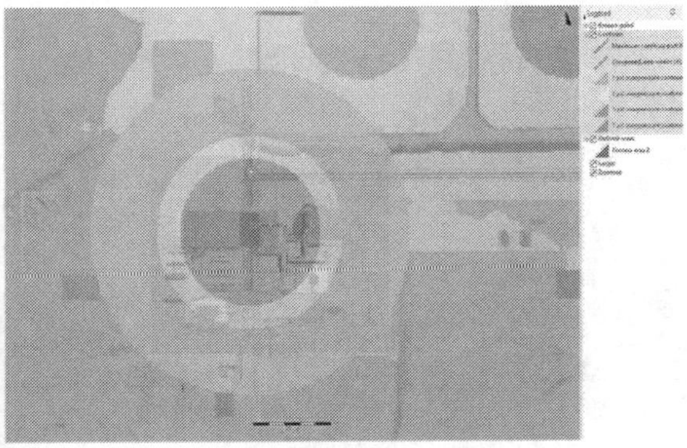

2D model Overpressure – H2 5Kg/s Release Rate

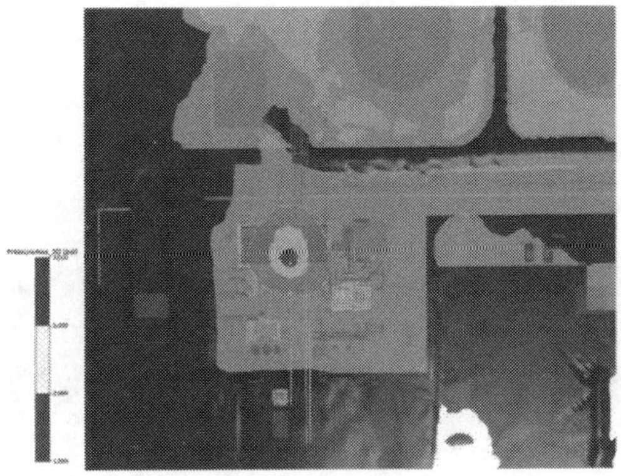

Cloud Volume from CFD Overpressure – H2 5 Kg/s Release Rate

GEXCON

Methodology – Consequence-Based - CFD

	Maximum Values					
	BDL	OV	Overpressure* (psi)	Thermal Radiation* (kW/m^2)	Recommended Mitigations	ROM Cost
Conventional	4	0.838	2.0	35	- The building should be replaced by blast&fire resistant building or - The building should be retrofitted to resist the blast overpressure and thermal radiation	500K-5M
Hybrid Facility Siting Approach	2	0.00001	< 1	Below 2.00	- Window glass and door and their frames should be replaced by blast resistant	5k-50K

GEXCON

Case Studies – Flare Location

GEXCON

Methodology – Consequence-Based - CFD

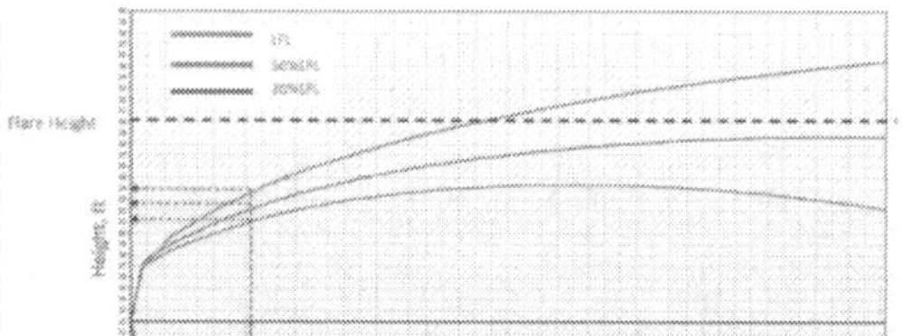

2D model Natural Gas Dispersion – 0.38 Kg/s 0.8 m/s

CFD Natural Gas Dispersion – 0.38 Kg/s 0.8 m.s

GEXCON

Methodology – Consequence-Based - CFD

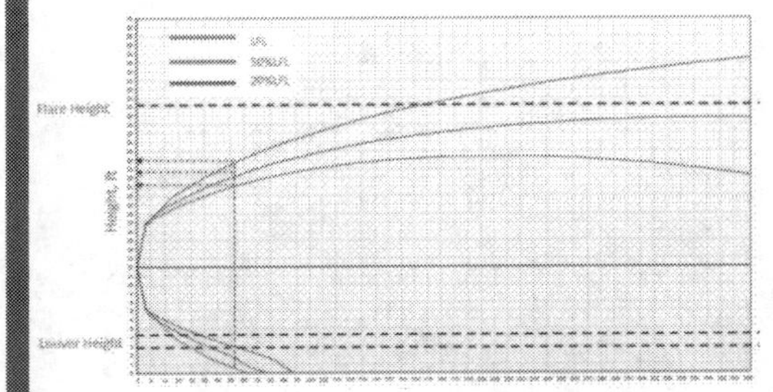

2D model Natural Gas Dispersion – 0.38 Kg/s no wind

CFD Natural Gas Dispersion – 0.38 m/s no wind

Gexcon.com

27 ▸▸

GEXCON

Methodology – Consequence-Based - CFD

	Recommendations	
	Recommended Mitigations	ROM Cost
Conventional	- The flare and louvre would receive a flammable concentration and the vent needs to be relocated or the flare would need to be moved.	100K-500K
Hybrid Facility Siting Approach	- The flare or louvre would not be reached and not modifications were needed.	

Gexcon.com

28 ▸▸

GEXCON

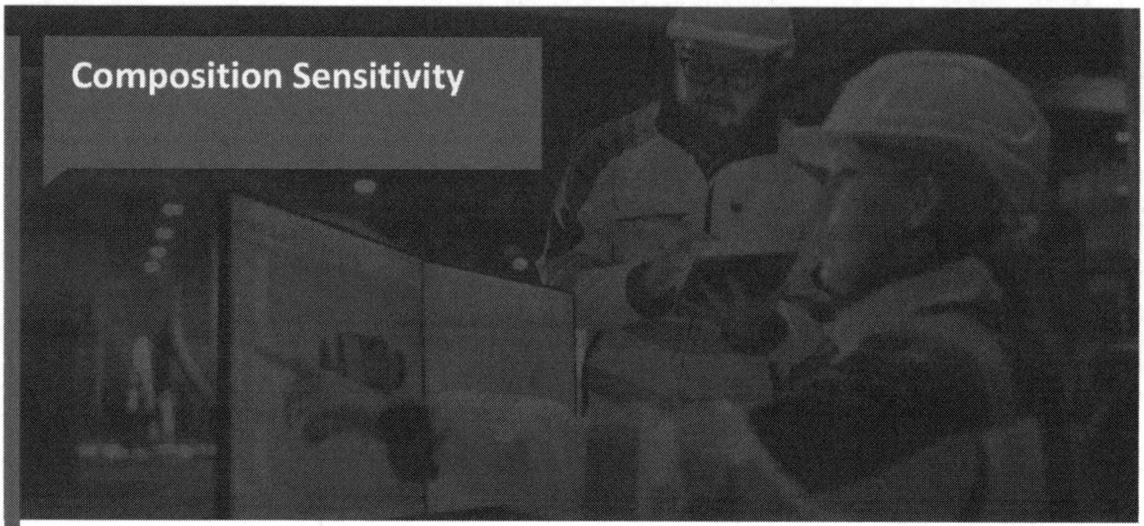

GEXCON

Natural Gas Composition Blending

- 1% H2 Blending
- 5% H2 Blending
- 10% H2 Blending
- 20% H2 Blending
- H2 Model

GEXCON

Natural Gas Composition Blending - Dispersion

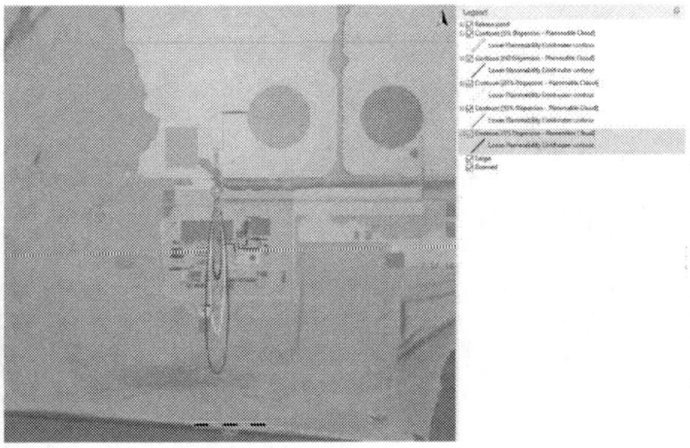

Dispersion – LFL contours

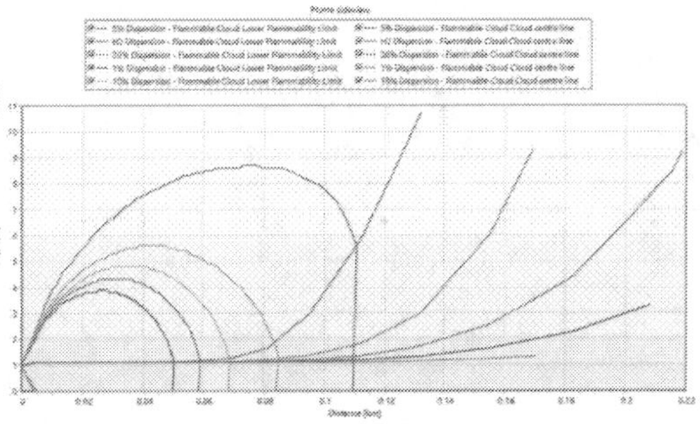

Dispersion - Sideview

GEXCON

Natural Gas Composition Blending - Dispersion

Contour maximum distances

Concentration contours distance [m]	1% Dispersion - Flammable Cloud	5% Dispersion - Flammable Cloud	10% Dispersion - Flammable Cloud	20% Dispersion - Flammable Cloud	H2 Dispersion - Flammable Cloud
Lower Flammability Limit outer contour	50	59	69	84	111
Lower Flammability Limit outer contour at 0 m	49	58	68	85	110

Concentration Results

Concentration contours distance [m]	1% Dispersion - Flammable Cloud	5% Dispersion - Flammable Cloud	10% Dispersion - Flammable Cloud	20% Dispersion - Flammable Cloud	H2 Dispersion - Flammable Cloud
Flammable Mass (Kg)	21.4	25.3	28.9	33.9	33.9
Cloud Volume (m3)	522	812	1202	2049	6756

GEXCON

Natural Gas Composition Blending - Radiation

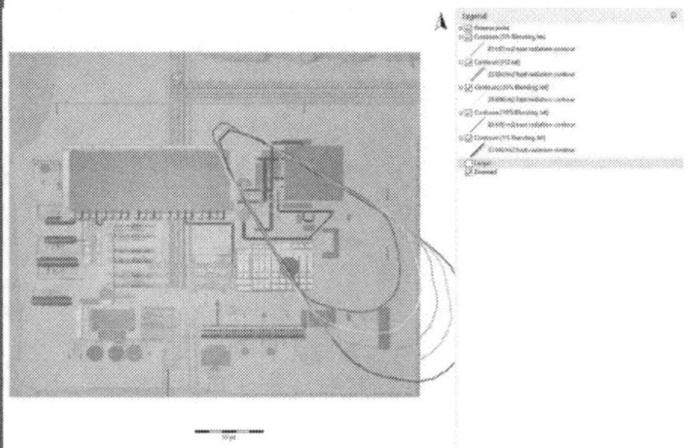

Radiation – 35 kW/m2 contours

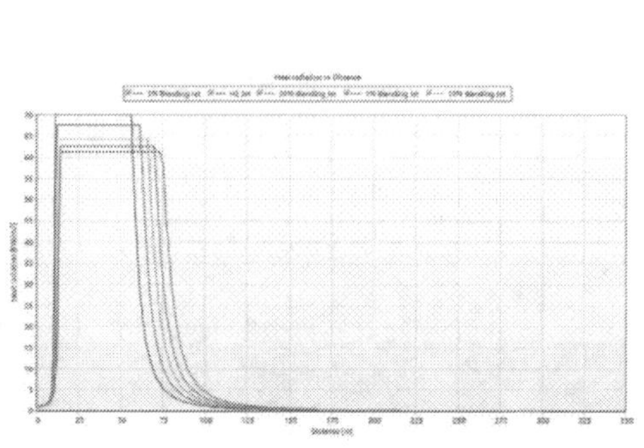

Jet Fire Radiation vs Distance

Process Safety Leadership in China

—Leading High-Quality Development for Enterprises

Haoshui Wang, Qiang Chen

Executive Vice President
China Occupational Safety and Health Association;
Associate Professor, China University of Petroleum

March 26, 2024

Content

1. **Definition of Process Safety Leadership (PSL)**
2. **Significance of PSL in the Context of China's Industrial Growth**
3. **Enhancement and Practice of PSL in Enterprises**

China's Chemical Industry Exhibited Rapid Expansion

- In 2022, China's total refining capacity reached 920 million tons, making it the world's largest refining nation.
- China's fast-growing chemical industry has been the largest in the world by revenue since 2011 and steadily grow every year.
- In 2023, China's GDP grew by 5.2% .

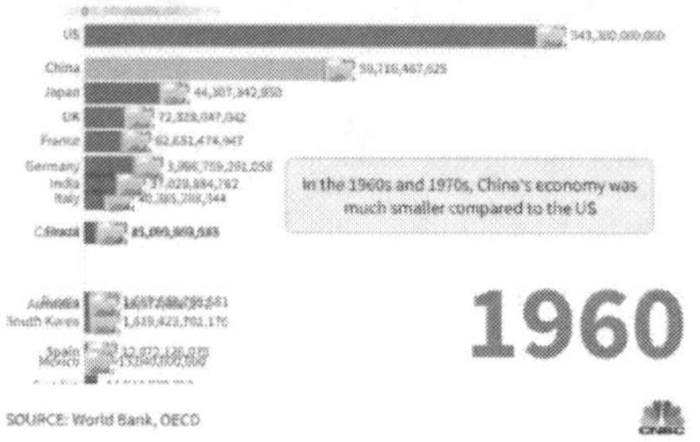

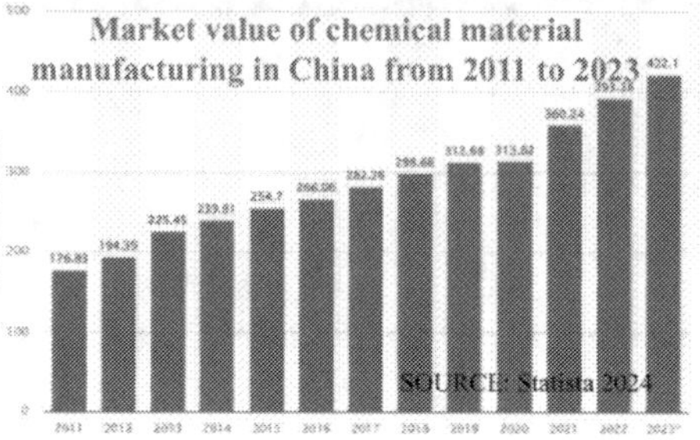

Several Major Chemical Safety Accidents Occurred in China

- **In recent years, China has encountered a series of significant safety incidents that have underscored the urgent need for enhanced safety measures.**

Aug 12, 2015; Tianjin Port Explosion, 173 dead

Mar 21, 2019; Xiangshui Explosion, 78 dead

Jan 12, 2024; Pingdingshan Coal Mine Accident, 16 dead

Jan 19, 2024; Chifeng Flash Explosion, 3 dead

May 3, 2023; Luxi Explosion, 9 dead

Oct 9, 2023; Ordos High Pressure gas, 10 dead

Safety Leadership

Accidents → Emphasizing the Need for Robust Process Safety

Inadequate Safety Leadership
Root causes of accidents

- ◆ The prevention of many accidents hinges on the extent to which leaders prioritize safety measures.
- ◆ If leaders demonstrate a keen awareness and commitment to ensuring implementation of safety protocols, the likelihood of avoiding mishaps increases substantially

CCPS Risk-Based 20 PSM Elements

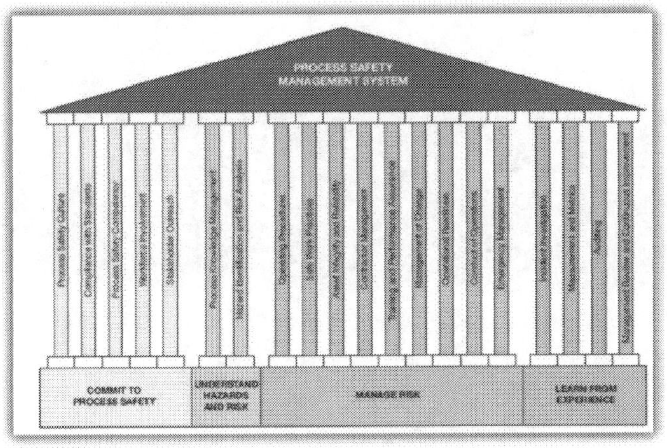

◆ **An invaluable roadmap for companies operating in the chemical industry, providing essential guidance for the development and maintenance of robust process safety management systems.**

China Safety Industry Standards: AQ/T 3034-2022

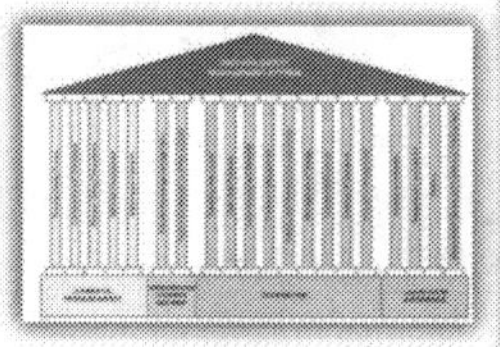

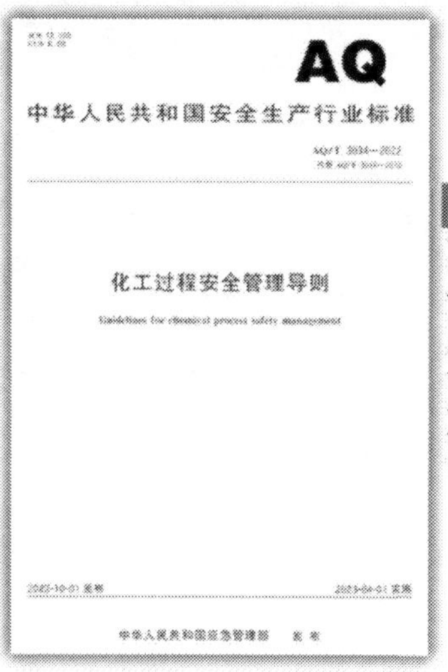

安全领导力是第一要素

Safety Leadership (first element)

1.安全领导力
2.全员安全生产责任制
3.安全生产合规性管理
4.安全生产信息管理
5.安全教育、培训和能力建设
6.风险管理
7.化工、危化品装置安全规划与设计
8.装置首次开车(投用）安全
9.安全操作
10.设备完好性管理
11.安全仪表管理
12.重大危险源(高后果场景）安全管理
13.作业安全管理
14.承包商(相关方）安全管理
15.变更管理
16.应急准备与响应
17.安全事件调查与管理
18.本质更安全
19.安全文化建设
20.持续改进

1. What is Safety Leadership ?

1. Definition of Process Safety Leadership

Mobilizing a group of individuals to handle difficult challenges

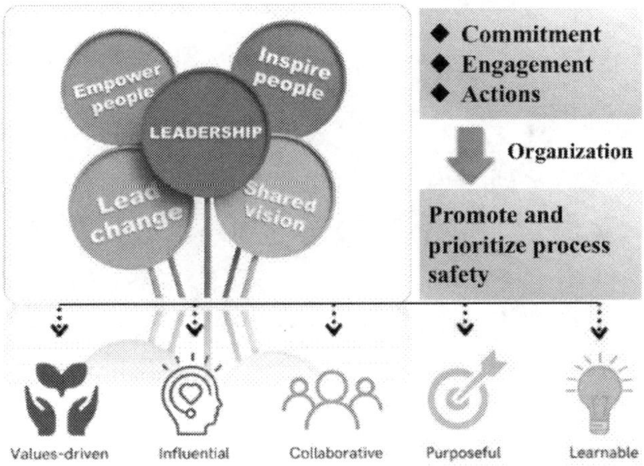

1. Definition of Process Safety Leadership

Process Safety Leadership

◆ **The ability to guide the sustainable safety development within an organization, involving leading a team in identifying and managing risks effectively**

PSL Components

- **Safety Care**: Well-being of individuals is a top priority; Genuine concern for the safety of members
- **Safety Control**: Effectively identify, assess, and control risks; Create a secure working environment
- **Safety Guidance**: Provide clear direction and support; Set standards and foster a collaborative approach

2. Significance of Process Safety Leadership

2. Significance of Process Safety Leadership

(1) Key to successful PSM implementation and thereby accident prevention

- **Top-tier leadership plays a central role in fostering an effective PSM culture**
- **Engaging the collective commitment and dedication of every member within an organization is essential for maintaining a steadfast and positive process safety culture across the workforce**

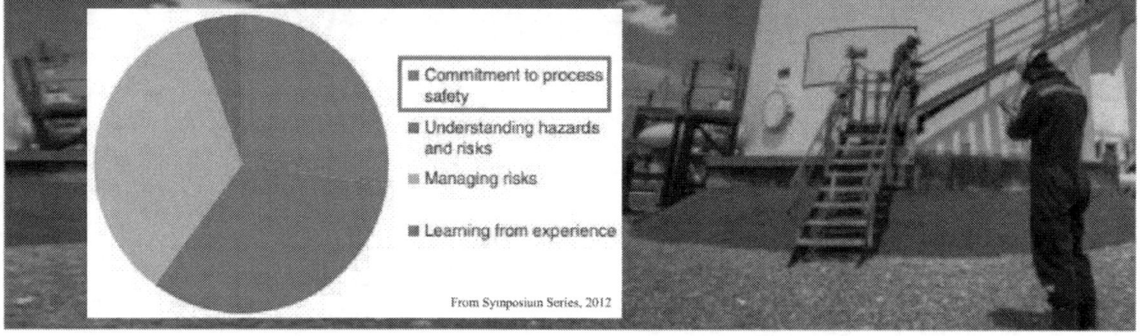

2. Significance of Process Safety Leadership

(2) Safety leadership underpins everything else we do in process safety

◆ **Leadership is not just one element among others in process safety but is foundational to the success of all safety-related efforts**

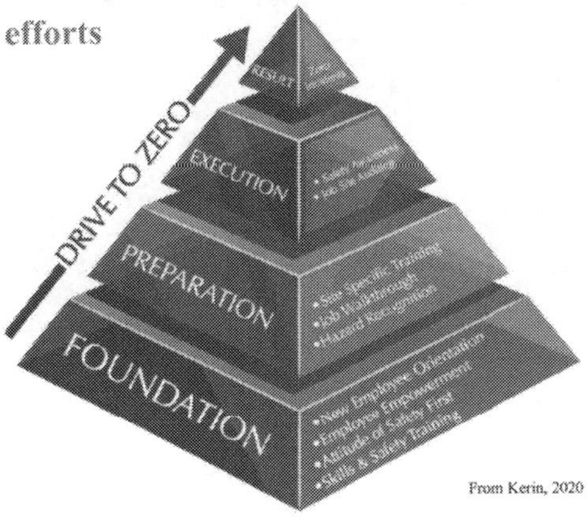

From Kerin, 2020

2. Significance of Process Safety Leadership

(3) Fit to Chinese cultural values

◆ **Leader's behavior and attitude are observed and followed by team members, setting the tone for the entire workforce**

A set of rules → (Leader's Commitment) → A shared value

◆ **Leadership accountability**

Fundamental Safety Improvement of Workplace → Enterprise Implementation → Leader

2. Significance of Process Safety Leadership

(4) Insufficient leadership in Chinese enterprises

DNV international safety assessments of five typical Chinese enterprises:
The leadership scores are not ideal, the majority scoring lower than the mean level

Enterprise	Safety Score	Leadership Score	Total Score
A	3	55.3%	58.20%
B	3	48.8%	56.9%
C	5	65.2%	64.8%
D	6	67.86%	68.9%
E	5	76.22%	67.02%

2. Significance of Process Safety Leadership

(4) Insufficient leadership in Chinese enterprises

The score of safety leadership elements of a typical central enterprise in China (46.73%)

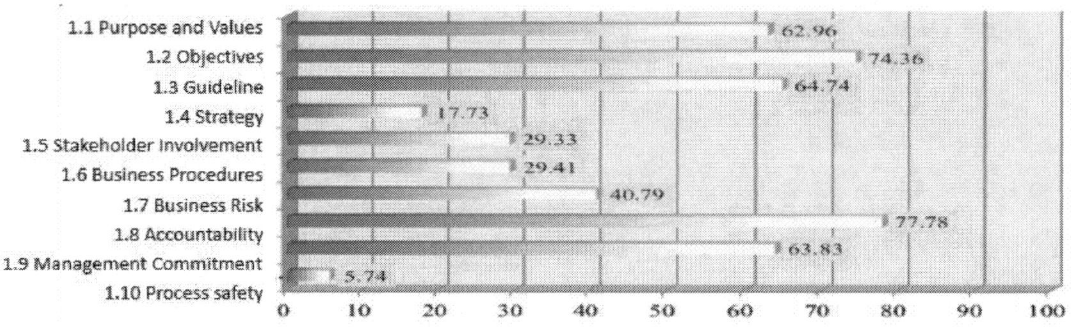

2. Significance of Process Safety Leadership

(4) Insufficient leadership in Chinese enterprises

◆ **Demise of leaders in accidents due to insufficient capabilities**

Learn the lesson before you suffer your own loss!

May 1st, 2023
Explosion occurred in the hydrogen peroxide production facility of Shandong Luxi Chemical Co., Ltd. claiming the lives of 9 individuals including plant managers

September 7th, 2023
a high-pressure gas release incident at Ordos resulted in the loss of 10 lives, including CEOs

3. Enhancement and Practice

3. Enhancement and Practice of Process Safety Leadership

(1) Four crucial skills in safety leadership (O'Dea and Flin)

Visibility	The extent to which leaders are present in the workplace, actively participating in job tasks, enforcing rules and company safety policies, and serving as safety role models.
Relationship	Establishing open, honest, and trust-based relationships through effective communication with employees, listening to their concerns, embracing their suggestions, and maintaining an open-door policy
Employee Engagement	Encouraging employees to actively participate in safety planning and decision-making, fostering a sense of ownership and responsibility among employees
Proactive Management	Creating an environment where incident reporting is encouraged and supported. Identify and address safety issues.

3. Enhancement and Practice of Process Safety Leadership

(2) Five components of process safety leadership

- ◆ Set a clear vision for safety
- ◆ Ensure safety as a top priority
- ◆ All accidents are preventable with proper measures

- ◆ Inspire and motivate others to prioritize safety
- ◆ Shape safety culture and practice throughout the organization

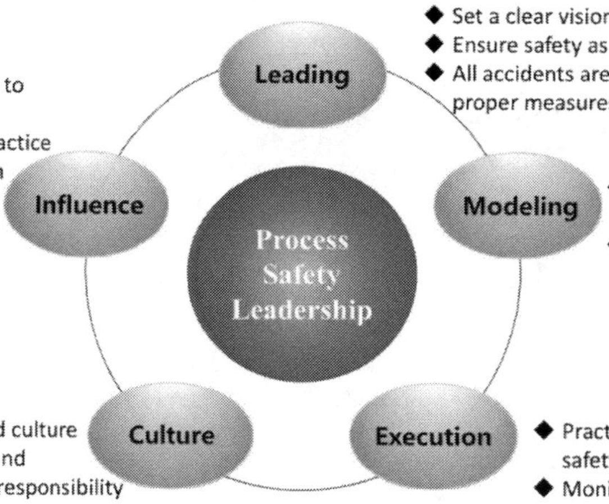

- ◆ Demonstrate a commitment to safety protocols
- ◆ Accountable for maintaining a safe workplace

- ◆ Foster a safety-oriented culture
- ◆ Open communication and encourage a collective responsibility for safety

- ◆ Practical implementation of safety measures and policies
- ◆ Monitor compliance

3. Enhancement and Practice of Process Safety Leadership

(3) Eight common practices for enhancing process safety leadership

◆Continuously elevate and solidify safety principles

◆Persist in learning; active in practice; rigorous in implementation; continuous in improving

3. Enhancement and Practice of Process Safety Leadership

(3) Eight common practices for enhancing process safety leadership

1) Visible and felt leadership to safety work

2) Lead by example, demonstrating through actual actions

3) High standards with rigorous requirements. Everyone is equal in the realm of safety

4) Establish a , set ambitious safety goals, leading the team in addressing safety issues

3. Enhancement and Practice of Process Safety Leadership

(3) Eight common practices for enhancing process safety leadership

5) Maintain effective communication with employees

6) Support employee training to foster safe behaviors and habits

7) Motivate and recognize excellent employees, regularly organize safety celebration or sharing events

8) Provide a [illegible] for employees to unleash their intelligence and contribute to enhancing safety

3. Enhancement and Practice of Process Safety Leadership

(4) Dynamic risk management

$$A(R)=IR_{(E)} \cdot DR_{(C,\ I,\ B)}$$

A: accident; R: risk; IR: inherent risk; DR: dynamic risk;
E: energy; C: control; I: information; B: behavior

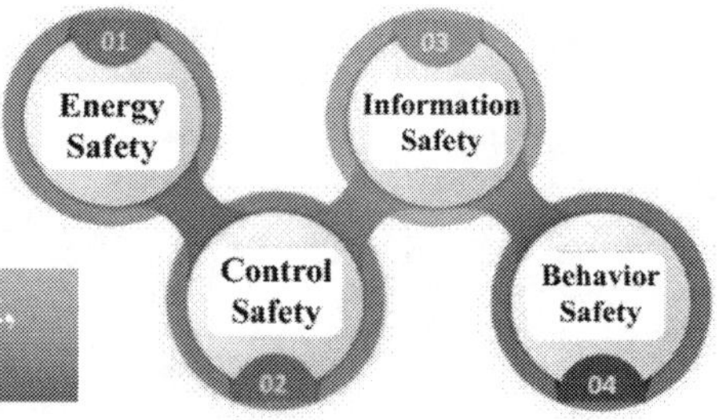

Managing safety is essentially managing risk, particularly dynamic risk.

1) Risk dynamic control technique, integrating heterogeneous data from multiple sources

◆ **For high-risk scenarios, dynamic quantification assessment based on the integration of heterogeneous data from multiple sources. Establish a model capable of dynamically monitoring the probability of incidents in high-risk scenarios.**

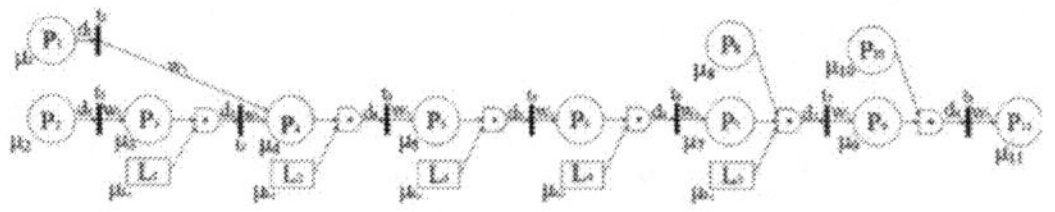

3. Enhancement and Practice of Process Safety Leadership

◆**Build an intelligent control system capable of analyzing dynamic risk changes when different protective layers fail.** Scientifically guiding daily inspections, preventive maintenance, and routine testing.

Examples of Dynamic Risk Management Platforms

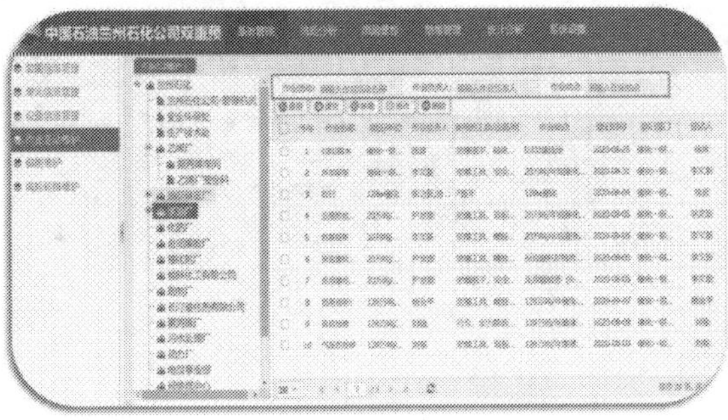

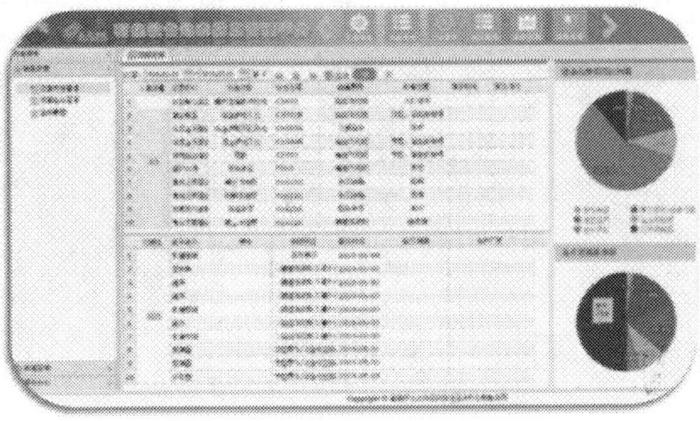

3. Enhancement and Practice of Process Safety Leadership

2) Centralized monitoring and early warning method for key parameters of chemical plant equipment based on real-time data

- ◆ A model for hierarchical management of alarm classification for major hazard sources of hazardous chemicals.
- ◆ Comprehensive lifecycle alarm management to assist leadership in decision-making.
- ◆ Deep learning-based abnormal operating condition monitoring and early warning model for
- ◆ Data-driven approach for

Multivariate Time Series Prediction Model

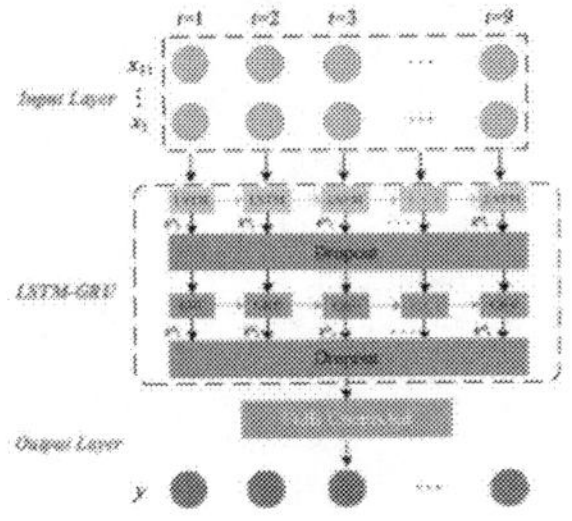

DCS data retrieval and processing

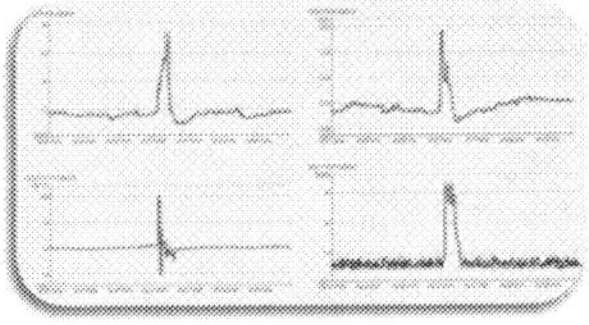

SRCC Correlation Analysis

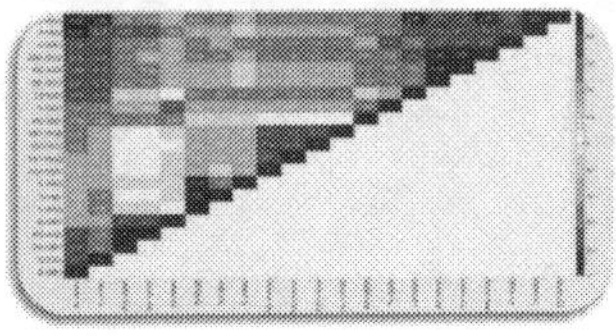

Data-Driven Network Model

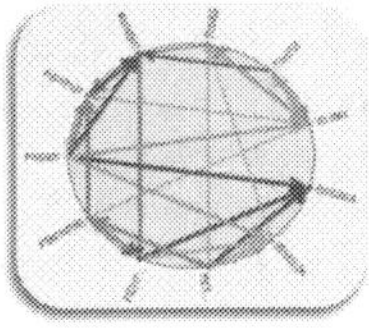

Predicted Results

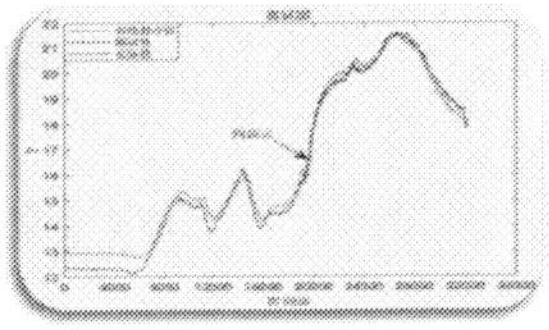

3. Enhancement and Practice of Process Safety Leadership

3) Accident scenario construction and emergency response techniques based on leakage unit division

- Method of accident scenario construction
- Emergency response techniques based on “Scenario-Response” approach , achieving the scientific and prompt deployment of emergency resources

Emergency Resource Demand Calculation

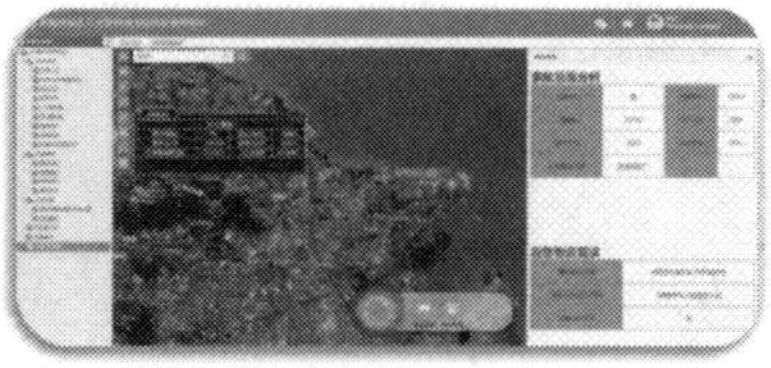

Examples of Leakage Unit Division

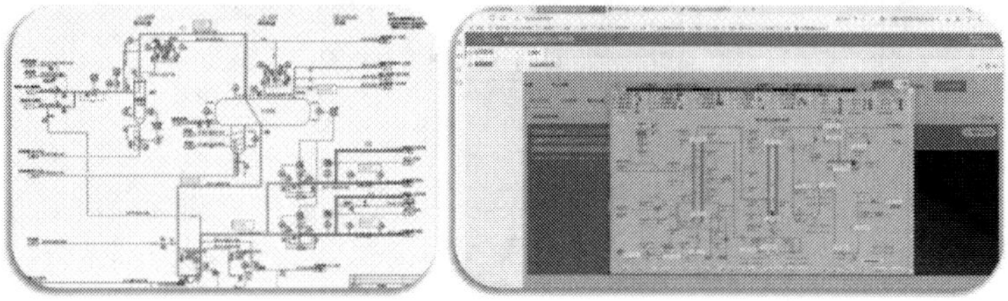

Emergency Resource Regional Coordination

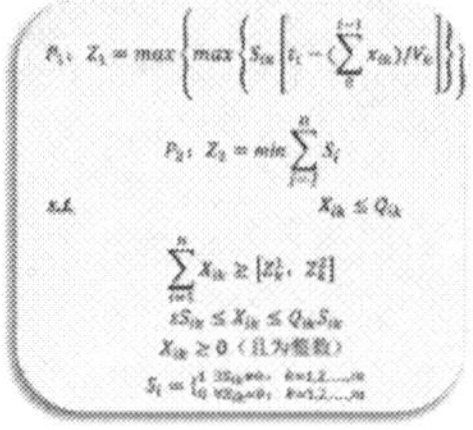

3. Enhancement and Practice of Process Safety Leadership

4) Four-tier element approach: Quantitative assessment method for evaluating process safety management

Completeness audit

Whether corresponding management procedures or systems have been established for the relevant management activities?

Effectiveness audit

Can existing management procedures and systems address all issues?

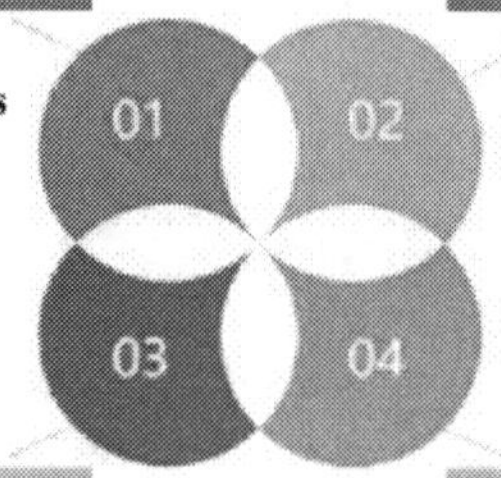

Execution audit

Whether work is being executed in accordance with relevant procedures and regulations?

Performance audit

How effective is the execution of on-site safety management activities?

Summary

Process Safety Leadership

4 crucial skills
visibility, relationship, employee involvement, proactive management

5 components
leading, modeling, execution, culture, influence

8 common practices
felt leadership, lead by example, rigorous requirements, safety vision, effective communication, employee training, motivate, platform

4 tier element approach
completeness, effectiveness, execution, performance

The 20th Global Congress on Process Safety

SPRING24 +20TH GCPS
A Joint AIChE and CCPS Meeting

Tell Me and I Forget; Teach Me and I May Remember; Involve Me and I Learn.

—— Chinese Proverb Xun Zi (translated by Benjamin Franklin)

UNITING
Process Safety Experts
SOLVING
Global Challenges
SPRING24
+20TH GCPS
A Joint AIChE and CCPS Meeting
March 24–28, 2024
New Orleans Ernest N. Morial
Convention Center
New Orleans, LA

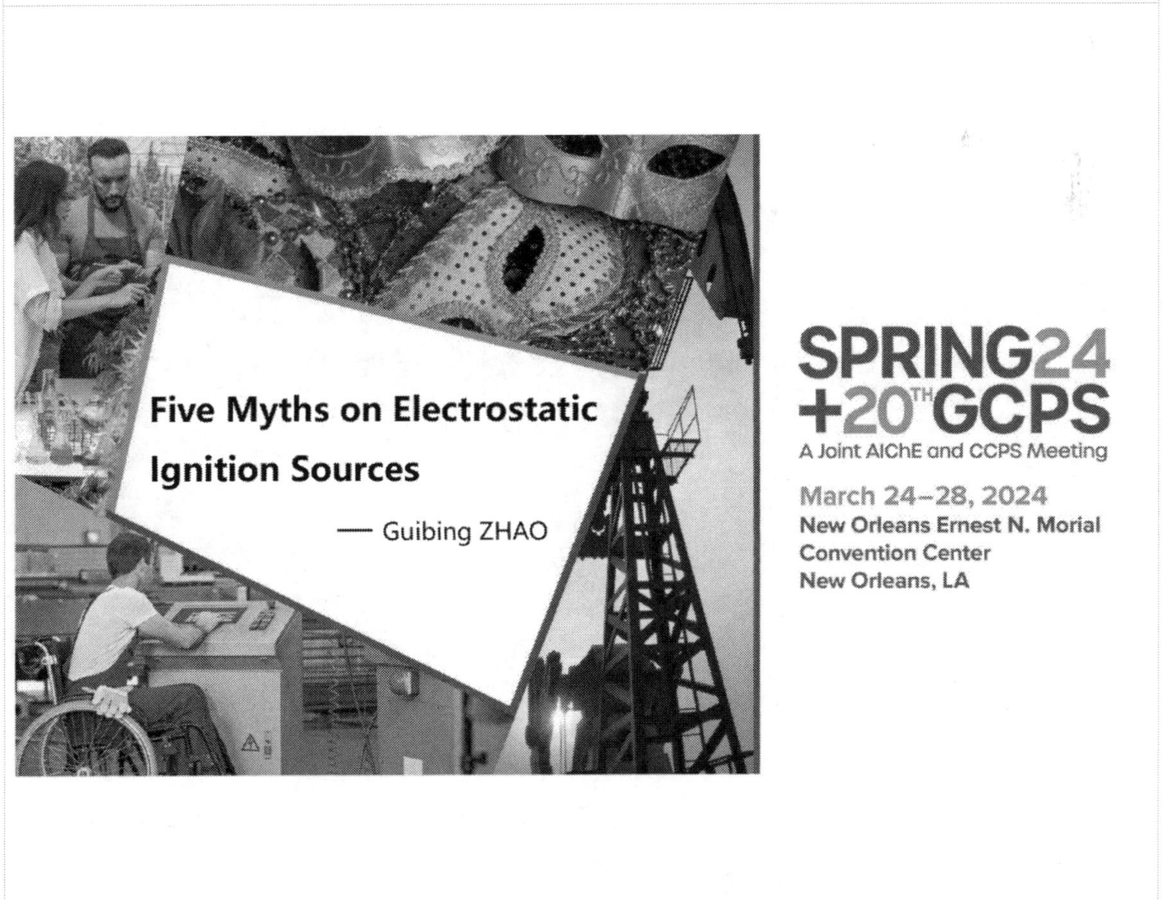
Five Myths on Electrostatic
Ignition Sources
— Guibing ZHAO
SPRING24
+20TH GCPS
A Joint AIChE and CCPS Meeting
March 24–28, 2024
New Orleans Ernest N. Morial
Convention Center
New Orleans, LA

SPRING24+20TH GCPS
A Joint AIChE and CCPS Meeting

March 24–28, 2024
New Orleans Ernest N. Morial Convention Center,
New Orleans, LA

Guibing ZHAO

Principal Engineer at Wanhua Chemical Group

Dr. Zhao is an eminent authority in the field of process safety, having spearheaded a wide array of projects. His proficiency includes carrying out Process Hazard Analyses, leading incident investigations, designing emergency relief systems using DIERS techniques, and assessing risks pertaining to dust explosions, electrostatic ignition hazard, flammable materials, and classifying hazardous zones. Dr. Zhao delivers exhaustive evaluations in process safety to his clients and is a widely respected trainer in the realms of process safety management and engineering. An author of more than 50 peer-reviewed papers across several multidisciplinary areas, he serves as a valued reviewer for prestigious journals such as the Journal of Loss Prevention in the Process Industries and the AIChE Journal.

SPRING24+20TH GCPS
A Joint AIChE and CCPS Meeting

March 24–28, 2024
New Orleans Hilton Riverside
New Orleans Convention Center

Five Myths on Electrostatic Ignition Sources

Guibing Zhao, PhD
Wanhua Chemical Group Co., LTD.

Content

SPRING24+20TH GCPS
A Joint AIChE and CCPS Meeting
March 24–28, 2024
New Orleans Ernest N. Morial
Convention Center,
New Orleans, LA

Part 01 Electrostatic Ignition Caused Incidents in Industry

- Numerous Incidents Occurred in Industry Due to ES Ignition

July 17, 2007, Kansas, USA

SPRING24+20TH GCPS
A Joint AIChE and CCPS Meeting

March 24–28, 2024
New Orleans Ernest N. Morial Convention Center,
New Orleans, LA

Part 01 **Electrostatic Ignition Caused Incidents in Industry**

- Numerous Incidents Occurred in Industry Due to ES Ignition

1 dead, 33 injured after explosions

Nov 20, 2017, Verla International in New Windsor, NY, USA

SPRING24+20TH GCPS
A Joint AIChE and CCPS Meeting

March 24–28, 2024
New Orleans Ernest N. Morial Convention Center,
New Orleans, LA

Part 01 **Electrostatic Ignition Caused Incidents in Industry**

- Numerous Incidents Occurred in Industry Due to ES Ignition

Jan 22, 2019, an industrial park in Guangdong Dongwan

SPRING24+20TH GCPS
A Joint AIChE and CCPS Meeting

March 24–28, 2024
New Orleans Ernest N. Morial
Convention Center,
New Orleans, LA

Part 01 **Electrostatic Ignition Caused Incidents in Industry**

- Numerous Incidents Occurred in Industry Due to ES Ignition

a worker dragging his foot along a factory floor sparked an explosion at a Texas chemical plant, Mar 15,2018

SPRING24+20TH GCPS
A Joint AIChE and CCPS Meeting

March 24–28, 2024
New Orleans Ernest N. Morial
Convention Center,
New Orleans, LA

Part 02 **Credible Static Ignition Source for Explosible Atmosphere**

- **Fire Triangle**
  - FUEL– Liquid (vapor or mist), gas, or solid capable of being oxidized. Combustion always occurs in the vapor phase; liquids are volatized and solids are decomposed into vapor prior to combustion
  - OXIDANT – A substance which supports combustion – Usually oxygen in air
  - IGNITION SOURCE – An energy source capable of initiating a combustion reaction

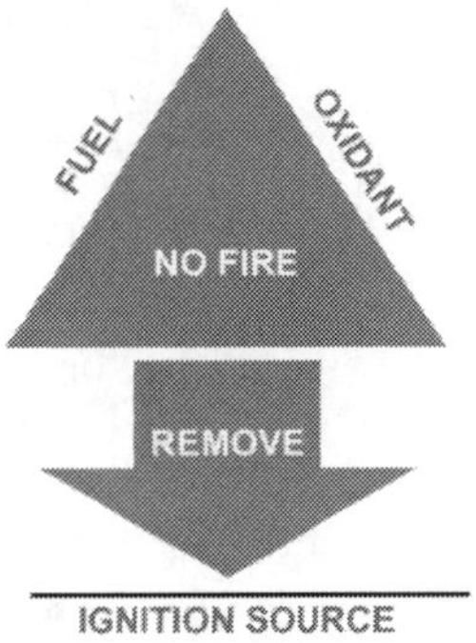

SPRING24+20TH GCPS
A Joint AIChE and CCPS Meeting

March 24–28, 2024
New Orleans Ernest N. Morial Convention Center,
New Orleans, LA

Part 02 Credible Static Ignition Source for Explosible Atmosphere

Different types of ignition sources for fires and explosions (from EN1127-1:2007)

Ignition sources	
1	Hot surfaces
2	Flames and hot gases (including hot particles)
3	Mechanically generated sparks
4	Electrical apparatus
5	Stray electrical currents, cathodic corrosion protection
6	Static electricity
7	Lightning
8	Radio frequency (RF) electromagnetic waves from 10^4 Hz to 3×10^{12} Hz
9	Electromagnetic waves from 3×10^{11} Hz to 3×10^{15} Hz
10	Ionising radiation
11	Ultrasonics
12	Adiabatic compression and shock waves
13	Chemical reactions, including self-ignition of dusts

SPRING24+20TH GCPS
A Joint AIChE and CCPS Meeting

March 24–28, 2024
New Orleans Ernest N. Morial Convention Center,
New Orleans, LA

Part 02 Credible Static Ignition Source for Explosible Atmosphere

- Minimum Ignition Energy (MIE) of Some Gases, Vapors, and Dust Clouds

Atmosphere	Material	Minimum Ignition Energy (mJ)
Vapor / Gas	ETHYLE ACETATE	0.460
	PROPANE	0.240
	METHANOL	0.140
	ACETYLENE	0.017
	HYDROGEN	0.016
	CARBON DISULPHIDE	0.009
Dust Cloud	PVC	1,500
	ZINC	200
	WHEAT FLOUR	50
	SUGAR	30
	ALUMINIUM	10
	ZIRCONIUM	5

SPRING24+20TH GCPS
A Joint AIChE and CCPS Meeting

March 24–28, 2024
New Orleans Ernest N. Morial Convention Center,
New Orleans, LA

Part 02 Credible Static Ignition Source for Explosible Atmosphere

- Analyze the credibility of electrostatic ignition risk need consider the following five questions:
  (1) Is there a flammable atmosphere?
  (2) Is there a mechanism to generate static charges?
  (3) Can charge accumulate?
  (4) Is there a discharge risk? What type of discharge is it?
  (5) Can static discharges generate enough energy to ignite the flammable atmosphere?

SPRING24+20TH GCPS
A Joint AIChE and CCPS Meeting

March 24–28, 2024
New Orleans Ernest N. Morial Convention Center,
New Orleans, LA

Part 03 Five Myths and Misunderstanding on ES Ignition Source

Myth 1: Bonding and grounding can solve all ESD issues

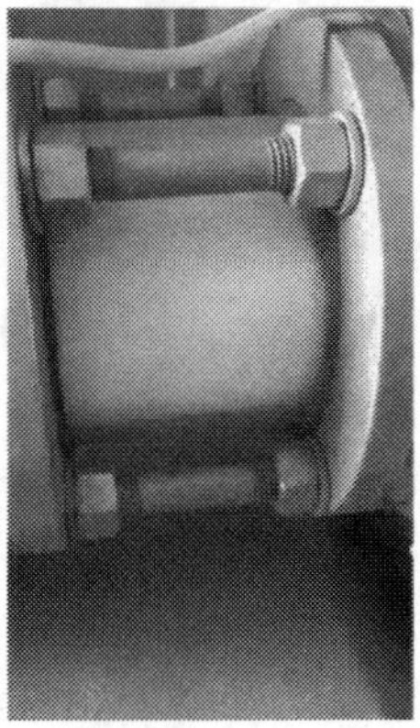

SPRING24+20TH GCPS
A Joint AIChE and CCPS Meeting

March 24–28, 2024
New Orleans Ernest N. Morial Convention Center,
New Orleans, LA

Part 03 Five Myths and Misunderstanding on ES Ignition Source

Myth 1: Bonding and grounding can solve all ESD issues

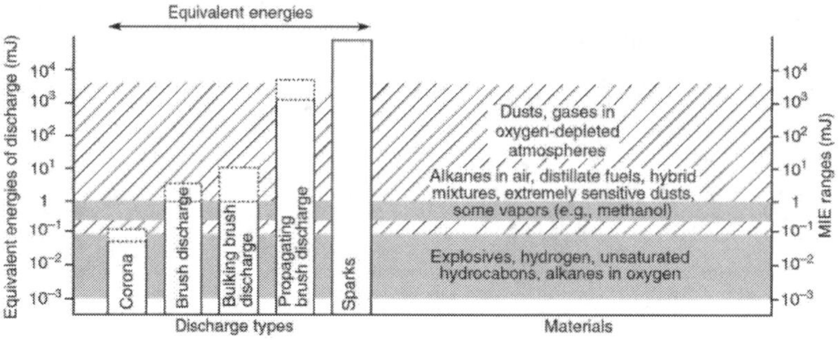

NFPA77, Figure 5.5.1, 2019 version

SPRING24+20TH GCPS
A Joint AIChE and CCPS Meeting

March 24–28, 2024
New Orleans Ernest N. Morial Convention Center,
New Orleans, LA

Part 03 Five Myths and Misunderstanding on ES Ignition Source

Myth 2: Lab-measured MIE can be directly applied to Industrial scenarios

$$MIE(T) = 10^{[-4.056+ (1.873-0.624\log T)\cdot(logMIE(T_0)+4.056)]}$$

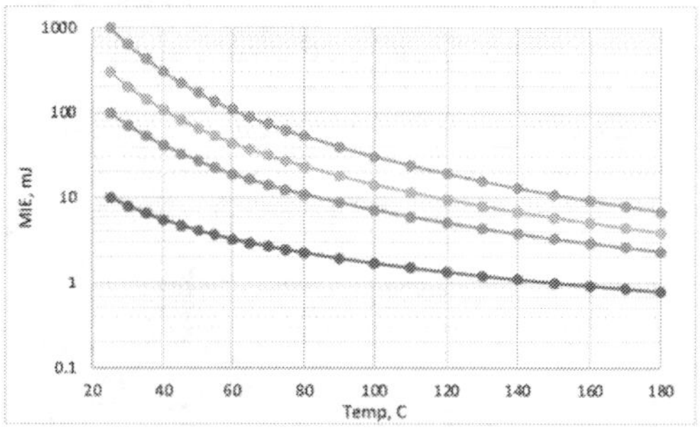

Siwek, Richard. "Determination of technical safety indices and factors influencing hazard evaluation of dusts." *Journal of loss prevention in the process industries* 9.1 (1996): 21-31.

SPRING24+20TH GCPS
A Joint AIChE and CCPS Meeting

March 24–28, 2024
New Orleans Ernest N. Morial Convention Center, New Orleans, LA

Part 03 **Five Myths and Misunderstanding on ES Ignition Source**

Myth 3: Dust explosion protection is not required if its MIE > 1000 mJ

Minimum Ignition Energy (MIE), (ASTM E 2019)

The minimum spark energy needed to ignite an optimum concentration of the material using a capacitive spark under ideal conditions

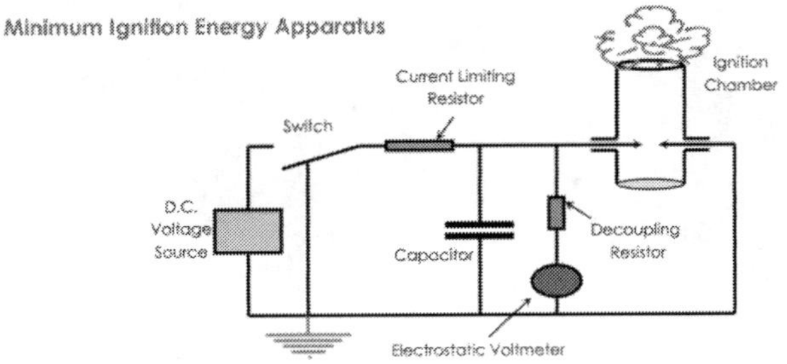

SPRING24+20TH GCPS
A Joint AIChE and CCPS Meeting

March 24–28, 2024
New Orleans Ernest N. Morial Convention Center, New Orleans, LA

Part 03 **Five Myths and Misunderstanding on ES Ignition Source**

Myth 3: Dust explosion protection is not required if its MIE > 1000 mJ

Typical High MIE Dust:

- Carbon Black
- Activated Carbon
- PVC
- ......

SPRING24+20TH GCPS
A Joint AIChE and CCPS Meeting

March 24–28, 2024
New Orleans Ernest N. Morial
Convention Center,
New Orleans, LA

Part 03 **Five Myths and Misunderstanding on ES Ignition Source**

Myth 3: Dust explosion protection is not required if its MIE > 1000 mJ

Case 1： Titan Tire Corporation Dust Explosion Incident

October 4, 1999, *"While running a batch of stock, a mixer ejected a large cloud of carbon black dust into the work area and up to the ceiling of the building. The dust ignited, causing a large flash fire that seriously burned two employees"*

Ignition Source: Electrical Device

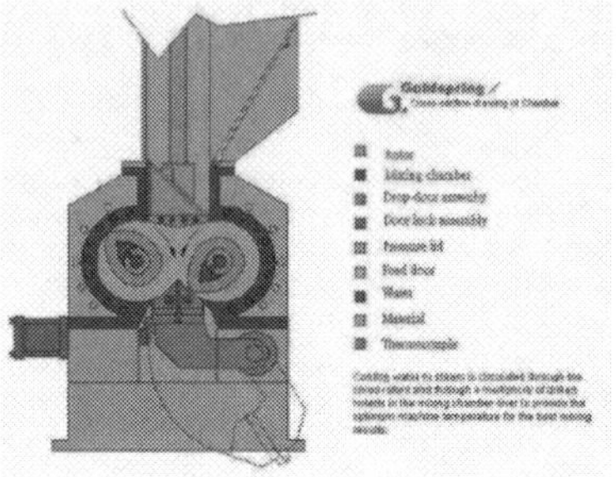

SPRING24+20TH GCPS
A Joint AIChE and CCPS Meeting

March 24–28, 2024
New Orleans Ernest N. Morial
Convention Center,
New Orleans, LA

Part 03 **Five Myths and Misunderstanding on ES Ignition Source**

Myth 3: Dust explosion protection is not required if its MIE > 1000 mJ

Case 1： Titan Tire Corporation Dust Explosion Incident

October 4, 1999, *"While running a batch of stock, a mixer ejected a large cloud of carbon black dust into the work area and up to the ceiling of the building. The dust ignited, causing a large flash fire that seriously burned two employees"*

Ignition Source: Electrical Device

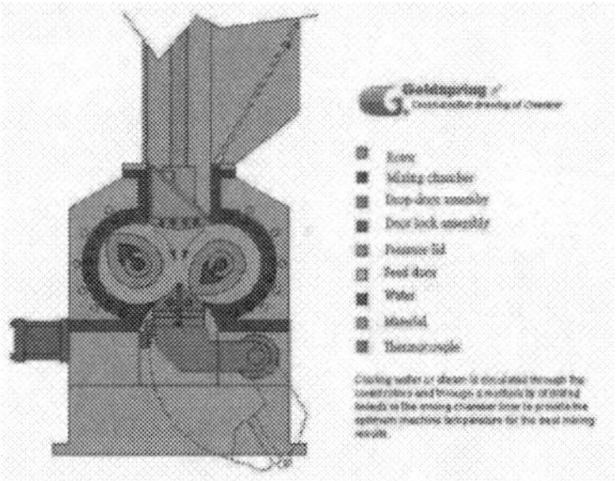

Part 03 Five Myths and Misunderstanding on ES Ignition Source

Myth 3: Dust explosion protection is not required if its MIE > 1000 mJ

Citation 313634842/01001

Inspection	Reporting ID	Open Date	SIC	Establishment Name
313634842	0419400	12/07/2010	3011	Cooper Tire & Rubber Company

Citation	Issuance Date	Hazard Category
01001	06/03/2011	Dust&Fumes

Section 5(a)(1) of the Occupational Safety and Health Act of 1970: The employer did not furnish employment and a place of employment which were free from recognized hazards that were causing or likely to cause death or serious physical harm to employees in that employees were exposed to Class II combustible dust explosion hazards due to the carbon black dust handling systems were not equipped with safety devices as to: On or about December 7-8, 2010 on the roof and 2nd floor of the mixing department, employees were exposed to combustible dust fire and explosion hazards due to the employer's failure to protect the carbon black dust storage silos, conveying systems, day bins, cyclones, and dust collectors with an adequate venting, suppression system, or suppression system. (a)The pneumatic/vacuum duct system used to transfer carbon black from the tank railcars and tank trailers was not provided with deflagration and explosion protection. (b)The eight (8) main storage silos located outside and away from the building were not provided with deflagration and explosion protection. (c)The eight (8) inch diameter pneumatic/vacuum conveying duct that transfers carbon black from the silos to the day bins on the roof of the building above the banbury mixers was not provided with deflagration and explosion protection. (d)The two day bins, "E" Day Bin for banbury mixer #4 and "D" Day Bin for banbury mixer 3 were four (4) compartment silos located on top of the roof above the master batch mixers. They were not provided with deflagration and explosion protection. Among others, one feasible and useful abatement method to correct these hazards is to follow the requirements found in NFPA 654 for; a) & c)Pneumatic/vacuum conveying ducts, refer to NFPA 654-2006, Chapter 7.1.4. b) & d)Storage silos and bins, refer to NFPA 654-2006, Chapter 7.2.3.3.3 and Chapter 7.1.2.1.

Part 03 Five Myths and Misunderstanding on ES Ignition Source

Myth 3: Dust explosion protection is not required if its MIE > 1000 mJ

An interesting Lawsuit:

- failing to determine the probability or likelihood that such a dust would ignite and explode in the actual workplace setting. It is important to note that the inherent properties of carbon black make it extremely difficult to ignite, and the Secretary did not have any data to establish that any of the identified ignition sources found in Cooper Tire's tire making plant could actually ignite the dust sample. Moreover, the entire tire making industry and the U.S. rubber manufacturers had never recognized carbon black to be a combustible dust hazard and, moreover, the NFPA had published the results of scientific studies showing that carbon black was one of the least likely carbonaceous dusts to ignite or explode and that even burning magnesium, at almost 3000 degrees, could not ignite carbon black.

SPRING24+20TH GCPS
A Joint AIChE and CCPS Meeting

March 24–28, 2024
New Orleans Ernest N. Morial Convention Center,
New Orleans, LA

Part 03 **Five Myths and Misunderstanding on ES Ignition Source**

Myth 3: Dust explosion protection is not required if its MIE > 1000 mJ

Case 2: US Ink/Sun Chemical Corporation: Ink Dust Explosion and Flash Fires in East Rutherford, New Jersey. October 9, 2012.

- Seven workers suffered burn injuries
- Mixing increases solution temperature to about 240°F. (116°C)
- Designed air flow velocity to DC: 21.5 m/s, actual air flow velocity was extremely low. Dust accumulation inside the ductwork was significant.
- A rotary airlock on the bottom of the hopper
- The employees were not wearing flame-resistant clothing (FRC).

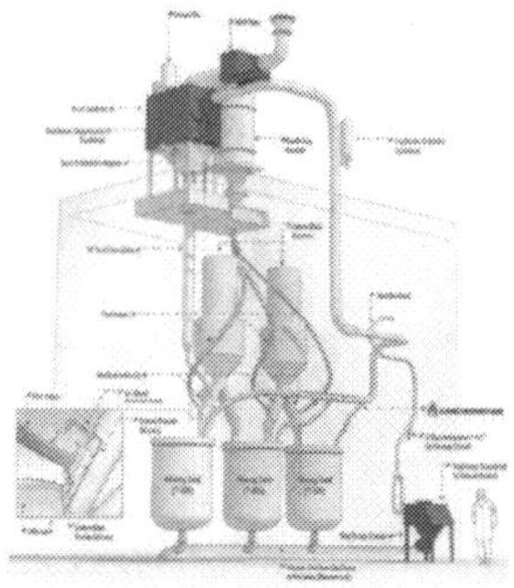

SPRING24+20TH GCPS
A Joint AIChE and CCPS Meeting

March 24–28, 2024
New Orleans Ernest N. Morial Convention Center,
New Orleans, LA

Part 03 **Five Myths and Misunderstanding on ES Ignition Source**

Myth 3: Dust explosion protection is not required if its MIE > 1000 mJ

Different types of ignition sources for fires and explosions (EN1127-1:2007)

Ignition sources	
1	Hot surfaces
	Flames and hot gases (including hot particles)
3	Mechanically generated sparks
	Electrical apparatus
5	Stray electrical currents, cathodic corrosion protection
6	Static electricity
	Lightning
8	Radio frequency (RF) electromagnetic waves from 10^4 Hz to 3×10^{12} Hz
9	Electromagnetic waves from 3×10^{11} Hz to 3×10^{15} Hz
10	Ionising radiation
11	Ultrasonics
12	Adiabatic compression and shock waves
	Chemical reactions, including self-ignition of dusts

Part 03 **Five Myths and Misunderstanding on ES Ignition Source**

Myth 4: Brush Discharge Can Ignite Dust with MIE < 4 mJ

SCHWENZFEUER, K.; GLOR, M. Tests to determine the ignition of dust by brush discharges. *Journal of electrostatics*, 1993, 30: 115-122.

The equivalent energy of brush discharges for the ignition of flammable gases has been determined in earlier work as 1 to 4 mJ. Although this lies in the region of the minimum ignition energy of several dusts, until today no dust ignitions due to brush discharges could be observed. The present work describes the successful ignition of sulphur by brush discharges with separation of the discharge zone and dust ignition. Estimation of the resultant energies is in good agreement with the determined equivalent energy.

SPRING24+20TH GCPS
A Joint AIChE and CCPS Meeting
March 24–28, 2024
New Orleans Ernest N. Morial Convention Center,
New Orleans, LA

Part 03 **Five Myths and Misunderstanding on ES Ignition Source**

Myth 4: Brush Discharge Can Ignite Dust with MIE < 4 mJ

GLOR, Martin; SCHWENZFEUER, Klaus. Direct ignition tests with brush discharges. *Journal of electrostatics*, 2005, 63.6-10: 463-468

To study the ignition behaviour of brush discharges, a new experimental approach was chosen. In a clear and easy manner it was possible to demonstrate the incendivity of brush discharges for flammable gases (what was already well known in the past) and the non-incendivity for powders, even in case of very low minimum ignition energies (MIE). Based on the investigations performed, it could be demonstrated, that the influence of the dust on the electric field and thus on the course of the brush discharge as well as the different wave form (power density) of the brush discharge compared with a spark discharge are responsible for the non-ignition of powders by brush discharges.

NFPA 77. 15.5.2. thereis no evidence suggesting that brush discharges are capable of igniting a dust cloud ....

SPRING24+20TH GCPS
A Joint AIChE and CCPS Meeting

March 24–28, 2024
New Orleans Ernest N. Morial
Convention Center,
New Orleans, LA

Part 03 **Five Myths and Misunderstanding on ES Ignition Source**

Myth 5: Gas flow alone generating static charges constitute an ES ignitions

NFPA 77. 5.2. The flow of clean gas over a solid surface produces negligible charging.

SPRING24+20TH GCPS
A Joint AIChE and CCPS Meeting

March 24–28, 2024
New Orleans Ernest N. Morial
Convention Center,
New Orleans, LA

Part 03 **Five Myths and Misunderstanding on ES Ignition Source**

Myth 5: Gas flow alone generating static charges constitute an ES ignitions

DRYER, Frederick L., et al. *Spontaneous ignition of pressurized releases of hydrogen and natural gas into air.* ***Combustion science and technology***, 2007, 179.4: 663-694.

- Certainly, the observed spontaneous ignition dependencies on pressure and geometry as well as their reproducibility are inconsistent with electrostatic discharge as ignition sources, frequently used to explain ignition of hydrogen releases into air.
- Hydrogen self-ignition mechanism under special working conditions has been extended by shock-induced ignition theory.

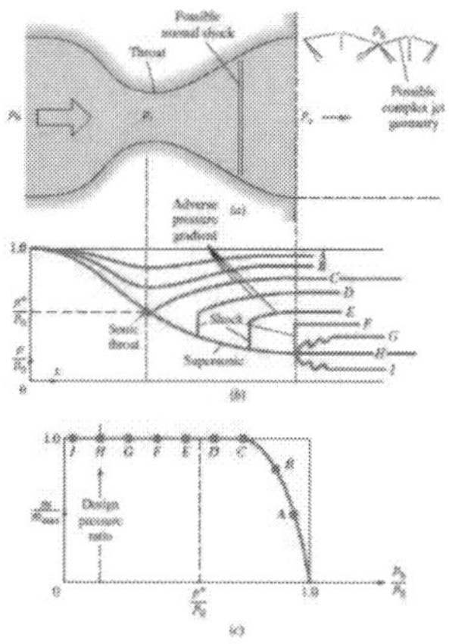

Part 04 **Summary**

- Key parameters to thoroughly understanding ESD hazard:
  (1) MIE of the flammable atmosphere
  (2) Volume resistivity/electrical conductivity
  (3) Surface resistivity
  (4) Chargeability
  (5) Types of ES discharges

- Do not guess ESD ignition sources if you cannot clearly explain it.

- Looking for expert support if any doubts are incurred.

SPRING24+20TH GCPS
A Joint AIChE and CCPS Meeting
March 24–28, 2024
New Orleans Ernest N. Morial Convention Center,
New Orleans, LA

Thank you for your time and participation!!!

Any questions?

Practical Sharing of Process Safety Management Systems Audit in China

Zhou Chao
Qingdao OASIS Science & Technology Development Group Co., Ltd.

SPRING24
+20TH GCPS
A Joint AIChE and CCPS Meeting
March 24–28, 2024
New Orleans Ernest N. Morial
Convention Center
New Orleans, LA

- **1. The Development of PSM in China**
- 2. PSM Systems Audit Application in China
- 3. Experience Sharing of PSM Systems Audit

ASIS

SPRING24+20TH GCPS
A Joint AIChE and CCPS Meeting

1. The development of PSM in China

2007
CCPS-CS Established

- PSM Training
- CCPS China PSM Conference
- CCPS Publication Translation
- Process Safety Beacon
- ……

2010
AQ/T3034-2010 Published

- China's first standard for PSM
- PSM theory research and training
- In 2012, CCPS-CS hosted the China branch of the Global Process Safety Conference for the first time
- ……

2013
Guidance on Strengthening the PSM of Chemical Enterprises

- The correct translation for process safety
- PSM auditing
- PSM improvement
- Cultivation of Chemical process safety personnel
- Construction of chemical safety academic subjects
- ……

2022年
AQ/T3034-2022 Published

- Government departments promote PSM auditing and improvement
- Published PSM auditing Guidelines
- Summarize experience and implement PSM good practices
- ……

ASIS

SPRING24+20TH GCPS
A Joint AIChE and CCPS Meeting

1. The development of PSM in China

- Chemical safety in China was initially implemented mainly through safety inspections focusing on the **"Yinhuan"** .
- In 2007,"Interim Provisions on the Investigation and Treatment of **Hidden Dangers** of Work Safety Accidents" (The Former SAWS No.16) was published.
- **Focus on compliance and integrity of on-site safety facilities.**

ASIS

SPRING24+20TH GCPS
A Joint AIChE and CCPS Meeting

1. The development of PSM in China

- In 2016, the "Opinions on implementing the Work Guidelines for Containing Major Accidents and the Construction of Double Prevention Mechanism" (Ministry of Emergency Management) was issued.
- **The key point of chemical safety management is to move forward from on-site safety inspection to risk identification and inspection of the key protection layers.**

ASIS

SPRING24+20TH GCPS
A Joint AIChE and CCPS Meeting

1. The development of PSM in China

- In 2022, "Guidelines for chemical process safety management" (AQ/T 3034-2022) was Published, which redefined **"Yinhuan"** as Unsafe-condition.
- *An unidentified risk, or a state in which the risk is not effectively controlled, beyond the acceptable level of risk. Including the unsafe behavior of people, the unsafe state of things, and management defects.*
- **The deficiencies in the system and the problems with the effectiveness on implementation need to be audited through systematic audit methods.**

- 1. The Development of PSM in China
- **2. PSM Systems Audit Application in China**
- 3. Experience Sharing of PSM Systems Audit

ASIS
SPRING24+20TH GCPS
A Joint AIChE and CCPS Meeting

2. PSM Systems Audit application in China

• Changing Mode of Safety Inspection

- Back to 2008, Qingdao Emergency Management Bureau promoted changes in safety supervision models with our company—Qingdao Oasis. And carried out the assessment **on Management of Change and Operating Procedures** in more than 30 enterprises.

Safety inspection mode based on on-site issues

- Based on site-specific issues;
- Difficult to achieve full coverage;
- Failure to implement improvements from management perspective;
- recurrence of issues.

→

Safety audit model based on management elements

- Based on systems thinking;
- Start with the root causes of issues;
- Rectification of the problem is based on management improvement;
- Effectively avoid the recurrence of issues.

- *The method of tracing safety management defects from on-site problems has been recognized by most enterprises and has achieved good results.*

SPRING24+20TH GCPS
A Joint AIChE and CCPS Meeting

2. PSM Systems Audit application in China

- **Expand the Scope of Auditing**
  - In 2020~2022, Qingdao expanded the scope of assessment management elements to 6 including **Hazard Identification and Risk Analysis, Management of Change, Operating Procedures, Permit to Work, Safety Instrument Management and Incident Investigation** in 23 companies, and quantified the auditing results.

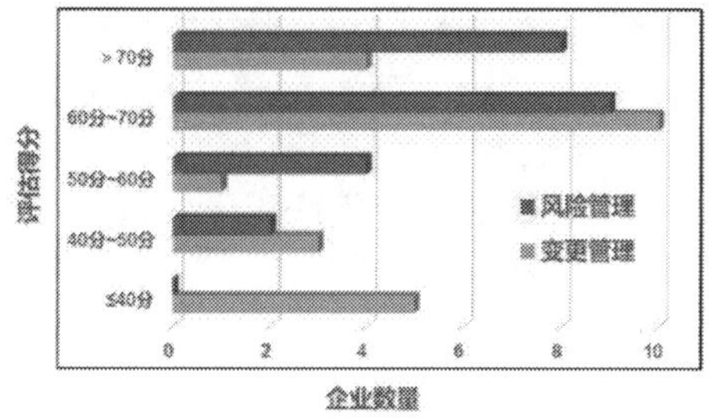

- *Through quantitative assessment results, the differences in the management level on elements of various enterprises in the region are measured.*

SPRING24+20TH GCPS
A Joint AIChE and CCPS Meeting

2. PSM Systems Audit application in China

- **Wide Range of Applications**
  - On the basis of fully investigation of the PSM audit in Qingdao, Shandong Province carried out the pilot audit of 44 enterprises in 2023, Qingdao Oasis was in charge of this work and published the "Audit Manual", which has been widely used in the chemical companies in the province.
  - Through the quantitative PSM audit of each enterprises, it is linked to the safety supervision method and frequency, and regular assessment is carried out to promote improvement of safety management.

ASIS

SPRING24+20TH GCPS
A Joint AIChE and CCPS Meeting

2. PSM Systems Audit application in China

- **Method of Auditing**
  - Based on the ***Guidelines for Auditing Process Safety Management Systems*** and the actual safety management status of the chemical enterprises in China, the auditing is carried out in four levels.

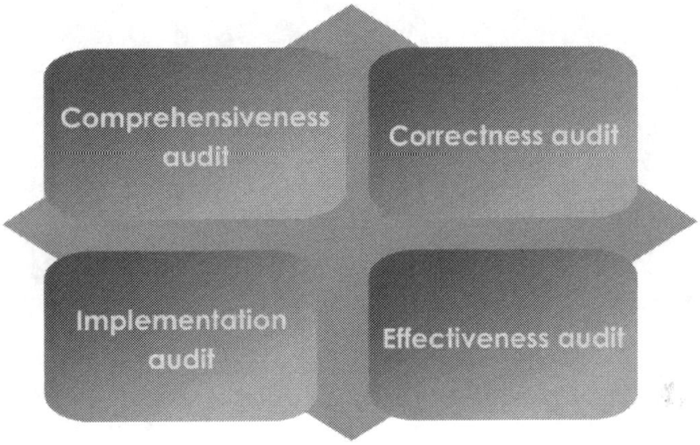

ASIS

SPRING24+20TH GCPS
A Joint AIChE and CCPS Meeting

2. PSM Systems Audit application in China

- **Comprehensiveness Audit**
  - Audit whether the written regulations can cover all activities that are elements of process safety management.

- *For example, in our auditing, chemical storage management in some companies lacked the requirement of the chemical reactive worksheet, so we saw the chemicals that can react with each other were mixed and stored together.*

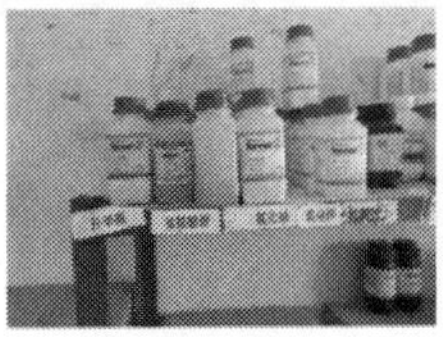

2. PSM Systems Audit application in China

- **Correctness audit**
  - Audit whether the written regulations are correct and easy to execute.

Cause of the leak	Quantity	Units	Quantity
Corroded	7	Alkylation unit	8
Poor welding	5	Aromatization unit	6
Isolation fails	5	Storage and transportation unit	6
Operational errors	4	Hydrogenation unit	3
Leakage of machine seals	2		

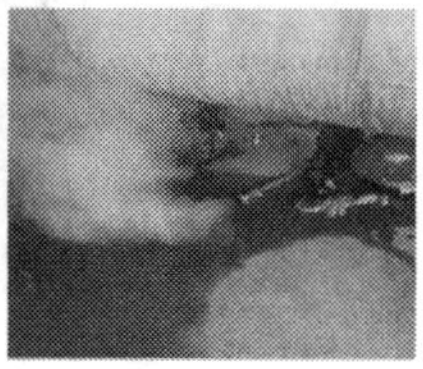

- *Due to the unclear definition of accident events in the written regulations in some companies, multiple leakage events were not managed as accident events, and finally a serious leakage accident occurred.*

ASIS
SPRING24+20TH GCPS
A Joint AIChE and CCPS Meeting

2. PSM Systems Audit application in China

- **Implementation audit**
  - Audit whether the management and production activities are strictly implemented in accordance with the relevant written regulations.

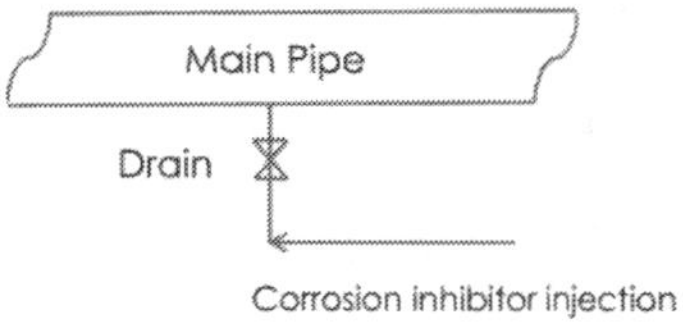

- *A company used the original design of drain line for corrosion inhibitor injection, did not implement the change management process, after a few years of scoured on the injection port, resulting in a leakage accident.*

2. PSM Systems Audit application in China

- **Effectiveness audit**
  - Audit whether the written regulations really plays a role in controlling risks after implementation.

- *Almost all companies have developed and implemented the permit regulation for interlock removal, but in some companies, interlocks were removed for the fear of unit shutdown due to interlock activation, especially in some high-risk situation.*
- *For example, removed the heating furnace fuel gas pressure low-low interlock when the fuel gas pressure fluctuates, and may result in a furnace flash explosion accident.*

Name	Frequency	SIF description	Ireason for interlock removal
Unit A	1	P5001A/B low low outlet flow	Switch the P5001A/B
	1	Circulating hydrogen flow low low	Switch the circulating hydrogen compressor
	3	P5003A/B low low outlet flow	Switch the P5001A/B
Unit B	4	F101 fuel gas pressure low low	Fuel gas pressure fluctuates
	4	F101 material flow low low	Material flow fluctuates

ASIS
SPRING24+20TH GCPS
A Joint AIChE and CCPS Meeting

2. PSM Systems Audit application in China

- After completing the audit, the team helped the company improve its safety performance by improving its key elements.

- *An company was helped to improve its incident investigation element by:*
- *Improved the incident management procedure;*
- *Than trained accident investigation experts.*
- *Formulated KPIs for incident management,*
- *And achieved quantitative improvement.*

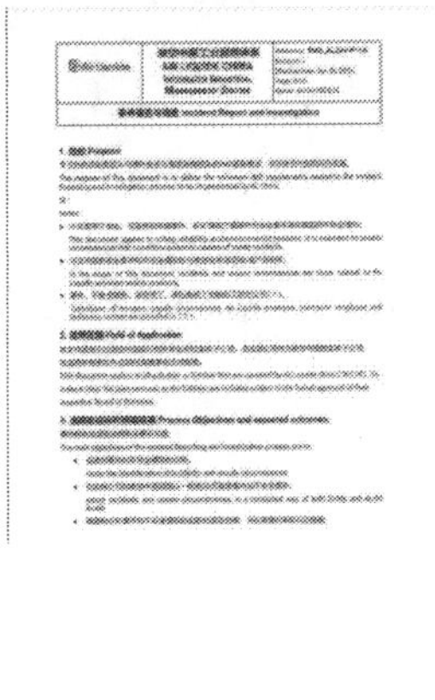

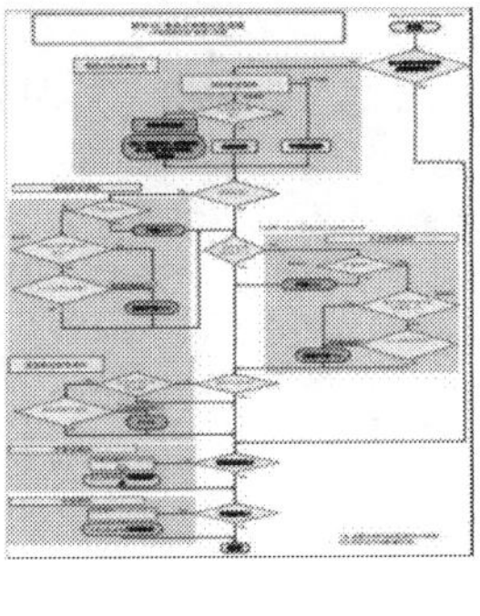

ASIS

SPRING24+20TH GCPS
A Joint AIChE and CCPS Meeting

2. PSM Systems Audit application in China

- Through PSM auditing and improvement, the overall process safety management level of chemical enterprises in that region shows a significant improvement trend.

Stage	Enterprises with very poor safety performance	Enterprises with poor safety performance	Enterprises with average safety performance	Enterprises with good safety performance
First Audit	**12**	**57**	**7**	1
Proportion	**15.58%**	**74.03%**	**9.09%**	1.30%
Second Audit	0	**48**	**28**	1
Proportion	0.00%	**62.34%**	**36.36%**	1.30%

- *33 of the 77 companies in some region showed a significant improvement displayed through the table.*

ASIS

SPRING24+20TH GCPS
A Joint AIChE and CCPS Meeting

- 1. The Development of PSM in China
- 2. PSM Systems Audit Application in China
- **3. Experience Sharing of PSM Systems Audit**

ASIS

SPRING24+20TH GCPS
A Joint AIChE and CCPS Meeting

3. Experience sharing of PSM Systems Audit

- Through the implementation of PSM systems audit and improvement in China, we summarized the experience of promoting process safety management from **the government level, the enterprise level and the social level.**

ASIS

SPRING24+20TH GCPS
A Joint AIChE and CCPS Meeting

3. Experience sharing of PSM Systems Audit

- **At the government level**
  - **The PSM audit cannot be well implemented in China without the promotion of government departments.**
- **Promote management audit**
  - Take management audit as the focus of safety supervision, and carry out quantitative audit of PSM elements.
- **Continuous Improvement**
  - Quantify the level of enterprise process safety management, set improvement goals, and continuously carry out re-audits.

ASIS

SPRING24+20TH GCPS
A Joint AIChE and CCPS Meeting

3. Experience sharing of PSM Systems Audit

- **Auditors Training**
- Establish a team of auditors to carry out process safety management audit.

- **Establishing Model Enterprises**
- Through the establishment of model enterprises, enhanced the confidence in safety management.

ASIS

SPRING24+20TH GCPS
A Joint AIChE and CCPS Meeting

3. Experience sharing of PSM Systems Audit

- **At the enterprise level**

As the main body of the implementation of process safety management, the implementation of process safety management at the enterprise level is mainly considered to do the following five aspects.

- Attitude
- ■Organizational Structure
- Based on Auditing Results
- Professional Management Group
- System of Assessment

ASIS

SPRING24+20TH GCPS
A Joint AIChE and CCPS Meeting

3. Experience sharing of PSM Systems Audit

1) Staff at various levels should have the correct attitude towards the improvement of process safety management. **Especially the leaders.**

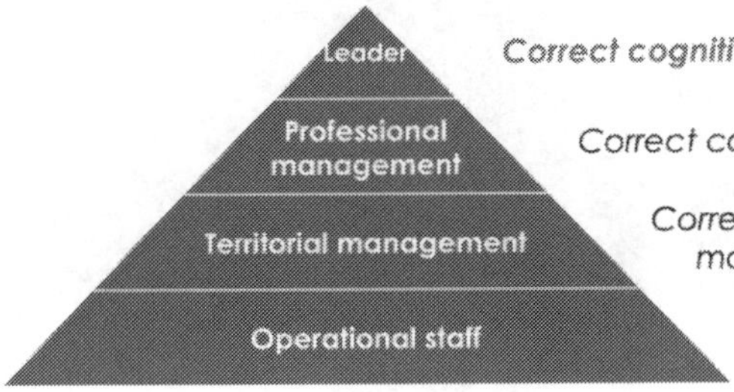

Correct cognition, sufficient attention, full support, leading demonstration

Correct cognition of professional management, with ability

Correctly recognize the responsibilities of the territory and mobilize employees to participate

Actively participate and execute correctly

ASIS

SPRING24+20TH GCPS
A Joint AIChE and CCPS Meeting

3. Experience sharing of PSM Systems Audit

2) Establish a complete organizational structure for process safety management improvement.

Promotion of the leadership group

It is usually the company's safety committee, which aims to control the progress of PSM improvement work, coordinate and solve key problems, make decisions on key issues, etc.

Promotion Office for PSM

A temporary organization, select capable personnel to serve, responsible for the organization, communication, assessment, and will become members of expert team in the later stage.

Groups for Promotion of Each Element.

In accordance with the PSM system, establish the RACI Model, responsible for the specific promotion of each element.

3. Experience sharing of PSM Systems Audit

3)Improvement plans should be based on auditing results of the PSM system, except for the Audit of Elements, but also the audit of the Foundation.

ASIS
SPRING24+20TH GCPS
A Joint AIChE and CCPS Meeting

3. Experience sharing of PSM Systems Audit

4) In the process of promotion, according to its own advantages and characteristics, establish corresponding professional management groups.

- Team of PHA
- Team of Internal Trainers
- Team of Equipment Reliability Experts
- Team for Management Review

3. Experience sharing of PSM Systems Audit

5) Establish an effective assessment mechanism for the improvement of process safety management including:

Assessment of Advancement Progress + **Assessment of System Implementation**

ASIS
SPRING24+20TH GCPS
A Joint AIChE and CCPS Meeting

3. Experience sharing of PSM Systems Audit

- **At the Social level**——Sharing of Best Practice

Academic Conferences	Seminars	Observation and Learning
Share academic presentations	Experience sharing and exchange of management elements to improve	Enterprises observe and learn on the spot

OASIS

SPRING24+20TH GCPS
A Joint AIChE and CCPS Meeting

路漫漫其脩遠兮
吾將上下而求索

OASIS

SPRING24+20TH GCPS
A Joint AIChE and CCPS Meeting

Thanks for Your Attention

ZHOU CHAO
Qingdao OASIS Science & Technology Development Group Co., Ltd.
(+86) 17606393740
zc@upchse.com

AUTHOR INDEX

AIChE
120 Wall Street, FL 23
New York, NY 10005-4020

ISBN 979-8-3313-1701-0

CURRAN ASSOCIATES INC.
proceedings.com